Analysis of Dynamic Psychological Systems

Volume 2

Methods and Applications

Analysis of Dynamic Psychological Systems
Volume 2
Methods and Applications

Edited by

Ralph L. Levine and
Hiram E. Fitzgerald

Michigan State University
East Lansing, Michigan

Plenum Press • New York and London

Library of Congress Cataloging in Publication Data

(Revised for volume 2)

Analysis of dynamic psychological systems.

  Includes bibliographical references and indexes.
  Contents: v. 1. Basic approaches to general systems, dynamic systems, and cybernetics—v. 2. Methods and applications.
  1. Psychology—Philosophy. 2. Psychology—Methodology. 3. System theory. I. Levine, Ralph L. II. Fitzgerald, Hiram E.
BF38.A46 1992   150′.1′1   91-44744
ISBN 0-306-43746-5

ISBN 0-306-43746-5

© 1992 Plenum Press, New York
A Division of Plenum Publishing Corporation
233 Spring Street, New York, N.Y. 10013

Printed in the United States of America

Contributors

David Chavis, School of Social Work, Rutgers University, New Brunswick, New Jersey 08903

Karl H. Clauset, Jr., Education Development Center, Inc., 55 Chapel Street, Newton, Massachusetts 02160

Edward L. Fink, Department of Speech Communication, University of Maryland, College Park, Maryland 20742-1221

Hiram E. Fitzgerald, Department of Psychology, Michigan State University, East Lansing, Michigan 48824

Paul Florin, Department of Psychology, University of Rhode Island, Kingston, Rhode Island 02881

Alan K. Gaynor, School of Education, Boston University, Boston, Massachusetts 02215

Irwin M. Greenberg, Department of Psychiatry, Harvard Medical School–Cambridge Hospital, 1493 Cambridge Street, Cambridge, Massachusetts 02139

Stan A. Kaplowitz, Department of Sociology, Michigan State University, East Lansing, Michigan 48824

Robert Langs, Nathan S. Kline Institute for Psychiatric Research, Building 37, Orangeburg Road, Orangeburg, New York 10962

Gilbert Levin, Department of Epidemiology and Social Medicine, and Department of Psychiatry, Albert Einstein College of Medicine, 1300 Morris Park Avenue, Bronx, New York 10461

Ralph L. Levine, Department of Psychology, Michigan State University, East Lansing, Michigan 48824

Weldon Lodwick, Department of Mathematics, University of Colorado, Denver, Denver, Colorado 80217-3364

James Grier Miller, Departments of Psychiatry and Psychology, University of California at Los Angeles, Los Angeles, California 90024, and Department of Psychiatry, University of California at San Diego, San Diego, California 92093

Jessie L. Miller, 1055 Torrey Pines Road, Suite 203, La Jolla, California 92037

William T. Powers, 73 Ridge Place, Durango, Colorado 81301

Richard Rich, Department of Political Science, Virginia Polytechnic Institute and State University, Blacksburg, Virginia 24061

Edward B. Roberts, Alfred Sloan School of Management, Massachusetts Institute of Technology, Cambridge, Massachusetts 02139

Mark W. Roosa, Department of Family Resources and Human Development, Arizona State University, Tempe, Arizona 85287

Gordon C. Ruscoe, School of Education, University of Louisville, Louisville, Kentucky 40292

Lois E. Tetrick, Department of Psychology, Wayne State University, Detroit, Michigan 48202

Abraham Wandersman, Department of Psychology, University of South Carolina, Columbia, South Carolina 29208

Preface

A system is a functional whole composed of a set of component parts (subsystems, units) that when coupled together, generate a level of organization that is fundamentally different from the level of organization represented in any individual or subset of the component parts. To understand the structure and function of a system, it is necessary to identify the component parts, their structural connections, and the properties of the system that are not evident in individual component parts of the system. Living systems are hierarchically organized, dynamic, self-regulating, holistic, purposeful, and open. As systems develop, their dynamic structure emerges via the organization of a network of feedback loops that regulate the growth, collapse, oscillation, and inhibition of the system. Thus, the study of systems is by definition concerned with change. In addition, the study of systems is by definition interdisciplinary, with historical roots in engineering, mathematics, computer science, biology, economics, and the management sciences. Authors of the various chapters in these volumes represent disciplines such as mathematics, medicine, psychology, sociology, economics, education, communications, political science, management, human development, industrial relations, neurophysiology, and cognitive science.

Although even the most casual students of the behavioral and social sciences are familiar with systems, their familiarity has been predominantly descriptive rather than analytic. The paradigmatic shift that has occurred in the behavioral and social sciences over the past generation, sometimes too narrowly referred to as the cognitive revolution, is gradually altering the way that behavioral and social scientists conceptualize the nature of human nature, the complexities of organism–environment transaction, and the emergent organization of such transactions.

Volume 1 focuses on approaches to systems analysis and is offered as a tutorial on systems theory; mathematical innovations that have changed modern physics and engineering; and nonlinear dynamical techniques for analysis of feedback processes, information flow, decision making, control theory, and modeling of human behavioral and social systems. Volume 2 focuses on methodological issues ranging from structural equation modeling to techniques for filtering random measurement error in time-series data, as well as on applications of structural models and systems approaches to the empirical analysis of a variety of behavioral and social problems.

Although we originally set out to assemble a collection of original chapters designed as tutorials for behavioral and social scientists, our intent is more than educational; it is missionary. Our mission is to stimulate a broader use of causal modeling and nonlinear dynamical modeling because we are convinced that such techniques will lead to a richer understanding of how the human system functions and how its functions are irrevocably and systemically bound to its environments.

RALPH L. LEVINE
HIRAM E. FITZGERALD

Contents

PART I. METHODS

PART II. APPLICATIONS OF SYSTEMS APPROACHES

Chapter 6. Applications of Living Systems Theory 153

James Grier Miller and Jessie L. Miller

Chapter 7. Complex Organizations and Living Systems Theory: The Case of the U.S. Army 181

Gordon C. Ruscoe

**Chapter 15.   Toward Building Psychoanalytically Based
Mathematical Models of Psychotherapeutic
Paradigms** 371

Robert Langs

1

Systems and Systems Analysis
Methods and Applications

Ralph L. Levine and Hiram E. Fitzgerald

Systems, Goal Seeking, and Feedback Mechanisms

A system can be defined as a set of components (subsystems, units), which when coupled together form a functional whole. This definition suggests three major themes to the study of systems: (1) identifying the subunits of the total system, (2) identifying the structural connections of subunits, and (3) assessing the properties that emerge when this collection of components are coupled together into a specific dynamic structure and allowed to change over time.

One fundamental concept used in living systems theory and in system dynamics is the idea of *feedback*. In a systems model, a network of feedback loops represents the dynamic structure of the system. Associated with that structure are one or more modes of behavior, such as growth, collapse, oscillation, and inhibition. Frequently the task of the systems scientist is to translate the proposed causes of behavior into a set of feedback loops. The advantage and power of this approach is that, through the analysis and simulation of this network of feedback loops, often it is not only possible to understand one's current data set, but also it is possible to predict qualitatively how the system should be-

Ralph L. Levine and **Hiram E. Fitzgerald** • Department of Psychology, Michigan State University, East Lansing, Michigan 48824.

Analysis of Dynamic Psychological Systems, Volume 2: Methods and Applications, edited by Ralph L. Levine and Hiram E. Fitzgerald. Plenum Press, New York, 1992.

have under conditions where there may not be much data available. A good model is one that can be used with confidence when conditions or policies change. This point is illustrated later in the introduction, when describing a model of a family's adjustment to diabetes. The relapse of the family's problem was predicted after the termination of therapy from observing the model's dynamic structure.

Although psychologists are quite knowledgeable about cognitive and information processes, they have not developed analytic tools for handling feedback mechanisms. This is an area where systems science can be helpful. Goal seeking is a very important facet of human behavior, and we suggest that knowing how to represent and analyze feedback mechanisms, which go hand-in-hand with goal-seeking behavior, should be given high priority. Placing more emphasis on feedback might have an effect on the use of causal modeling techniques. An informal survey of psychological journals indicates that in the majority of articles using causal modeling techniques, the arrows go in one direction. Nonrecursive causal models, which are models that contain reciprocal causation, are rarely found in the psychological literature. That is unfortunate, for once one studies the techniques developed by systems analysis for thinking in terms of closed loop structures, with practice, it becomes easy to see that there are feedback mechanisms embedded in most psychological theories that deal with dynamics, that is, change in the individual or group over time. Feedback is inherent in numerous psychological theories, ranging from the James–Lange theory of emotion (Lange, 1967) to the latest version of attribution theory.

Developmental approaches to the organization of behavior have been quick to embrace general system theory, largely through the influence of Urie Bronfenbrenner (1977). Recently much attention has been directed to the use of dynamical system modeling for representing developmental phenomenon (e.g., Fogel & Thelen, 1987). General system theory and dynamical system modeling are especially appropriate for developmental approaches, which, by definition, seek to identify and explain continuities (stability) and discontinuities (change) in the organization of behavior. Such disciplines include, among others, developmental neurobiology, developmental psychobiology, developmental–behavioral pediatrics, human ecology, and embryology. In developmental psychology, Sameroff and Fiese (in press) have been leading spokespersons for adoption of transactional models, which conceptualize developmental outcome as flowing from reciprocal interactions between the organism and its environmental contexts. Sameroff's concept of *environtype* focuses attention to environmental regulatory systems at the level of family, neighborhood, and culture and to the utility of systems approaches to preventive intervention. His approach historically derives from general system theory and from Bronfenbrenner's (1977) description of micro-, meso-, exo-, and macrosystem influences on the organization of individual and family systems. Although Sameroff and Bronfenbrenner did not propose specific dynamical system models, other developmentalists have advocated dynamical systems approaches to the analysis of expressive and communicative actions during the first year of life (Fogel & Thelen, 1987), the newborn's adaptation to its environment (Wolff, 1987), to the organization of motor behavior during infancy (cf. Thelen, 1989), and to the organization of cerebral lateral specializations during infancy (Fitzgerald *et al.*, in press). The rapidity with which developmentalists are embracing systems theory and systems modeling is illustrated by publication of an issue of *Child Development* (Connell & Tanaka, 1987), presented as a tutorial on structural equation modeling; the Twenty-Second Minnesota Symposium on Child Psychology (Gunnar & Thelen, 1989), devoted to papers on systems and development; and to various innovators who have used systems concepts to transform and revitalize traditional theoretical explanations of development (e.g., Emde, 1987), and/or to examine development from a perspective that stresses the organism's active engagement of its multimodal environments as a means of impelling differentiation and integration during adaptation (e.g., Gibson, 1979; Neisser, 1976). If we are ever to understand the organization of living systems, we must begin to use models that help to explain how systems emerge and become transformed into more complex (more organized) systems as a result of various endogenous and exogenous influences.

Let us differentiate between dynamic structure, which is the configuration of feedback loops, and the behavioral output of the system. The dynamic structure determines the behavior of the system. To see this, compare Figures 1a and 1b. This is a stripped-down version of a model of classroom dynamics developed by Gaynor and Clauset (1983) to account for the differences between effective and noneffective schools. In Figure 1a, which represents the simple dynamics of the teacher's compensation for poor performance, it is assumed that there is no direct causal link between performance and expectations (standards). On the other hand, Figure 1b represents the dynamics of teaching in noneffective schools. Here although the variables are the same, there is a causal link between performance and expectations.

Thus both causal loop diagrams have exactly the same variables. They only differ for one small connection between variables. This small difference in structure generates huge *qualitative* differences in behavior. In Figure 2a, when there is a large gap between the teacher's expectations of achievement and actual achievement, the teacher will in-

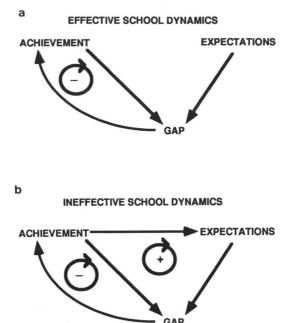

Figure 1. Differences between the dynamic structures of effective and ineffective schools.

crease the intensity of instructions to eliminate the gap. The behavior of the achievement variables is shown in Figure 2a. There is a smooth change in achievement up to the teacher's expectation level over time.

In Figure 2b, which describes the behavior found in ineffective schools, only one new connection has been added to Figure 1. Here the teacher's expectations are influenced by the child's low achievement level. The structure is now different, even though no new variables have been added. In this case, there is an erosion of expectations *down* to much lower achievement levels.

Applications of Systems Science to Psychology

The Gaynor and Clauset (1983) model illustrates the relationship between dynamic structure and the behavior output of the system. However, there is a very different application of systems analysis to psychology. This application extends systems methodology to clinical aspects of family dynamics. One of the major approaches to family therapy is to view family problems from a systems perspective. Family

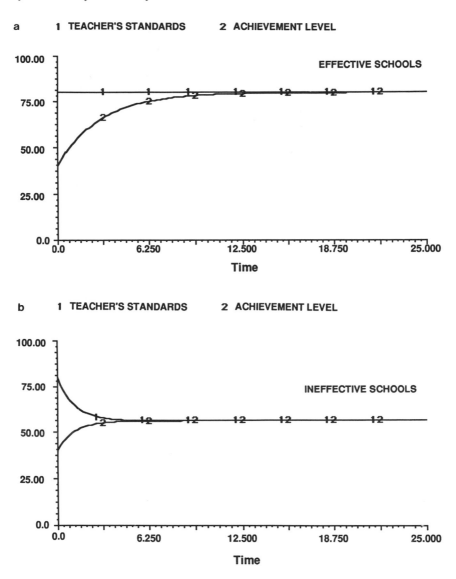

Figure 2. Differences in behavior patterns of effective and ineffective schools.

systems analysis was pioneered by such workers as Bowen (1978), Bateson (1972), Minuchin (1974), Papp (1983), and Selvini Palazzoli, Boscsolo, Cecchin, and Prata (1978). The systems approach has been used effectively in working with families that deal with problems of alcoholism and other drug abuse (see Wegscheider, 1981; Bepko, 1985).

In addition, John Bradshaw gave a series of public lectures on family problems (also see Bradshaw, 1988) that popularized the notion of viewing the family as a system. He pointed out that changes in one member of the family system can have differential reverberations across the whole family. The adjustments and roles taken by family members can lead to compulsive behavior and to making the problem worse.

A team effort by the first editor (RL) and two family therapists, Jane Pearson and Nicholous Ialongo, attempted to model the dynamics of a specific clinical case. The Levine, Pearson, and Ialongo approach to family therapy fits nicely into the family systems framework, although they place more stress on formal representation of the feedback processes associated with the family interaction patterns. The major goal of the project was to ascertain the effectiveness of modeling a specific case history of a family that had a terrible problem of adjusting to the child's chronic diabetes. Levine *et al.* built a model of the dynamics of the family's problem. It was a first attempt to show the power of viewing dynamics in terms of cybernetic feedback loops. The model describes the dynamics of a family system under stress. The approach was to focus on a single case history, model the dynamics underlying the family problem, and then use the model to follow the impacts of the therapeutic approach used in the case. They also used the model to understand why other suggested interventions might fail if applied to this situation. Eventually, one would hope to use this modeling methodology "on line," so to speak, as a more general therapeutic tool.

Briefly, this family had a problem adjusting to the son's noncompliance to his diabetes regimen, his recent hospitalization for ketoacidosis, frequent absence from school, and the like. Here are the most important facts in the case: (1) The mother spent large amounts of time working with the child. (2) The father refused to help her maintain the child's diabetes regime. (3) The father denied the potential risks associated with the disease. (4) The mother grew angry and frustrated as time passed. (5) The father became defensive and angry, which increased his wife's anger level. (6) The boy reacted to the amount of strife between his parents by becoming worse as his parents fought.

With the help of the two clinical members of the team, RL built the model of the dynamics of the family's problem in three stages. First, we modeled each component of the family system; the mother, father, and child separately. Then we coupled the three sectors together to get the emerging properties that occur when the whole system evolves over time using computer simulation techniques provided by a commercial program called STELLA™. Next, we assessed the importance of each major feedback loop in the model and the effect of changing the model's input parameters. This last stage embodied an analysis of the impact of various therapeutic interventions, their intensity and timing.

The reactions of the family to the father's refusal to help the child suggested that at least part of the dynamics of the problem could be handled by using an equity theory framework. Equity theory, which has been around for many years (see, e.g., Walster, Walster, & Berscheid, 1978), and has a substantial empirical foundation, states that, when a person perceives that he or she has been treated unfairly, a set of processes begin to come into play, to attempt to change the situation. The mother felt that the father was not living up to his responsibilities. The records show that her anger mounted very quickly and reached such a height that she threatened to leave him and the child.

The other aspect of equity theory deals with the reaction of the father to his lack of participation. Equity theory assumes that he should feel some degree of discomfort, which in turn can lead to one or more actions or rationalizations to decrease this discomfort. Indeed, the application of equity theory to the dynamics of the family's problem seemed to be a good starting place (see Figure 3). However, in this complex family system, there was much more going on than just one process. Levine *et al.* hypothesized other social mechanisms also coming into play. Each theoretical process was then represented as one or more cybernetic loop structures, our basic building blocks. When the loops representing equity theory were embedded into the larger network, it became evident that other processes inhibited the impacts of equity theory on the overall behavior of the system. In this situation, for example, the father's increased anger inhibited the growth of his discomfort (see Figure 3), so that the classical results of equity theory might not be observed without therapeutic intervention.

Once the major mechanisms have been worked out, the next stage is to translate this proposed dynamic structure to a computer environment by using a simulation program like STELLA™ to simulate the systems model. In this case, we developed a working model of the system, composed of six state variables, like the level of each parent's anger, in about 1½ hours.

The next figure shows the results of two computer runs (see Figure 4). These are theoretical curves of the emotional variables associated with this system, namely the mother's level of anger, their father's level of anger, and the father's level of distress (guilt). In the first run, we looked at what would happen without any therapeutic intervention. Here one would predict that the levels of tension and anger would remain at a maximum over the period of observation. The lower graph shows the prediction of what would happen with the therapeutic inter-

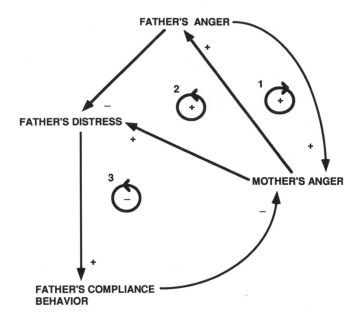

Figure 3. The proposed dynamic structure of the ongoing interaction between the parents of a diabetic child that corresponds to an equity process, inhibited by the husband's anger.

vention used in this case. In the actual situation, the co-therapists first attempted to lower the mother's anger to prevent her from walking out on the situation. Second, they increased the father's distress level by reminding him of his responsibilities and by indicating appropriate behaviors for aiding his son.

This figure shows the predictions of the model, assuming that the therapeutic interventions described took place starting at Week 4 and ending on the twenty-eighth week. According to the model, there is a definite timing of emotional responses, which were verified in the actual case by comparing the qualitative predictions of the model with clinical notes about the real case. The exact sequence generated by the model—increase in mother's anger, followed by the growth of the father's anger, and the like—exactly matched what occurred over sessions in the actual family system.

The model was formulated from the details given to the model builder by the co-therapists. However, there are always bound to be some details not thought important enough to relate. A test of a good model is whether or not the model can generate predictions about the data that were there to begin with, but nobody noticed them until the

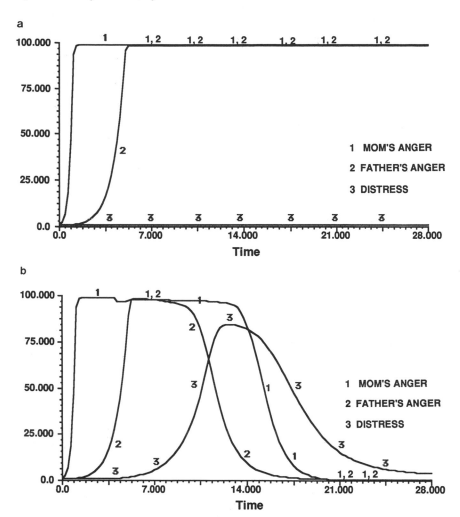

Figure 4. Output of the model showing what would happen (a) without the therapeutic intervention and (b) with the therapeutic intervention.

model pointed out their existence. This is called the "surprise behavior test."

An analysis of this configuration of variables indicated the strong possibility of an erosion of the father's behavior and a recycling of anger some time after therapy was terminated.

To be more specific, the modeler reran the model under identical conditions, the only difference being that he followed the family system for 60 weeks, instead of 28 weeks, the time at which therapy was

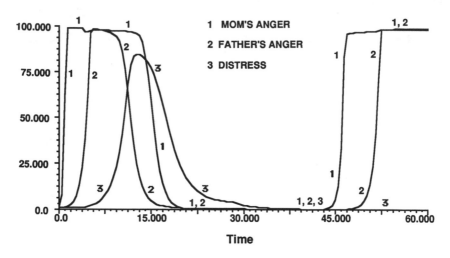

Figure 5. Prediction of the model when the time horizon is lengthened to 60 weeks, indicating the recycling of the anger.

terminated. The next figure (see Figure 5) shows the results of this run. There is a period where all of the emotional variables were "silent," so to speak, followed by a recycling of anger. One would predict, then, that there should be a relapse in about 4 to 6 months after termination of therapy. After consulting with the therapists on the team, the modeler was told that this is exactly what happened. The therapists had to see the family again for a few brief sessions where they worked on helping the family maintain the father's motivation for helping his son.

This application of systems analysis to the dynamics of psychopathology is a prototype of what we hope to do in the future. We must make it clear that these are theoretical curves and remind you that in this case we have no specific *quantitative* measures of the variables. This model only provided qualitative information concerning the relationship among the variables and described behavior in general terms. Nevertheless, it provided a cybernetic framework to understand complex family behavior patterns. As we have pointed out, one can make predictions that can be either verified or discarded.

In future modeling efforts, we would plan to use more traditional clinical research methodology, obtaining measurements of the state variables from ratings of tapes and fitting the model to these measures in a more formal way. Nevertheless, we would be looking for common cybernetic structures that run across families with similar problems.

Organization of Volume 2

This volume addresses several technical problems that might be of interest to the reader who wants to apply the systems approach to psychological problems. Chapter 2 deals with the application of structural models to psychological problems from a systems point of view. A growing number of psychologists are using structural equation techniques to analyze data. Lois Tetrick's chapter goes beyond describing how to analyze structural models by stressing how to carefully apply the models for policy analysis. This is a relatively new application of structural modeling in psychology, although the structural modeling approach is used extensively in the economic forecasting field.

Chapter 3, by Levine and Lodwick, describes the process of finding which aspects of the system are most sensitive to small changes in either input (e.g., changing parameters) or changes in cybernetic loop structures. This chapter is particularly important for assessing the impact of interventions and the initiation of new policies in applied settings. The chapter reviews the more technical aspects of sensitivity analysis first described in Chapter 4, Volume 1. Sensitivity analysis, among other things, helps to locate so-called *policy levers*, which may be used to guide the system to desired goals through new management policies.

System analysis is very strong on generating a theoretical framework for understanding system problems. Quantitatively, however, there is a need to develop the other side of the systems approach, namely empirical validation of the system model to real psychological data, especially if the model has been formulated mathematically as a set of differential or difference equations, which generate time series. Psychological variables are usually measured on interval data scales, and sometimes the scales themselves may only have relatively low reliabilities. Research psychologists are well versed in statistical analysis and have developed a set of standard techniques for measuring and validating psychological concepts. In Chapter 4, Levine and Lodwick suggest several procedures that might be done before one fits a systems model to a given data set. Because most psychological variables are operationally defined in terms of interval scales, it may be important to see if the level of the measurement makes a difference in what the theoretical curve looks like. Levine and Lodwick present an example of a nonlinear attitude change model that assumes that the underlying *theoretical* variable, the latent trait, has ratio scale properties. That was on the theoretical side. On the empirical side, the researcher might find it convenient to measure the variable by operationally defining the variable in terms of an interval scale composed of responses to a set of items. They show that the same model may generate two qualitatively different theoretical time curves, depending on whether or not attitudes were measured on an interval or ratio scale. These authors point out that the researcher might immediately reject the model because he or she had the wrong measurement version of the theoretical model. They resolve this problem by showing how to generate theoretical predictions of interval scaled time series. These new theoretical curves provide the basis for fitting the model to the real data. Although there might be some costs involved with using interval scaled variables, in general the authors lay the foundation for using standard interval variables instead of measuring all the variables in the model in terms of ratio scales.

Levine and Lodwick also discuss use of filtering techniques to screen out random measurement error inherent in one's empirical time series data before estimating parameters. They briefly review some of the engineering literature and show what happens when attempting to filter out noise in a time series variable that had a reliability of .6 versus a reliability of .9.

Chapter 4 describes how to prepare the data prior to estimating the parameter values and assessing the degree of fit. Chapter 5 deals with ways to obtain estimates of parameter values of one's dynamic model from filtered data. In Chapter 5, Levine and Lodwick describe several types of estimation procedures and identify a software package that is available for this task. The last part of the chapter discusses the assessment of not only fit but how to decompose the residual variance into systematic and nonsystematic model error for each state variable. This assessment procedure pin points where the model falls down and what is the nature of errors being made.

Part II of the volume is composed of short papers or vignettes that describe specific applications of the system approach to psychology and related fields. In organizing Volume 2, we intended to present as many concrete examples as possible to give the reader an idea about how the systems approach can actually be applied to active areas of psychological research. Volume 1 presented the more theoretical and abstract aspects of the systems approach. Part 1 of Volume 2 moved toward more technical aspects of dealing with systems but nevertheless dwelled on methodological material. Part II is divided into several subsections that are organized along different approaches to systems analysis. Table 1 shows the title of each chapter in each volume and describes which of three basic approaches is represented by the chapter: general systems theory (living systems theory), cybernetics, or dynamical processes (system dynamics).

Table 1. Approaches to Systems Analysis Represented in Each Volume
Organized by Chapter

Chapter and author(s)	Topic/approach
Volume 2	
1. Levine & Fitzgerald	Living systems, dynamical systems, cybernetics
2. Tetrick	Structural modeling/general systems theory
3. Levine & Lodwick	Sensitivity/system dynamics
4. Levine & Lodwick	Scaling and filtering/system dynamics
5. Levine & Lodwick	Estimation and evaluation/system dynamics
6. Miller & Miller	L.S.T. applications/living systems theory
7. Ruscoe	Organizations/living systems theory
8. Greenberg	Psychiatric disorders/general systems theory
9. Florin et al.	Grassroots organizations/general systems theory
10. Fitzgerald & Roosa	Teenage child rearing/G.S.T./structural modeling
11. Levin & Edwards	Service delivery systems/system dynamics
12. Clauset & Gaynor	School dynamics/system dynamics
13. Powers	Cognitive control/cybernetics
14. Kaplowitz & Fink	Attitude change/dynamical processes
15. Langs	Dynamics of therapy/dynamical processes
Volume 1	
1. Levine & Fitzgerald	Living systems, dynamical systems, cybernetics
2. Miller & Miller	Living systems theory
3. Abraham, Abraham, & Shaw	Dynamical processes
4. Levine, Van Sell, & Rubin	System dynamics/cybernetics
5. Levine	Cybernetics

General Systems Theory

Chapter 6, written by James and Jessie Miller, describes research applications of living systems theory at a variety of levels of organization.
Chapter 7, written by Gordon Ruscoe, follows the Millers' chapter with
a review of a specific study utilizing living systems theory as a framework. These two chapters should give the reader a better idea about
how one can apply living systems theory to specific research problems.
In Chapter 8, Irwin Greenberg uses another type of general systems
theory to diagnose psychiatric disorders. Greenberg shows how to carefully solve the mystery of piecing together evidence about a number of
physiological and psychological systems in order to generate a correct
diagnosis. In a sense, Greenberg is his own expert system, integrating
across many levels of organization in order to understand the patient's
problem.

At a different level of behavior, Florin, Chavis, Wandersman, and Rich (Chapter 9) deal with a systems approach to understanding and enhancing grassroots organizations. Their approach uses the analysis and assessment of resources, organizational structure, social climate, and information processing. It is another fine example of a general systems approach that stresses the functionality of the *open system* concept, whereby viable social systems expend time and energy, bringing new members into the organization and utilizing outside resources.

Chapter 10, written by Fitzgerald and Roosa, is the last chapter in the series of applications that might be categorized as an application of general system theory. These authors use a causal modeling framework to organize their data. One should also note that in Chapters 6 to 10, which constitute general system theory applications, empirical results frequently address policy questions. Perhaps this is a natural consequence of viewing a problem in terms of many interrelated variables.

The Cybernetic Approach

The next three chapters (11 to 13) illustrate how the cybernetics approach to systems apply feedback principles to understanding specific behaviors. Chapter 11, written by Levin and Roberts, analyzes the generic *loop structures* associated with the dynamics of service delivery systems. Their system dynamic model deals with the instability of the patient's standard of functioning, and so forth. Chapter 12, written by Clauset and Gaynor, deals with problems in the classroom, showing generic loop structures associated with poor classroom performance. Once those generic structures are identified, a systems model can be used to study the effects of proposed new policies and to trace the impacts of new sources of information flow.

Chapter 13, written by William Powers, is another fine example of a cybernetic approach to behavior, this time in the domain of cognitive psychology. Powers makes the point about how internally the subject has control over defining the goals. He applies this internal framework to explain behavior in a perceptual-motor task. His emphasis on internal processing and internal states is very compatible with the systems dynamics approach to modeling behavioral problems. All three chapters (11, 12, 13) illustrate how to analyze the role of feedback involved in the problem under study.

Oscillatory Behavior and Dynamical Systems

The last two chapters (14, 15) deal with applications of mathematical dynamical theory, which frequently explains oscillatory behavior. In

Chapter 14, Kaplowitz and Fink describe an oscillatory model of attitude change, whereas in Chapter 15, written by Robert Langs, emphasis is on the nature of the interaction of therapist and client during therapy sessions. Here Langs is breaking new theoretical ground, making an elegant argument for viewing the therapy process as one that displays complicated twists and changes that can only be captured by using a nonlinear dynamical systems approach. Recently Langs has been collaborating with a number of mathematicians formulating models of the therapy process and fitting these models to time series data obtained from therapy sessions.

References

Bateson, G. (1972). *Steps toward an ecology of mind.* New York: Ballantine Books.

Bepko, C. (1985). *The responsibility trap: A blueprint for treating the alcoholic family.* New York: The Free Press.

Bowen, M. (1978). *Family therapy in clinical practice.* New York: Jason Aronson.

Bradshaw, J. (1988). *Bradshaw on: The Family.* Deerfield Beach, FL: Health Communications, Inc.

Bronfenbrenner, U. (1977). Toward an experimental ecology of human development. *American Psychologist, 32,* 513–531.

Connell, J. P., & Tanaka, J. S. (Eds.). (1987). Special section on structural equation modeling. *Child Development, 58,* 1–175.

Emde, R. (1987). Infant mental health: Clinical dilemmas, the expansion of meaning, and opportunities. In J. D. Osofsky (Ed.), *Handbook of infant development* (pp. 1297–1320). New York: Wiley Interscience.

Fitzgerald, H. E., Harris, L. J., Barnes, C. L., Wang, X., Cornwell, K. S., Kamptner, N. L., Dagenbach, D., & Carlson, D. (1991). The organization of lateralized behavior during infancy. In H. E. Fitzgerald, B. M. Lester, & M. W. Yogman (Eds.), *Theory and research in behavioral pediatrics* (Vol. 5; pp. 155–184). New York: Plenum Press.

Fogel, A., & Thelen, E. (1987). Development of early expressive and communicative action: Reinterpreting the evidence from a dynamic systems perspective. *Developmental Psychology, 23,* 747–761.

Gaynor, A. K., & Clauset, K. H. (1983). Implementing effective school policies: A system dynamics policy analysis. *Proceedings of the 1983 International System Dynamics Conference, 1,* 307–314.

Gibson, J. J. (1979). *The ecological approach to perception.* Boston: Houghton-Mifflin.

Gunnar, M., & Thelen, E. (Eds.). (1989). *Systems and development: The Minnesota Symposium on Child Psychology* (Vol. 22). Hillsdale, NJ: Erlbaum.

Lange, C. G. (1967). The emotions: A physiological study. In C. G. Lange & W. James (Eds.), *The emotions* (pp. 33–90). New York: Hafner (Originally published 1885).

Minuchin, S. (1974). *Families and family therapy.* Cambridge, MA: Harvard University Press.

Neisser, U. (1976). *Cognition and reality.* San Francisco: W. H. Freeman.

Papp, P. (1983). *The process of change.* New York: Guilford Press.

Sameroff, A. J., & Fiese, B. H. (in press). Transactional regulation and early intervention. In S. J. Meisels & J. P. Shonkoff (Eds.), *Early intervention: A handbook of theory, practice, and analysis.* New York: Cambridge University Press.

Selvini Palazzoli, M., Boscolo, L., Cecchin, G., & Prata, G. (1978). *Paradox and counterparadox*. New York: Jason Aronson.

Thelen, E. (1989). The (re)discovery of motor development: Learning new things from an old field. *Developmental Psychology, 25,* 946–949.

Walster, E. H., Walster, G. W., & Berscheid, E. (1978). *Equity: Theory and research*. Boston: Allyn & Bacon.

Wegscheider, S. (1981). *Another chance: Hope and health for the alcoholic family*. Palo Alto, CA: Science and Behavior Books.

Wolff, P. H. (1987). *The development of behavioral states and the expression of emotion in early infancy*. Chicago: University of Chicago Press.

I

**Technical and Methodological
Aspects of Systems Analysis**

2

The Study of Dynamic Systems Using Structural Equations Analysis

Lois E. Tetrick

Structural equations modeling involves laying out the relationships among events or phenomena along a time horizon based on theory. The model or schematic representation of these relationships then can be expressed in a system of structural equations. Solving for the parameters in these structural equations allows one to determine the usefulness of the theory, which guided the development of the model.

This approach is especially useful to researchers investigating complex phenomena in natural settings in which traditional experimental designs are not feasible or in instances where the variables of interest cannot be directly observed. Structural equations analysis, therefore, can be valuable in evaluating the performance of systems or programs within organizations for policy decisions. For example, this analytic technique has been used to describe aspects of selection systems in organizations (Powell, 1984; Schmidt, Hunter, & Outerbridge, 1986), career transitions within organizations (Latack, 1984), organizational withdrawal or turnover (Michaels & Spector, 1982; Price & Mueller, 1981; Williams & Hazer, 1986), and organizational decision making (Singh, 1986; Walker & Weber, 1984). However, this statistical technique does not automatically justify inferring causation from correlational designs.

Lois E. Tetrick • Department of Psychology, Wayne State University, Detroit, Michigan 48202.

Analysis of Dynamic Psychological Systems, Volume 2: Methods and Applications, edited by Ralph L. Levine and Hiram E. Fitzgerald. Plenum Press, New York, 1992.

Modeling of hypothesized relationships among constructs, as in any research design, requires that the researcher develop the model based on theory and then examine how well the model explains or accounts for the observed covariance among the data collected after model development. The general considerations in developing and testing a model using structural equations analysis are the same as in any empirical study. The purpose of this chapter is to present the reader with the logic and design considerations as well as the statistical techniques for developing and testing an hypothesized model.

Developing a Model Appropriate for Causal Analysis

James, Mulaik, and Brett (1982) enumerated seven conditions that a model reasonably must satisfy in order for one to submit a model to causal analysis using structural equation methods. These conditions parallel traditional standards for research designs. The first condition is a formal statement of theory in terms of a structural model; that is, one must delineate the relationships among the constructs as specified by theory. Secondly, the researcher must be able to specify a theoretical rationale for each causal hypothesis included in the model. The third and fourth conditions are that in specifying the theoretical relationships, the researcher also must specify the causal order and causal direction of the hypothesized relationships among the constructs included in the model.

Causal order refers to the sequence of cause and effect that carries with it an implicit causal interval, the length of time elapsing between a change in a purported cause, and the associated change or effect on an endogenous variable. (An endogenous variable is any variable in a model or system of equations for which antecedents are hypothesized.) Traditionally, the causal interval in psychological, laboratory experiments has been assumed to be the length of the experimental task. This condition in confirmatory analysis requires researchers to explicitly consider the time it takes for one variable to effect a change in another variable.

Causal direction introduces alternative directions over those traditionally explored or tested in laboratory experiments. Typically, researchers have assumed, in their designs, unidirectional causation; however, many psychological theories postulate nonrecursive relationships. Nonrecursive relationships are bidirectional, either cyclical or reciprocal. That is, two variables may mutually and simultaneously influence each other in the case of a reciprocal relationship, or a later occurring variable may influence a prior variable through feedback loops. Self-loops in which a variables essentially causes itself are handled in structural equations analysis by either specifying a cyclical relationship (i.e., A causes B that in turn causes A) or using a longitudinal time series design.

One distinction, then, between reciprocal and cyclical relationships is the length of the causal interval theorized between two endogenous variables. This explains the need to relabel and extend the common distinction between independent and dependent variables, because in a structural model a given variable can be both a cause and an effect. Hence, any variable in a system of equations for which antecedents are hypothesized is referred to as an endogenous variable, and any variable for which the model does not attempt to explain its occurrence, no antecedents are hypothesized, is referred to as an exogenous variable.

The fifth condition enumerated by James *et al.* (1982) states that the functional equations specified in the model must be self-contained. To meet this condition, it is necessary to include all relevant causes of the endogenous variables included in the hypothesized model. A relevant cause is one that has at least a moderate and unique effect on an endogenous variable and is correlated with the other causes of that endogenous variable that have been included in the model. Failure to meet this condition results in biased estimates of the structural parameters because the effects of the unmeasured, relevant causes are included in the disturbance term for that equation (James, 1980).

In experimental work, the condition of self-containment is assumed to have been met by random assignment of subjects to treatments. Thus any extraneous effects from other variables are assumed to be nullified (Cook & Campbell, 1979). This condition is probably the most frequently overlooked by researchers in using structural equation analysis in nonrandom assignment designs. Models appearing in the literature that contain only a few variables and where subjects were not randomly assigned to condition rarely can be considered to have met the condition of self-containment. Of course, one cannot meet this condition in an absolute sense unless one knows all the possible causes of the endogenous variables, and if one had this knowledge, there would be no need to conduct the test of the structural model. Thus, in developing a structural model, one should attempt to reasonably meet this condition based on theory and prior empirical investigations.

In conjunction with trying to insure that the model meets the condition of self-containment, it is necessary to consider the identification of the model. Identification, simply defined, is whether there is enough information in the system of equations to estimate the structural coefficients. If one specifies a model in which all constructs of a lower causal order influence directly all subsequent constructs, the model will be, at best, just identified. If the model is just identified, the model will fit the data perfectly. In essence, all possible degrees of freedom have been given up, and the model is not testable. Therefore, to develop a testable model, certain variables must be hypothesized not to directly influence some of the endogenous variables. This results in an overidentified model and allows one to test the goodness of fit of the model to the data.

Although determining identification is an extremely complex process, several relatively simple techniques have been offered to determine whether a model or system of equations is identified (see Kenny, 1979; Blalock, 1971; Duncan, 1975; Goldberger & Duncan, 1973; Joreskog & Sorbom, 1983; Bentler, 1985; James et al., 1982). A general guide one might adopt is that each endogenous variable should have at least one unique cause within the model unless the endogenous variable is involved in a reciprocal relationship with another endogenous variable.

In order for a nonrecursive model, a system containing reciprocal, feedback relationships, to be identified, it is necessary to include "at least one instrument for each endogenous variable involved in a reciprocal relationship" (James et al., 1982, pp. 40–41). An instrument is an exogenous variable that affects only one of the endogenous variables involved in the reciprocal relationship (James & Singh, 1978). With only one instrument per endogenous variable, the system of equations for the two endogenous variables involved in the reciprocal relationship would be just-identified. To overidentify the model and therefore be able to test the hypothesized relationships, one then would need to include at least two instruments for each of the endogenous variables involved in the reciprocal relationship.

None of these methods are necessary and sufficient conditions for identification, but they may be the only reasonable approach for most researchers. Further discussion of issues of identification are beyond the scope of this chapter. The reader is referred to the preceding references, as well as more advanced discussions of identification for the mathematically sophisticated as found in Johnston (1972) and Judge, Griffiths, Hill, Lutkepohl, and Lee (1985).

The sixth condition for developing a model appropriate for confirmatory analysis is specifying the context within which the functional relations are expected to generalize. This condition is equivalent to consideration of the generalizability of findings of any experimental investigation. Kenny (1979) indicated that one has met this condition if there is linearity in the parameters and variables and the functional equations are not contingent on other variables.

Kenny and Judd (1984) and Rindskopf (1984) have proposed techniques for handling nonlinear and interactive effects in structural equation analyses. The logic underlying the treatment of nonlinear and interactive effects is analogous to curvilinear and moderated hierarchical regression analysis (Pedhazur, 1982). One creates the appropriate polynomial or cross-product terms among the exogenous variables and introduces these terms into the system of structural equations. Moderator effects also can be handled using subgroup analysis where appropriate (Magidson, 1979).

The last condition for consideration in developing a model appropriate for confirmatory analysis is the stability of the structural model. This implies that the values of the structural parameters are stationary over the specified time horizon and that the variables are in equilibrium. Thus, given the causal interval specified in considering the causal order and direction of the hypothesized relationships, one must allow sufficient time for a change in one variable to work its way throughout the system of structural equations when collecting data. Depending on the proposed time horizon, this may necessitate longitudinal designs or, if one assumes that changes have a very short causal interval such as in reciprocal relationships, cross-sectional designs may be sufficient (James et al., 1982). Again, the importance of the time required for a change in one variable to complete its influence on all subsequent hypothesized relationships is apparent. Treatment of longitudinal designs such as time series or panel designs requires special consideration of disturbance terms, instrumentation effects, and stability (Arminger, 1986; Gollob & Reichardt, 1987; Hertzog & Nesselroade, 1987; Joreskog, 1979; Joreskog & Sorbom, 1977; Sorbom, 1979).

Presentation of the Structural Model

By convention, one usually draws a pictoral representation of the hypothesized relationships as one develops the theoretical, structural model. The constructs are connected by arrows representing the proposed direct influence of one construct on another. For example, if one proposed that X, an exogenous variable, causes Y, an endogenous variable, which in turn causes Z, another endogenous variable, the structural diagram would look as follows: $X \rightarrow Y \rightarrow Z$. Each arrow represents a hypothesized direct effect. Implicit in this model also is the hypothesized lack of a direct effect of X on Z (there is no arrow from X to Z). This model proposes that any covariation between X and Z is a result of the indirect effect of X on Z as mediated by Y. If one hypothesized a bidirectional, reciprocal relationship between two variables, two single-headed arrows would be used (e.g., $X \rightleftarrows Y$). Correlation is depicted by double-headed arrows on a curved line and typically reflects covariance among exogenous variables, those variables included in the model for which no causal explanation is being attempted.

Associated with each endogenous variable, all variables with an arrow pointing toward them other than double-headed curved arrows, is a disturbance term. The disturbance term includes all other causes of the variable that are not explicitly included in the model as well as random shock. If one reasonably has met the condition of self-containment, the disturbance is assumed to be uncorrelated with the antecedents of the endogenous variable included in the model. However, one must consider whether the disturbance terms for different endogenous variables are correlated. This depends on substantative and research design considerations. For example, the disturbance terms for two variables that are reciprocally related would be correlated. Likewise, if the two endogenous variables are the same construct measured at two different points in time, the disturbance terms unless consisting only of random shock would be hypothesized to be correlated. Because a researcher cannot directly measure the disturbance term, hypothesizing covariance among disturbance terms increases the likelihood of identification problems.

When one completes the schematic representation of the hypothesized relationships, it is wise to review the model to verify that the previously stated conditions have been met. If one is satisfied that the model reasonably meets the conditions for submitting a model to confirmatory analysis, one is ready to operationalize the constructs reflected in the model.

Figure 1 presents a schematic representation of a hypothetical model among six constructs. The double-headed curved arrows among Construct A, Construct B, Construct C, and Construct D indicate that these four exogenous variables are correlated, but the source of this covariance is not explained by the model. Construct E is influenced by Constructs F, A, and B, whereas Construct F is influenced by Constructs E, C, and D. All other causes of Constructs E and F are represented by d_E and d_F (the disturbance terms). Constructs E and F are hypothesized to be reciprocally related; these two constructs mutually and simultaneously influence each other.

Assuming that we have a formal theory to explain the relationships depicted in this hypothetical model and a theoretical rationale explaining the mediating mechanisms by which Construct A influences Construct E, and so forth, the first two conditions may be considered to have been met. For example, if Construct E represents perceived role stress, a cognition, and Construct F represents job tension or anxiety, an affective response, one might use the following as the theoretical ra-

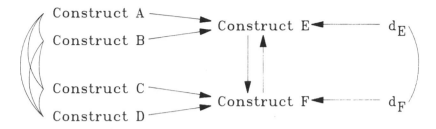

Figure 1. Schematic representation of a hypothetical structural model.

tionale for the reciprocal relationship between these two constructs: The relationship between affect and cognition has been demonstrated to be one of reciprocal causation (James & Jones, 1980; James & Tetrick, 1986; Lazarus, 1982; Zajonc, 1980). The more apprehensive one is concerning his/her organizational well-being, that is, the more job tension one is experiencing, the less efficient one is in processing information (Organ, 1981). The less efficient one is in processing information, the greater the perception of stress would be. Therefore, it is hypothesized that not only does stress cause job tension, but job tension causes the perception of stress.

The condition of self-containment is not apt to have been reasonably satisfied in such a simple model as reflected in Figure 1. Depending on what Construct E and F represent, there are probably other variables that have a major, unique effect on these two constructs and also are correlated with the causes depicted in the model, Constructs A, B, C, and D. That is, if there is another variable known to influence Construct E that is correlated with Constructs A, B, or F, the equation is not self-contained, and hence the model is not self-contained. Any such extraneous variables should be included in the model at this stage, before the model is operationalized, data collected, and the fit of the model tested. For purposes of this chapter, it will be assumed that this simple model as shown in Figure 1 is self-contained as is.

The remaining consideration, at this point, is the question of identification. Because the model contains a reciprocal relationship between Constructs E and F, it is necessary to have at least two instruments for each endogenous variable involved in the reciprocal relationship to overidentify these equations and therefore be able to test the relationship hypothesized between Constructs E and F. Constructs A and B represent instruments for Construct E in that they are hypothesized to have a direct effect on Construct E but are not hypothesized to have a direct effect on Construct F. Likewise, Constructs C and D are hypothesized to be instruments for Construct F. Therefore, the model is overidentified and subject to test.

The model as presented assumes linear, additive relationships. It also is necessary to assume that the system is in equilibrium. The data collection procedures must be aligned with the theoretical causal interval such that given a change in the exogenous variables, the effect of such change has progressed through the model to the point where the variables are again stationary. Theory and prior research must be relied upon to justify meeting these two conditions.

At this point, one can write the functional equations for the relationships in the hypothetical model. Because there are two endogenous variables in the model shown in Figure 1, there will be two functional equations:

$$E = b_E + b_{EF} + b_{EA} A + b_{EB} B + b_{EC} C + b_{ED} D + d_E$$
$$F = b_F + b_{FE} E + b_{FA} A + b_{FB} B + b_{FC} C + b_{FD} D + d_F$$

where the capital letters refer to the corresponding construct in the model, the d's refer to the disturbance terms, and the b's refer to the structural parameters.

One should note that in the first equation b_{EC} and b_{ED} are hypothesized to be zero as are b_{FA} and b_{FB} in the second equation. Thus for the model in Figure 1, there is a system of two functional equations with two parameters in each equation to be estimated, b_{EA} and b_{EB} in the first equation and b_{FC} and b_{FD} in the second equation. If one then takes expectations and derives the normal equations for the model, one reduces the functional equations to equations for covariances among the variables. Unique solutions for the unknown parameters, b's, can be obtained based on the overidentifying restrictions in the model, those parameters hypothesized to be zero. The fit of the model is based on how well these reproduced covariances with restrictions approximate the observed covariances (see Duncan, 1975; James et al., 1982; Kenny, 1979, for presentation of the mathematics involved in deriving the normal equations).

Operationalization of the Model

At this stage, it is necessary to decide how one is going to measure the constructs in the model. As in all research it is important to select reliable, valid measures. In addition, one has the option of using a latent variable approach, a manifest variable approach, or a combination approach using both latent and manifest variables. In the latent variable approach, one selects multiple indicators or measures of each

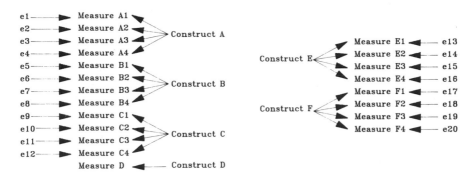

Figure 2. Hypothetical measurement model for Figure 1.

construct in the model. The variances and covariances among the latent variables are then computed based on the common variance among the multiple indicators for each construct. In the manifest variable approach, one selects only one observed variable to represent the construct and then treats it as a perfectly reliable and valid measure of the theoretical construct. In the combination approach, some constructs have multiple indicators and others have only a single indicator.

Figure 2 is a graphic representation of the measurement model for the hypothesized structural model in Figure 1. All latent, theoretical constructs are allowed to be correlated with each other, although the curved arrows have been omitted in the figure for clarity. Each construct has four observed, measured variables indicating the construct with the exception of Construct D. Construct D is treated as a manifest variable; there is a direct 1 to 1 correspondence between the actual observed variable and the underlying construct. Figure 2 also reflects a unique component for each measure, except measure D, and that these unique components are orthogonal.

With some visual integration of Figures 1 and 2, it can be seen that using multiple indicators of theoretical constructs increases the overidentifying conditions for the full model (structural model plus measurement model). However, these overidentifying conditions serve to test the measurement model and do not satisfy the need for overidentification in the structural model to be able to test the hypothesized theoretical relationships among the constructs of interest.

Test of the Measurement Model

As in any research study, after one obtains the data one must check the adequacy of the measurement model. In a fully manifest variable approach, one obtains estimates of reliability but does not typically have the ability to test the adequacy of the measurement model in total. Use of the latent variable approach allows one to test the overall adequacy of the measurement model by determining whether the manifest indicators are representative of the specified underlying structure of common factors as hypothesized, giving one some evidence of construct validity as well as reliability. Although it is possible to test the measurement model simultaneously with the structural model, it has been recommended that one test the measurement model separately and then proceed with the test of the structural model using the covariances or correlations among the latent variables and any manifest exogenous variables (Geraci, 1977; Hsiao, 1976; Hunter & Gerbing, 1982). This procedure allows one to test the structural model without the influence of random measurement error on the structural parameters associated with the manifest indicators and without excessive use of computer resources (Boardman, Hui, & Wold, 1981; Hausman, 1977; Hsiao, 1976; Hunter & Gerbing, 1982).

Testing the measurement model is done by performing a confirmatory factor analysis of the data. As Long (1983) indicates, confirmatory factor analysis allows one to specify which observed variables (manifest indicators) are related to which hypothesized latent variables or common factors and which common factors are correlated with other common factors. Also, one can specify which observed variables are affected by a unique factor and which unique factors are correlated.

The manner in which one specifies the constraints, that is, which observed variables are a function of which latent constructs depends on the hypothesized measurement model and the statistical program by which one conducts the confirmatory factor analysis. Basically, the latent, theoretical constructs in the structural model are allowed to covary; the covariance among the theoretical constructs is addressed in the test of the structural model. Constraints are placed on the hypothesized covariance matrix such that each manifest indicator is a function of one latent, theoretical construct as hypothesized in the operationalization of the model and a unique component. For constructs that had only one manifest indicator (a manifest variable), one specifies that this observed variable was measured without error (there is no unique component). In addition, the parameter linking the latent construct to the observed variable is fixed to be equal to 1.0 indicating that there is a perfect correspondence between the latent construct and the observed variable.

If one has measured several of the variables using a single method, such as a questionnaire or test, one may specify that the error terms of the manifest indicators are correlated. Or, one can specify an additional latent variable reflecting method variance (Rindskopf, 1984; Royce, 1963; Schmitt & Stults, 1986). This approach then specifies all observed variables measured by the same method to be complex variables influenced by the hypothesized latent theoretical construct and a phantom, latent method variable. Both of these procedures result in a loss of degrees of freedom and increase the likelihood that the measurement model will be underidentified, as one allows more parameters to be estimated without increasing the number of observed variables, or will result in improper solutions (Dillon, Kumar, & Mulani, 1987).

It has been suggested that one have at least four manifest indicators for each latent construct in order to fully test the measurement model (James et al., 1982). If one cannot obtain four indicators for each latent construct, it has been suggested that additional constraints be placed on some of the parameters in the model (Bentler, 1985; Jöreskog & Sörbom, 1988). That is, the parameter linking a latent construct with one of the manifest indicators might be fixed to 1.0. This gains a degree of freedom and sets the metric on the latent variable to that of the manifest indicator whose parameter is fixed to 1.0. The choice of which of the manifest indicators should be selected as the one with closest correspondence to the latent construct is a matter of judgment and knowledge of the psychometric properties of the measures used.

The outcome of the confirmatory factor analysis testing the measurement model provides several indexes of the adequacy of the operationalization of the model depending on the computer program used. The goodness of fit indicators commonly employed are chi-square, Jöreskog and Sörbom's (1988) goodness of fit index and adjusted goodness of fit index, Bentler and Bonett's (1980) normed fit index, and James et al.'s (1982) parsimonious fit index. Although it is likely given the power of chi-square for large samples that the chi-square value will be statistically significant, it has been suggested that one gauge the adequacy of the measurement model based on the chi-square to degrees of freedom ratio or one of the other goodness of fit statistics. The distributional properties of the goodness of fit indexes have not been determined, and therefore the absolute acceptable value of the different fit indexes cannot be stated at this time. As a rule of thumb, the ratio of chi-square to degrees of freedom should not be greater than 5:1. Bentler and Bonett (1980) suggest that if the normed fit index is not at least .90, based on a null model of complete independence, more of the covariance in the data can be accounted for by modifying the model, although it should be noted that they were not referring specifically to tests of measurement models.

If the measurement model does not adequately account for the covariance among the observed variables, it has been suggested that additional parameters other than those based on the hypothesized measurement model should be relaxed, freed to be estimated, in order to improve the fit of the measurement model. One must be careful in so doing as one may actually change the meaning of the theoretical constructs. To continue on then with the test of the structural relations after, in effect, changing the dimensionality of the measurement model is nonsensical. One may no longer be talking about the same constructs, and the theoretical rationale for the structural relations may no longer apply. Therefore, as a rule of thumb, if the adjusted goodness of fit or normed fit indexes of the hypothesized measurement model are not close to .80 or higher and improvement in fit of the measurement model can be obtained only by allowing certain manifest indicators to be a function of more than the one hypothesized latent construct, testing of the structural relationships among the hypothesized constructs should not proceed.

Test of the Structural Relations

Once one has obtained the covariances among the latent variables and any manifest variables, one is in a position to empirically test the predictions based on the causal hypotheses reflected in the structural model. Implicit in the system of structural equations for the model are two types of hypotheses: (1) that certain structural parameters of the model are statistically different from zero and (2) that all other possible structural parameters are not statistically different from zero. The first type of hypotheses is reflected by arrows linking constructs in the model, and the second type is reflected by the absence of arrows directly linking two constructs. Testing the structural model involves evaluating both types of hypotheses. If any parameter hypothesized to be nonzero is found not to be statistically significant or if any path between two variables hypothesized to be zero is found to be statistically significant, the model is disconfirmed. Any theory trimming or revisions to the model to improve the fit of the model must be recognized as exploratory. Until the revised model is tested with new data, the usefulness of the revised, best-fitting model in explaining the relationships among the constructs is undetermined.

Two general approaches may be used to test the structural model. One method is to use a simple regression or path analytic approach, computing a regression equation for each structural equation in the model at a time. This provides estimates of the hypothesized nonzero parameters, and statistical significance of the regression coefficients supports the hypothesized relationships. Then one can regress the disturbance terms as estimated from the first set of regression analyses against the variables excluded from each equation to test whether these paths are zero as hypothesized (James & Singh, 1978). If one has hypothesized reciprocal relations among endogenous variables, ordinary least squares regression cannot be used to estimate the parameters, and one must use two-stage least squares to adjust for the fact that the disturbance terms are correlated.

The advantage of this procedure is that misspecification errors influence only those equations directly involved in the misspecified relations since the parameters are estimated one equation at a time. However, this approach does not readily give one an indication of the overall goodness of fit of the model, and the parameter estimates are sensitive to multicollinearity, the presence of strong relationships among explanatory variables (Dutta, 1975; Intriligator, 1978; Johnston, 1972). Multicollinearity is especially troublesome if one has tried to include all relevant causes of the endogenous variables to satisfy the condition of self-containment.

The full-information or simultaneous estimation procedure as used in LISREL (Jöreskog & Sörbom, 1983) and EQS (Bentler, 1989) is the other method for testing the hypothesized structural relationships reflected in the model. The parameter estimates are based on the entire system of equations, and the fit function is minimized conditionally on all a priori identifying restrictions or constraints specified by the model being tested.

The full information procedures have the advantage over the preceding approach in that one obtains an overall goodness of fit statistic by which to gauge how well the hypothesized system of equations explains the sample covariance matrix. By the use of full information procedures, multicollinearity is less of a problem. It also is possible to place restrictions on the error structure in ways not possible with the regression approach (Intriligator, 1978; Johnston, 1972). This feature can be extremely important if one encounters problems with identification (Jöreskog and Sörbom, 1988). Full information maximum likelihood also has the advantage over two-stage least squares that the estimates are asymptotically efficient as well as consistent. The disadvantage of full information estimation procedures is that they are extremely sensitive to misspecification error. If a relationship is misspecified, all parameter estimates may be affected unlike the regression approach that confines specification error to the equations involved in the misspecification.

Testing the support of the model using full information methods of estimation can be done by comparing the goodness of fit among an a priori nested sequence of models, each model being more restricted than the previous model. The progression of nested models from the just-identified model to the hypothesized model constitutes an empirical confirmation of predictions that certain parameters are zero, whereas the sequence of nested models from the structural model to the null model constitutes a test that the parameters hypothesized to be nonzero are in fact significantly different from zero. The test of the difference between each pair of models in the sequence is given by the goodness of fit function of the more restricted model less the goodness of fit function for the less restricted model (Bentler & Bonett, 1980; James et al., 1982; Roy, 1958). Because the comparison between nested models is a test of the overidentifying restrictions and the sequence is terminated when a null hypothesis has been rejected, it is important to arrange the testing of hypotheses about parameters in the order of their relative importance.

In addition to the a priori nested sequence test, it is advisable to compute a normed fit index (Bentler & Bonett, 1980) or a parsimonious fit index (James et al., 1982). These indexes give an indication of the relative decrease in lack of fit between two nested models (Mulaik, James, Van Alstine, Bennett, Lind, & Stillwell, 1989). Typically one computes the fit index comparing the model of interest with the null model, that model that constrains all constructs to be orthogonal or unrelated. However, the researcher may use a somewhat less restricted model other than the null model as the standard to compare other models. For example, if one has included several exogenous variables in the model that are known to be correlated, it may be more appropriate not to restrict the correlations among these variables in the null model.

Table 1 reflects the indicators of fit for a series of nested models starting with an hypothesized model and freeing one additional parameter in each successive model. In this system of equations there were five endogenous variables labeled eta_1, eta_2, eta_3, eta_4, and eta_5 as contrasted to only two for the model shown in Figure 1. The series of nested models reflects freeing additional parameters, allowing additional exogenous variables to directly influence endogenous variables other than those originally hypothesized; thus Model I differs from the proposed model by freeing GA(2,9) to be estimated thereby allowing the exogenous variable X_9 to directly affect eta_2. By freeing up this parameter, one loses one degree of freedom, and the chi-square value drops from 432.22 to 347.33. This difference chi-square is statistically significant indicating the parameter is significantly different from zero contrary to what was hypothesized. The goodness of fit index given by LISREL increased from .960 to .967, and the adjusted goodness of fit index increased from .800 to .833. The normed fit index, based on a

Table 1. Sequential Tests of Goodness of Fit

		df	χ^2	Goodness-of-fit indices			
				GFI	AGFI	NFI	PFI
Proposed model		81	432.22	.960	.800	.630	.444
Less restricted models							
Model I	Free GA(2,9)	80	347.33	.967	.833	.703	.489
Model II	Free GA(2,15)	79	306.56	.971	.850	.738	.507
Model III	Free GA(1,19)	78	279.01	.973	.859	.761	.516
Model IV	Free GA(1,8)	77	251.82	.975	.870	.785	.525
Model V	Free GA(1,11)	76	224.13	.978	.881	.808	.534
Model VI	Free GA(4,12)	75	208.52	.979	.888	.822	.536
Model VII	Free GA(1,13)	74	195.85	.981	.893	.833	.536
Model VIII	Free GA(5,6)	73	183.22	.982	.898	.843	.535
Model IX	Free GA(3,2)	72	169.83	.983	.904	.855	.535
Model X	Free GA(3,14)	71	156.52	.984	.909	.866	.535
Model XI	Free GA(1,28)	70	144.95	.985	.915	.876	.531
Model XII	Free GA(1,22)	69	134.22	.986	.920	.885	.531
Model XIII	Free GA(3,24)	68	124.73	.987	.924	.893	.528
Model XIV	Free GA(3,16)	67	117.32	.988	.927	.900	.524
Model XV	Free GA(3,18)	66	111.95	.989	.929	.904	.519
Model XVI	Free GA(4,25)	65	106.60	.989	.932	.909	.514
Model XVII	Free GA(4,21)	64	101.29	.990	.934	.913	.508
Model XVIII	Free GA(2,17)	63	96.61	.990	.936	.917	.503
Model XIX	Free GA(5,16)	62	92.14	.991	.938	.921	.497
Model XX	Free GA(2,11)	61	87.79	.991	.940	.925	.491

null model allowing the exogenous variables to be correlated rather than the complete independence model, rose from .630 to .703, and the parsimonious fit index rose from .444 to .489.

As additional parameters are freed to be estimated, chi-square continues to decrease significantly. All the fit indexes continue to improve except the parsimonious fit index. The parsimonious fit index levels off around Model VI and VII and then begins to decrease. This suggests that freeing additional parameters beyond this point does not improve the fit of the model relative to the loss in degrees of freedom. Thus the most parsimonious, best-fitting model would be Model VI. It should be noted that the first step from the proposed model to Model I results in a disconfirmation of the model because a path hypothesized to be zero was found to be significantly different from zero. All subsequent revisions to the model, theory trimming, is exploratory, and such specification searches may or may not lead to discovery of the correct population model (MacCallum, 1986).

Computer Programs for Testing the Model

LISREL 7 (Jöreskog & Sörbom, 1988) is presently the most widely used program for testing systems of linear structural equations. This program has been publicly available for several years and currently is marketed by SPSS. LISREL is a general purpose program allowing the user to perform simple multiple regression, confirmatory factor analysis, and analysis of complex systems of structural equations (Jöreskog & Sörbom, 1988; Magidson, 1979).

The fundamental equation of LISREL is one which partitions the covariance matrix into functions of the elements in eight parameter matrices: lambda Y, lambda X, beta, gamma, phi, psi, theta epsilon, and theta delta. Lambda Y and Lambda X are the parameters relating the observed variables to the hypothesized latent constructs. If one is using a manifest variable approach or has extracted the covariance among the latent variables in a test of the measurement model and is then testing the structural relations separately, lambda X and lambda Y become identity matrices. Similarly, theta epsilon and theta delta represent the unique components of the observed variables. If one has a fully manifest model, implying a 1 to 1 correspondence between the observed variable and the latent construct, these matrices are zero. If one specifies the unique components are uncorrelated as in Figure 2, these matrices are diagonal matrices. The modeler fixes the off-diagonal elements representing covariance among these unique components to be equal to zero.

The phi matrix represents the correlation among the exogenous variables. Gamma represents the effects of exogenous variables on endogenous variables, and beta represents the effects of endogenous variables on exogenous variables. Psi represents unaccounted-for variance in the endogenous variables on the diagonal, and the off-diagonal represents covariance among the residuals. Whether any or all of the elements in these matrices are constrained or fixed to be 0, 1, or some other value depends on the model developed.

For example, based on the structural relations depicted in Figure 1, beta would be a 2×2 matrix in which the diagonal elements are zero indicating a variable does not cause itself. Both off-diagonal elements would be free to be estimated because it is hypothesized that Construct E causes Construct F and Construct F causes Construct E. The gamma matrix would be of order 2×4. The two rows of the gamma matrix represent the hypothesized effects of the exogenous variables on the two endogenous variables. Consistent with the hypothesized relationships, the first two elements in Row 1 associated with Constructs A and B, respectively, would be free to be estimated, whereas the last two elements would be fixed to zero indicating that there is no hypothesized, direct effect of Constructs C and D on Construct E. The second row would be just the opposite. The first two elements would be fixed to zero, and the last two elements would be free to be estimated. Psi would be symmetric and free to be estimated because the disturbance terms for the two endogenous variables would be correlated.

Beyond specifying which elements in these eight parameter matrices are to be estimated, constrained to be equal to one another, or fixed to some value, the user of LISREL specifies one of five different methods of parameter estimation: instrumental variable approach, two-stage least squares, unweighted least squares, generalized least squares, and maximum likelihood. The first two methods are noniterative and therefore require less computer time than the last three methods. Additional estimation procedures, as well as tests of multivariate normality are available through PRE-LIS. The reader is referred to the LISREL manual (Jöreskog & Sörbom, 1988) as well as other sources such as Judge et al. (1985) for detailed discussion as to which estimation procedures may be most appropriate given the data and model being tested.

LISREL automatically produces chi-square, goodness of fit indexes, and coefficients of determination to aid the user in assessing the goodness of fit of a model. In addition, the program generates the matrices of parameter estimates, standardized solutions for the parameter estimates, matrices of modification indexes that signify the reduction of chi-square if a parameter were freed to be estimated in a subsequent run, t values and standard errors for the parameter estimates, the residual covariance matrix (the sample covariance matrix less the estimated covariance matrix based on the model), factor score regression, covariances among the parameter estimates, and total effects (the sum of the direct and indirect effects) of exogenous variables on endogenous variables.

EQS (Bentler, 1985) recently has been publicly released through BMDP. The model specification used in this program is based on the Bentler–Weeks model (Bentler & Weeks, 1980). In this model there are three parameter matrices. Beta contains the coefficients reflecting the effect of endogenous variables on other endogenous variables. Gamma elements reflect the effects of exogenous variables on endogenous variables, and phi represents the covariance matrix of the exogenous variables.

The user of EQS specifies all the equations contained in the system of equations to the program. The program then orders the vectors of endogenous and exogenous variables. As in LISREL, the user must specify which parameters are free to be estimated or constrained (fixed to zero, some other value than zero, or constrained to be equal to other parameters or a linear combination of other parameters). This is done as one specifies the equations.

EQS offers seven parameter estimation methods. These are least squares, generalized least squares, maximum likelihood, elliptical least squares, elliptical generalized least squares, elliptical reweighted least squares, and arbitrary distribution generalized least squares. The elliptical estimation procedures automatically generate estimates based on multinormal theory as well thus allowing the user to compare the estimates based on distributional assumptions. The reader is referred to the EQS manual (Bentler, 1989) and the references contained therein as well as those mentioned previously for a more detailed discussion of these parameter estimation methods.

EQS produces much the same information as LISREL. The parameter estimates are shown in equation form rather than in matrices. The residual covariance matrix is produced automatically. In addition to the chi-square indicator of fit, Bentler and Bonett's (1980) normed and nonnormed fit indexes are provided. Coefficients of determination are not computed but can be calculated from the information given for recursive models. The standardized solution provided by EQS is different from that provided by LISREL, as this program also standardizes measured variables, errors in variables, and disturbances in equations (Bentler, 1989). EQS provides information concerning multivariate normality and identification of outliers if one inputs the raw data matrix. A path diagram of the model also can be requested.

Interpretation

Interpretation of models submitted to structural equation analysis is dependent on two primary considerations. If one has reasonably satisfied the conditions for causal analysis and the model is empirically supported, causal inferences may be made. Empirical support means that all hypothesized relationships, nonzero and zero relationships, are supported. If a model received support, it does not, however, indicate that this is the only model that could account for the data. To the extent the model is disconfirmed, one or more hypothesized relationships are not supported, and the results can be interpreted only in correlational terms. The theory trimming resulting in a revised, most parsimonious, best-fitting model must be validated on another sample before one can draw causal inferences. If the model does not meet the conditions for confirmatory analysis or was not supported, the model still may be helpful in theory development or policy evaluation.

Interpretation of the parameter estimates is much like interpretation of beta weights in regression analysis. Depending on the size and sign of the parameter estimate, one can conclude the relative importance or impact of variables on the endogenous variables and whether a change in one variable will increase or decrease another variable. The size of the effect is dependent on the absolute magnitude of the parameter estimates and the total amount of the variance accounted for in a given endogenous variable by the model.

If one found that the model depicted in Figure 1 was supported, the following conclusions could be drawn:

1. Construct E and Construct F have a mutual and simultaneous effect on each other. If the signs of the parameter estimates are positive, then a change in Construct E results in an increase in Construct F, whereas Construct F simultaneously increases Construct E until equilibrium is reached.

2. In addition to the effect of Construct F, Construct E is directly affected by Constructs A and B. If Construct A and/or B change, Construct E will increase or decrease depending on the sign of the parameters linking Constructs A and B to Construct E. Because Construct E changes, Construct F also will change. Constructs A and B are said to have indirect effects on Construct F; however, the effects of Constructs A and B are completely mediated by Construct E.

3. Similarly, Construct F is directly effected by Constructs C, D, and E. Changes in any of these three constructs result in a change in Construct F. Constructs C and D indirectly affect Construct E, their effect being mediated by Construct F.

Let, for illustrative purposes, Construct A be the leader's upward influence in the organization, Construct B be the leader's controlling behaviors toward subordinates, Construct C be the degree of standardization of personnel policies, Construct D be the subordinates' locus of control, Construct E be perceived leader support, and Construct F be perceived role stress. This hypothetical model is reflected in Figure 3. (Note the double-headed arrows indicating that the exogenous variables are correlated have been omitted for clarity.) One could then interpret the model as confirming that perceptions of leader support are reciprocally related to perceived role stress. Changes in either perceptions of support or role stress will mutually and simultaneously influence each other, thus maintaining cognitive consistency. In addition to perceived role stress, the leaders' upward influence in the organization and controlling behaviors directed at subordinates influence the perception of leader support. The degree of standardization of personnel

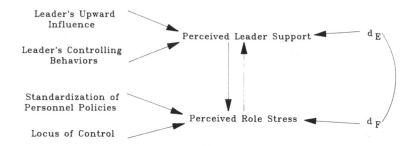

Figure 3. Hypothetical structural model for illustration purposes.

policies and the subordinates' locus of control directly affect perceptions of stress in addition to perceptions of leader support. The leaders' upward influence and controlling behaviors affect stress only as mediated by perceptions of leader support. Standardization of personnel policies and locus of control indirectly affect perceptions of leader support as mediated by perceptions of stress.

By assessing the relative size of the standardized parameter estimates, the sign of the parameter estimates, and the ability to manipulate the construct, policymakers or change agents can use the model to formulate intervention programs. Considering the hypothetical example given, leaders' controlling behaviors toward subordinates and the degree of standardization of personnel policies might be considered to be more easily manipulated than an individual's locus of control or a leader's upward influence. Standardization of personnel policies has a direct effect on stress, whereas leaders' controlling behaviors only have an indirect effect. Therefore, the model suggests to the policymaker or change agent interested in reducing stress that increasing the degree of standardization of personnel policies (assuming the sign of the parameter linking standardization to stress is positive) would result in a more direct impact on role stress than changing leaders' behaviors.

The degree of impact of a given intervention is dependent on the size of the total effect for the variable that is manipulated on the variable of interest. The total effect of a variable is the sum of the direct effects, if any, and the indirect effects. In this hypothetical example, leaders' controlling behaviors toward subordinates do not have any direct effect on perceptions of stress. However, leaders' controlling behaviors do have an indirect effect. If the parameter estimates for the indirect paths (the path from leaders' controlling behaviors to perceptions of leader support and the path from perceptions of leader support to perceptions of stress) are quite large compared to the direct path for standardization, the policymaker may be advised to focus on changing leaders' behaviors. By examining the direct and indirect effects in this way, policymakers and change agents can simulate alternative plans of action.

Summary

This chapter presented the logic and design considerations that should be followed in developing and testing causal models. If one can develop a model based on theory that meets the conditions outlined, one then selects measures of the theoretical constructs contained in the model and obtains sample data upon which to test the goodness of fit of the model. Testing the model is a two-step process. First, one must demonstrate the adequacy of the measurement model. Then, if the measurement model is adequate, one tests the structural relationships. This step includes empirical verification that the parameters hypothesized to be nonzero are significantly different from zero and that the parameters hypothesized to be zero are not significantly different from zero. If all hypothesized relationships are empirically supported, the model is supported. If any hypothesized relationship is not empirically supported, the model is disconfirmed.

References

Arminger, G. (1986). Linear stochastic differential equation models for panel data with unobserved variables. In N. B. Tuma (Ed.), *Sociological methodology, Vol. 16* (pp. 187–212).

Bentler, P. M. (1989). *EQS: Structural equations program manual.* Los Angeles: BMDP Statiscal Software, Inc.

Bentler, M. P., & Bonett, D. G. (1980). Significance tests and goodness of fit in the analysis of covariance structures. *Psychological Bulletin, 88,* 588–606.

Bentler, P. M., & Weeks, D. G. (1980). Linear structural equations with latent variables. *Psychometricka, 45,* 289–308.

Blalock, H. M. (Ed.). (1971). *Causal models in the social sciences.* Chicago: Aldine-Atherton.

Boardman, A. E., Hui, B. S., & Wold, H. (1981). The partial least squares-fit point. *Communication in Statistics: Theory and Methods, A10,* 613–639.

Cook, T. D., & Campbell, D. T. (1979). *Quasi-experimentation: Design and analysis issues for field settings.* Boston: Houghton-Mifflin.

Dillon, W. R., Kumar, A., & Mulani, N. (1987). Offending estimates in covariance structure analysis: Comments on the causes of and solutions to Heywood cases. *Psychological Bulletin, 101,* 126–135.

Duncan, O. D. (1975). *Introduction to structural equation models.* New York: Academic Press.

Dutta, M. (1975). *Econometric methods.* Cincinnati: South-Western Publishing Co.

Geraci, V. J. (1977). Estimation of simultaneous equation models with measurement error. *Econometrica, 45,* 1243–1255.

Goldberger, A. S., & Duncan, O. D. (1973). *Structural equation models in the social sciences.* New York: Seminar Press.

Gollob, H. F., & Reichardt, C. S. (1987). Taking account of time lags in causal models. *Child Development, 58,* 80–92.

Hausman, J. A. (1977). Errors in variables in simultaneous equation models. *Journal of Econometrics,* 389–401.

Hertzog, C., & Nesselroade, J. R. (1987). Beyond autoregressive models: Some implications of the trait-state distinction for the structural modeling of developmental change. *Child Development, 58,* 93–109.

Hsiao, C. (1976). Identification and estimation of simultaneous equation models with measurement error. *International Economic Review, 17,* 319–339.

Hunter, J. E., & Gerbing, D. W. (1982). Unidimensional measurement, second order factor analysis, and causal models. In B. M. Staw & L. L. Cummings (Eds.), *Research in Organizational Behavior,* Vol. 4 (pp. 267–320).

Intriligator, M. D. (1978). *Econometric models, techniques, and applications.* Englewood Cliffs, NJ: Prentice-Hall.

James, L. R. (1980). The unmeasured variables problem in path analysis. *Journal of Applied Psychology, 65,* 415–421.

James, L. R., & Jones, A. P. (1980). Perceived job characteristics and job satisfaction: An examination of reciprocal causation. *Personnel Psychology, 33,* 97–135.

James, L. R., & Singh, B. K. (1978). An introduction to the logic, assumptions, and basic analytic procedures of two-stage least squares. *Psychological Bulletin, 85,* 1104–1122.

James, L. R., & Tetrick, L. E. (1986). Confirmatory analytic tests of three causal models relating job perceptions to job satisfaction. *Journal of Applied Psychology, 71,* 77–82.

James, L. R., Mulaik, S. A., & Brett, J. M. (1982). *Causal analysis: Assumptions, models, and data.* Beverly Hills: Sage Publications.

Johnston, J. (1972). *Econometric methods* (2nd ed.). New York: McGraw-Hill Book Company.

Jöreskog, K. G. (1979). Statistical estimation of structural models in longitudinal developmental investigations. In J. R. Nesselroade & P. G. Baltes (Eds.), *Longitudinal research in the study of behavior and development* (pp. 303–351). New York: Academic Press.

Jöreskog, K. G., & Sörbom, D. (1977). Statistical models and methods for analysis of longitudinal data. In D. J. Aigner & A. S. Goldberger (Eds.), *Latent variables in socio-economic models* (pp. 285–325). Amsterdam: North-Holland.

Jöreskog, K. G., & Sörbom, D. (1988). *LISREL 7: A guide to the program and applications.* Chicago: SPSS, Inc.

Judge, G. G., Griffiths, W. E., Hill, R. C., Lutkepohl, H., & Lee, T. (1985). *The theory and practice of econometrics.* New York: John Wiley & Sons.

Kenny, D. A. (1979). *Correlation and causality.* New York: John Wiley & Sons.

Kenny, D. A., & Judd, C. M. (1984). Estimating the nonlinear and interactive effects of latent variables. *Psychological Bulletin, 96,* 201–210.

Latack, J. C. (1984). Career transitions within organizations: An exploratory story of work, nonwork, and coping strategies. *Organizational Behavior and Human Performance, 34,* 296–322.

Lazarus, R. S. (1982). Thoughts on the relations between emotion and cognition. *American Psychologist, 37,* 1019–1024.

Long, J. S. (1983). *Confirmatory factor analysis.* Beverly Hills: Sage Publications, Inc.

MacCallum, R. (1986). Specification searches in covariance structure modeling. *Psychological Bulletin, 100,* 107–120.

Magidson, J. (1979). *Advances in factor analysis and structural equation models.* Lanham: Abt Associates, Inc.

Michaels, C. E., & Spector, P. E. (1982). Causes of employee turnover: A test of the Mobley, Griffeth, Hand, and Meglino model. *Journal of Applied Psychology, 67,* 53–59.

Mulaik, S. A., James, L. R., Van Alstine, J., Bennett, N., Lind, S., & Stillwell, C. D. (1989). Evaluation of goodness-of-fit indices for structural equation models. *Psychological Bulletin, 105,* 430–445.

Organ, D. W. (1981). Direct, indirect, and trace effects of personality variables on role adjustment. *Human Relations, 34,* 573–587.

Pedhazur, E. J. (1982). *Multiple regression in behavioral research* (2nd ed.). New York: Holt, Rinehart & Winston.

Powell, G. N. (1984). Effects of job attributes and recruiting practices on applicant decisions: A comparison. *Personnel Psychology, 37,* 721–732.

Price, J. L., & Mueller, C. W. (1981). A causal model of turnover for nurses. *Academy of Management Journal, 24,* 543–565.

Rindskopf, D. (1984). Using phantom and imaginary latent variables to parameterize constraints in linear structural models. *Psychometrika, 49,* 37–47.

Roy, J. (1958). Step-down procedure in multivariate analysis. *Annals of Mathematical Statistics, 29,* 1177–1187.

Royce, J. R. (1963). Factors as theoretical constructs. *American Psychologist, 18,* 522–528.

Schmidt, F. L., Hunter, J. E., & Outerbridge, A. N. (1986). Impact of job experience and ability on job knowledge, work sample performance and supervisory ratings of job performance. *Journal of Applied Psychology, 71,* 432–439.

Schmitt, N., & Stults, D. M. (1986). Methodology review: Analysis of multitrait-multimethod matrices. *Applied Psychological Measurement, 10,* 1–22.

Singh, J. V. (1986). Performance, slack, and risk taking in organizational decision making. *Academy of Management Journal, 29,* 562–585.

Sorbom, D. (1979). Detection of correlated errors in longitudinal data. In J. Magidson (Ed.), *Advances in factor analysis and structural equation models* (pp. 171–184). Lanham: Abt Associates, Inc.

Walker, G., & Weber, D. (1984). A transaction cost approach to make-or-buy decisions. *Administrative Science Quarterly, 29,* 373–391.

Williams, L. J., & Hazer, J. T. (1986). Antecedents and consequences of satisfaction and commitment in turnover models: A reanalysis using latent variable structural equation methods. *Journal of Applied Psychology, 71,* 219–231.

Zajonc, R. B. (1980). Feeling and thinking: Preferences need no inferences. *American Psychologist, 35,* 151–175.

3

Sensitivity of Parameters and Loop Structures

Ralph L. Levine and Weldon Lodwick

Classical Sensitivity Analysis

Sensitivity analysis, which can be performed even before embarking on an extensive empirical time series study, provides a powerful technique for assessing the impact of specific initial values of the state variables, table functions, and the impact of changing parameters of the model on the behavior of the system. The assessment of initial values and table functions has been described elsewhere, especially in Chapters 4 and 5 in Volume 1. The current chapter will emphasize assessing the sensitivity of parameter values that can serve as leverage points for changing policies. Indeed, once leverage points are found, the modeler will then know which parameters must be estimated with a great deal of precision, as opposed to other parameter values that have little effect on the behavior of the system and therefore may not require extensive time and effort in obtaining precise estimates. Note that performing a sensitivity analysis on the model provides information about how robust the system is to changes in policy. If a particular policy is found to positively affect the problem under study, one can use the Policy

Ralph L. Levine • Department of Psychology, Michigan State University, East Lansing, Michigan 48824.   Weldon Lodwick • Department of Mathematics, University of Colorado, Denver, Denver, Colorado 80217-3364.

Analysis of Dynamic Psychological Systems, Volume 2: Methods and Applications, edited by Ralph L. Levine and Hiram E. Fitzgerald. Plenum Press, New York, 1992.

Parameter Sensitivity test to see the range of movement one has in making the change.

Classical approaches to sensitivity analysis attempt to discover the efficacy of changing parameter values around normal operating ranges in which the phenomenon under study has been observed. Sensitivity analysis has been described in detail in the systems literature (e.g., Tank-Nielsen, 1980). In psychological realms, dynamic behavioral models in general will have parameters related to individual differences among subjects, groups, and so forth. Whatever level of aggregation, these parameters are related to the time constants of the system, which in effect are reaction times associated with change in the system variables. For example, if one is interested in how people adjust to being laid off, there is a natural lag involved in processing the impact of not coming to work and dealing with the changes in status, and so forth that must be captured in the model. The lags themselves are usually represented by one or more constants that must be estimated by a numerical or statistical process.

Time Constants and Individual Differences

Sometimes parameters represent the role of individual differences in a model. For example, in a model of attitude change, some individuals might be more persuasible than others to the influence of the same message. Differences in persuasiveness might be captured by a parameter whose units represent the time to adjust to an incoming message, the time constant related to change in attitudes. Thus, for example, in a simple linear attitude change model, the change in a person's attitude, ATT, might be

$$CH\_ATT = PERSUAS*(MESS - ATT), \tag{1}$$

where
CH_ATT is the rate of change in an attitude,
PERSUAS is a persuasiveness parameter,
MESS is the value of the message in terms of attitude units,
and ATT is the person's attitude value at a given time

PERSUAS is an individual difference parameter dealing with how fast that individual adjusts to incoming attitude messages. In this modeling context, PERSUAS is inversely related to time, so that the more persuasible the person becomes, the more quickly he/she changes attitude values. A simple assumption is that PERSUAS is the inverse of the time constant, that is, in terms of units, PERSUAS would be thought of as an adjustment fraction equal to 1/time, where time refers to the time to adjust to the message.

Another illustration of using individual difference parameters is found in a model of burnout among workers as described in Chapter 4 of Volume 1. In this situation, supervisors might vary in terms of their sensitivity to the needs of their workers. In the burnout situation, supervisors may not be monitoring the performance of their workers to give them support, guidance, and structure in ambiguous situations. Insensitive supervisors may aid in bringing on the rapid progression of burnout. In terms of the model, supervisor's SENSITIVITY was a parameter defined from zero, completely insensitive to 1.0, extremely sensitive. It was found that this parameter, SENSITIVITY, certainly contributed to either slowing up or quickening the burnout process, but, even when SENSITIVITY was set to 1.0 in order to simulate an extremely sensitive supervisor, burnout still occurred when other parts of the model remained in effect. For example, if workers continued to have a large discrepancy between the level of performance and their high expectations, burnout progressed even when the sensitive supervisor initiated structure and gave the workers support. To play on words, a sensitivity analysis of the SENSITIVITY parameter showed that SENSITIVITY was insensitive to changing (reversing) system behavior.

The Method of Single Parameter Changes

In sensitivity analysis, the modeler changes a single parameter at a time, while keeping all other parameters, initial values of the state variables, and table functions the same from one simulation run to the other. Usually, one might vary each parameter a small percentage point away from its normal value to see if small changes generate shifts in behavior modes. For example, one might find that a given parameter, when varied by only as much as 2% in one direction, had the effect of changing the original behavior of the system from an oscillatory system to a system whose state variables moved *directly* to a stable equilibrium point. To illustrate, let us take a very simple model that is similar to a Box-Jenkens, ARIMA autoregressive model (see McCain & McCleary, 1979). Suppose one was modeling changes of mood, representing the feeling of well-being, as a state variable, *WELL*. An extremely simple model might take the form of the following difference equation:

$$WELL_{K+1} = A(WELL_K), \tag{2}$$

where

WELL_K is the feeling of well-being at the Kth time period, and

A is the self-loop time constant.

This would be a very elementary model based on discrete time. Nevertheless, for our purposes, it is an excellent example of a model in which small changes in the A parameter will make vast changes in behavior. Suppose originally A was set at .8 and the initial value of WELL was 3.0. Figure 1a indicates the wellness variable converges to an equilibrium point at zero, indicating that the system is stable. On the other hand, suppose one raised the value of A slightly to just above 1.0, say 1.05. This seemingly minor change generates a very different behavior mode, so that WELL goes off to infinity (see Figure 1b). On the other hand, suppose value of A was set to −.8. In this case, Figure 1c indicates that the system displays dampened oscillations. The variable, WELL, eventually will converge to a stable equilibrium point, damping to a value of 0.0. Finally, if the value of A is set slightly lower, say at −1.05, then the oscillations become divergent. This model is an extreme example of how one can generate major changes in the behavior

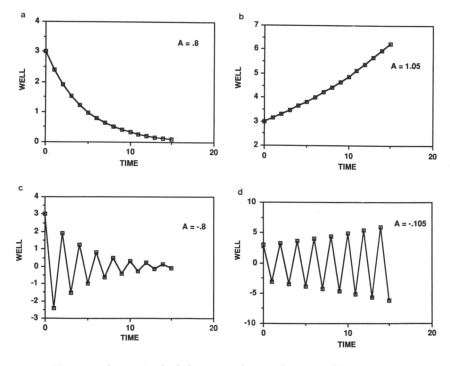

Figure 1. Changes in the behavior modes as a function of Parameter A.

of the system, with relatively small changes in parameters. Sensitivity analysis is the method by which we discover which parameters, such as A, act as leverage points to generate different qualitative behavior modes.

Insensitivity of Parameters and the Evaluation of Policies

Single parameter sensitivity analysis can be a valuable tool for assessing the impact of new proposed policies, laws, and other social interventions. However, sometimes it is also important to find which of the potentially many model parameters are not sensitive to change, to save time, effort, or money on policies that ultimately will not work. An excellent example is presented by Barry Richmond who has written an interesting modeling exercise for his students at Dartmouth College in a course on system dynamics. He presents them with a small version of an urban dynamic model first proposed by Forrester (1969) and further elaborated by a number of Forrester's students and colleagues (see Mass, 1974; Schroeder, Sweeney, & Alfeld, 1975; Alfeld & Graham, 1976). We shall use this urban dynamic example to point out the application of sensitivity analysis to assessing the impact of newly proposed policies that may initially seem reasonable in light of one's past experiences with city problems. In this case we shall show the relationship between changing policies and changing the values of a particular parameter of the model.

Figure 2 shows a typical pattern of urban growth noted in many cities across the world. The urban models developed by Forrester (1969) and his colleagues attempted to capture not only demographic variables such as urban population size, job availability, and the like, but also some of the physical characteristics of the city, such as the amount of land available and the amount of actual structures available for business to take place. For example, currently there are many cities in the United States with a surplus of partially unoccupied office build- ings. This situation exists because building construction outpaced demand for filling the building with businesses. The major characteristic of these urban dynamic models is the concentration on linking business construction and population growth with job availability.

As seen in Figure 2, this urban dynamic model is characterized by such variables as POPULATION and BUSINESS_STRUCTURES, which indicates the general size of the business community in the city being under study. The model also includes the LABOR_JOBS_RATIO, which is an index of the attractiveness of the city for obtaining jobs. In terms of relative numbers, cities with LABOR_JOBS_RATIOs much over 1.0 can be characterized as having too many people, relative to the number of

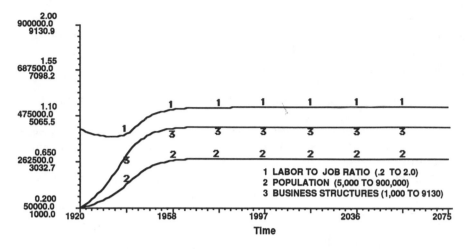

Figure 2. Typical growth pattern as produced by a baseline model of urban dynamics.

jobs and thus are relatively unattractive to outsiders looking for employment. Figure 2 indicates that at the beginning of the growth period, there was an opportunity for rapid expansion of BUSINESS_STRUCTURES and economically productive facilities. In addition, because of economies of scale for production and the fact that BUSINESS_STRUCTURES could increase rapidly due to the availability of land, the positive loops associated with the growth of BUSINESS_STRUCTURES initially dominated the situation. As BUSINESS_STRUCTURES increased, JOB_AVAILABILITY increased, thus lowering the LABOR _JOBS_RATIO because there were more jobs than people to fill them. When the LABOR_JOBS_RATIO decreased, making the city attractive to outside workers, a second major process came into the situation, namely the influx of men and women to fill the jobs. Given the natural birth rate of the population and the migration of people into the city because of job availability, the growth in BUSINESS_STRUCTURES was overpowered by the influx of the population. Moreover, as more and more BUSINESS_STRUCTURES came into being, land was used up, thereby constraining new construction. After a number of years, POPULATION began to level off because of the lack of attractiveness of employment opportunities, as seen by the high LABOR_JOBS_RATIO. In general, all variables level off because there is no more land to expand business, POPULATION immigration equals those who are leaving the city, and in general the employment picture looks relatively bleak. This picture may be typical of many modern cities.

In Richmond's laboratory exercise, the student is asked to evaluate the long-term impacts of several different policies, which include such policies as (1) generating a vigorous business construction policy, (2) building skyscrapers so that land per structure is less, (3) expanding the land area by land annexing, (4) decreasing the labor force by encouraging one of the marriage partners to stay at home, and (5) creating more jobs.

In the interest of brevity, we will only review the impact of policy 5, the creation of more jobs in the community. The student was told that he or she was a powerful labor leader and economist, who was lobbying for the president's attention. The leader states that what the country needs is jobs. According to the lobbyist, we have followed policies in the past that have encouraged the substitution of capital for labor. The lobbyist wants to reverse the process by encouraging every place of business to double its employment capacity.

In order to test the impact of creating more jobs, one way is to relate this initiative to changing one of the parameters of the model. In particular, the parameter, JOBS_PER_STRUCTURE, gives the average number of jobs associated with each structure. In the model, JOBS_PER_STRUCTURE is arbitrarily set at 18 jobs per structure, which can be thought of as an average figure. The labor leader suggests doubling that figure, so that one would want to test the model when the parameter was increased to 36 JOBS_PER_STRUCTURE, while keeping all other model parameters constant. The results of doubling the value of this parameter are shown in Figure 3. In this simulation run, the

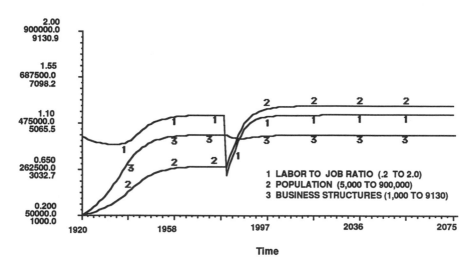

Figure 3. Response of the urban system when the value of the parameter, JOBS/STRUCTURE, was changed from 18 to 36.

system was allowed to run as in the previous baseline run, but the value of JOBS_PER_STRUCTURE was changed from a value of 18 to 36 in 1980. As one can see, the sudden doubling of jobs generated a rapid drop in the LABOR_JOB_RATIO, making the community very attractive. The figure also shows a very rapid increase in the POPULATION. However, this change in policy had little effect on BUSINESS_STRUC-TURES. This state variable decreased briefly but after a few years returned to the same level as before. More important, the desirable rapid drop in the LABOR_JOB_RATIO was offset by the rapid increase in POPULATION. The net effect was that within a few years, the city became as unattractive as before. The suggested policy proved to be only temporally effective and did not provide a permanent solution to this urban problem.

Another advantage of sensitivity analysis is that one can ask whether or not a particular policy would work under different conditions. For example, instead of attempting to create jobs during the time when the city was in equilibrium, which reflects severe problems, one might ask whether job creation would be helpful at times when the city is growing. In the next simulation run, JOBS_PER_STRUCTURE was doubled in the year 1930, during the city's growth phase. Figure 4 indicates that doubling the parameter at an earlier stage had little effect on the final equilibrium values of the three variables. POPULATION just began to increase a bit earlier.

Politically, the creation of jobs has been a favorite policy direction to follow, if jobs can be created. In terms of what we have been able to

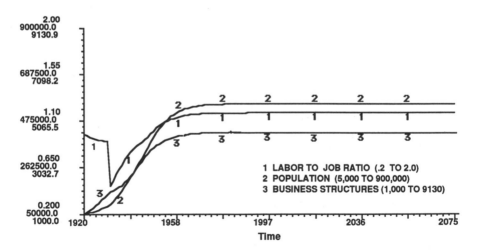

Figure 4. Projected pattern of urban growth, when the change in JOBS/STRUCTURE was initiated earlier in the twentieth century.

glean from this example thus far, the model's output indicates little evidence that increasing jobs *per se* will provide a permanent solution to the problem of unemployment. However, from a *short-term view*, the LABOR_JOB_RATIO did go in the right direction to a lower figure. In terms of a political time horizon, creating jobs does have some very positive benefits, but they appear to be quite transitory. An advantage of sensitivity analysis is that one cannot only discover parameters that radically affect the system's behavior, but also one can discover parameters that do not seem to be policy levers, like the JOBS_PER_STRUCTURE parameter.

To gain further insight for the reason why putting effort into changing JOBS_PER_STRUCTURE will not be rewarding, we suggest following the *causal structure* underlying the suggestion of creating more jobs. Figure 5 shows a causal diagram representing the closed loop nature of this portion of the urban system. The labor leader's policy

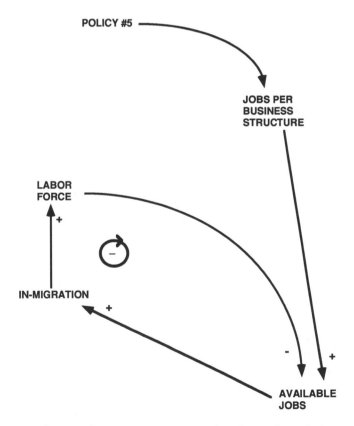

Figure 5. Basic dynamic loop structure associated with a policy of changing JOBS/STRUCTURE.

intervention should somehow increase JOBS_PER_STRUCTURE. Increasing this parameter value would generate more jobs. As one can see from the figure, although this may be true, unfortunately the availability of jobs also eventually makes the city more attractive. When attractiveness increases, IN_MIGRATION goes up expanding the size of the LABOR_FORCE and eventually closing off AVAILABLE_JOBS.

The labor leader was probably using *open loop thinking*, discussed in Chapter 4 in Volume 1, when suggesting that doubling JOBS_ PER _STRUCTURE would eventually lead to an increase in JOB_AVAILABILITY. In this situation, any policy suggested to solve the problem has to contend with the negative loop shown in Figure 5. In general, changing parameters that do not directly slow up IN_MIGRATION will have little permanent impact on the system. This powerful negative loop has to be considered in any policy initiative dealing with urban problems.

Thus far we have found that changing the value of the parameter, JOBS_PER_STRUCTURE, as a manifestation of a policy that stresses job creation, had little permanent impact on solving this urban problem. This also tells us that when it comes to estimating this parameter with data, that there is no real necessity to obtain exact values of this JOBS_PER_STRUCTURE parameter, especially if the cost of obtaining the data and costs of estimating the value of this parameter are factors. On the other hand, suppose one found a parameter that was extremely sensitive to even the slightest change in either direction. Indeed this might be the policy lever one is waiting for. As shown in Chapter 5 of this volume, when a parameter serves as a policy lever, it is quite important to spend resources to estimate as accurately as possible. Usually when a sensitive parameter is found, one attempts to control it and to model potential structures that can modify the value of the parameter. Thus the parameter becomes a variable in its own right.

Single Parameter Changes, Chaos, and Bifurcations

We have been describing the use of sensitivity analysis for policy purposes that stresses changing one parameter or initial value at a time. Usually those parameter changes are performed around the normal value of the parameter, because in most cases an intervention in a social system will not change the value of a parameter very much, so policy levers must be quite responsive to minute changes. Moreover, being able to control the social process associated with the change in the parameter usually is quite difficult in social settings. As a result, most system dynamicists have found that such software packages as DYNA-

MO™ and STELLA™ are adequate for performing a limited classical sensitivity analysis of a model. However, recently system dynamics, like many other systems approaches that work with nonlinear dynamical models have found that some models display bifurcations (qualitative behavior modes) and even chaos. Recently, for example, a whole special issue of the *System Dynamics Review* was devoted to chaotic social systems (Mosekilde, Aracil, & Allen, 1988). It is clear that there is a need for augmenting the present software to including subroutines that will systematically vary each parameter and initial value along its total range, store intermediate calculations, and then present the results of this total sensitivity analysis to the modeler.

There are several software packages that have been developed recently that perform this more sophisticated type of sensitivity analysis. One such package is called INSITE, which has been developed for the IBM PC family of microcomputers (see Parker & Chua, 1987). Among other things, this package will calculate projected time series, as will most simulation packages. However, in addition, INSITE programs will calculate and display bifurcations, plot out phase portraits, and plot the behavior of a state variable as a function of the parameter which was varied. This package takes the modeler a long way in analyzing the complete behavior of the system. It explores all aspects of the model quite thoroughly, giving the modeler an excellent picture of the potential behavior modes of the system.

Interval Approaches to Sensitivity Analysis

The classical method of sensitivity analysis emphasizes changing one parameter while holding all other parameters constant. Presumably one can find those parameters that when changed slightly will greatly affect the behavior of the system. However, the classical approach does have some disadvantages. First there is the problem of attempting to work with numerous parameters associated with large models. Second, the classical method of sensitivity analysis assumes independence of the effects of the parameters, *per se*. However, in some complex models, there may be an interaction among parameters, so that the effect of changing Parameter A might be dependent on the value of Parameter B.

A second approach to sensitivity analysis is to think of parameters in terms of a band or interval of values, so that, instead of treating the value of the parameter as a single value, such as .8, let us say, one specifies a reasonable range of values for the parameter, such as .5 to .95. The material to follow will describe the advantages of working simultaneously with specified intervals of parameter values, especially when one can use intervals to work with interactions among parameters.

Frequently in working with models, we ask decision makers to give assistance in specifying ball park figures of the parameter values so that we can begin to simulate the system under study with relatively realistic numbers. For each parameter, it is an easier task for decision makers to specify a *range* of realistic values. They feel more confident in dealing with intervals than with specifying one particular value.

These interval methods follow two major pathways. The first type of interval method for sensitivity analysis allows the modeler to obtain as much knowledge about the potential behavior modes of the system with a minimal number of simulation runs, perhaps only one or two. The second method also obtains maximum amounts of information concerning the potential behavior patterns of the system under study, but it does this through a very complex sampling procedure in which values of each parameter are appropriately sampled over many simulation runs. This latter method provides a sophisticated statistical exploration of the parameter space, just as one might perform an analysis and exploration of an experimental empirical situation using factorial designs and the analysis of variance. Both types of interval analyses provide information concerning a band of potential outputs over time, and in one case, this band can be considered as a confidence interval detailing what might happen over time as one projects into the future.

The Interval Arithmetic Approach

In this section we present the reader with another approach to sensitivity analysis, one that computes all small changes or perturbations between a lower and an upper bound using a form of mathematics that deals with interval arithmetic and interval analysis. Interval arithmetic is an arithmetic much like complex arithmetic in that it uses two values for its computations instead of one as is true in our usual arithmetic of numbers. As its name implies, the primary unit of analysis is the interval that is the set of real numbers of the form,

$$A^I = \{x | a^- \leq x \leq a^+\} = [a^-, a^+\}, \tag{3}$$

where $|$ means "such that."

One can think of the interval as a "receptacle" for all the perturbations. In a special case, an interval is called a real number of a *degenerate interval* when $a^- = a^+$. The interval is called a *point interval* (our usual real number). Point intervals are used in traditional sensitivity analysis. As such, interval numbers and their arithmetic are extensions of real numbers and arithmetic used in everyday calculations. The superscript "I" on a letter denotes an interval number but when the context is clear, we shall use unsuperscripted letters in capitals.

The power of representation (3) is that it is a *set* and a *number* at the same time. That is, an interval is a collection of numbers (more than one point in general) and at the same time can be operated upon by an arithmetic as if it were a single entity. A third aspect of interval entities that we do not use here is their induced three-valued logic, which is different from a traditional dichotomous logic (e.g., true/false) or with the 2 continuum-valued logic associated with fuzzy set theory (see Zadeh, 1973).

We now can develop methods to exploit the dual nature of interval arithmetic when confronted with *qualitative* modeling, modeling in the presence of *uncertainty*, or performing a *sensitivity analysis*. However, it is the use of interval methods for this latter case that will be explored next. Given the dual nature of our interval representation (3), an entire range of values, a continuum of values, can be analyzed as a whole. Moreover, in a dynamical model, any or all parameters of the model can be analyzed with a single framework. That is, if a dynamical model is couched in terms of interval numbers and interval analytic techniques are applied, then such a model can in principle represent all perturbations of parameters, be they a product of the simplifying assumptions (specification errors), data uncertainty (measurement error), ambiguity in the way variables are defined, numerical truncation errors generated by the computational methods used, or induced perturbations typically used in sensitivity analysis.

Traditional sensitivity analysis not only takes one real number to represent the perturbation of the interest, it usually only subjects one variable or parameter to that perturbation. On the other hand, the interval arithmetic approach allows us to represent a continuum of perturbations and the overall accumulation of all the different types of error to be determined. The interactive effects of the band of perturbations can thus be seen. In addition, if some of the parameters are always exact, for example, 0 or 1, they can be represented as degenerate intervals and carried in the model without major changes. To be sure, an additional complexity arises, and extra care must be taken to insure picking meaningful bounds on the parameters. Nevertheless, this approach provides a tool for the researcher and practitioner to handle the continuum of perturbations.

We first present how the arithmetic works, and then we will exploit it in sensitivity analysis of dynamical models. The reader who wishes to explore interval arithmetic and its use in sensitivity analysis further can consult Moore (1979), Alefeld and Herzbergfer (1983), Lodwick and Levine (1985), Deif (1986), and Dinkel and Tretter (1987).

Operations on Intervals

Here are the rules of interval arithmetic: Let the arithmetic operations of addition, subtraction, multiplication, and division be denoted by * where any of these operations defined on intervals A and B are given by

$$A * B = \{x|x = a * b, a \in A, b \in B\} \tag{4}$$

Specifically, Moore (1966) shows that for our usual algebraic operations (4) is equivalent to the following:

$$A + B = [a^- + b^-, a^+ + b^+] \tag{5}$$

$$A - B = [a^- - b^+, a^+ - b^-] \tag{6}$$

$$A * B = [c^-, c^+], \text{ where} \tag{7}$$
$$c^- = \min \{a^-b^-, a^-b^+, a^+b^-, a^+b^+\},$$
$$c^+ = \max\{a^-b^-, a^-b^+, a^+b^-, a^+b^+\}$$

$$A/B = [c^-, c^+], \tag{8}$$

where $0 \notin B$ and
$$c^- = \min \{a^-/b^-, a^-/b^+, a^+/b^-, a^+/b^+\}$$
$$c^+ = \max \{a^-/b^-, a^-/b^+, a^+/b^-, a^+/b^+\}.$$

These are the rules of interval arithmetic. An example of their use can be seen in Figure 6, which shows the operations of addition, subtraction, multiplication, and division on two intervals, A = [1, 3] and B = [2, 4]. A computer can be programmed to perform these operations quite easily. Also there are commercial extensions of PASCAL and FORTRAN available that handle intervals as data types. The interested reader is directed to Moore (1988).

Interval Sensitivity Analysis on Linear Models

We begin with the study of how to perform sensitivity analysis on linear dynamical models whose numbers are intervals instead of points on a line, first because the salient issues can be more readily seen with linear models and secondly because many nonlinear dynamical models are "linearized" in practice. Then we touch upon the more difficult task of applying interval arithmetic to nonlinear models.

Sensitivity analysis imposes perturbations or minute changes to

1. ADDITION

$$A + B = [\bar{a} + \bar{b}, \ a^+ + b^+]$$

$$[1, 3] + [2, 4] = [3, 7]$$

2. SUBTRACTION

$$A - B = [a^- - b^+, \ a^+ - \bar{b}]$$

$$[1, 3] - [2, 4] = [1 - 4, 3 - 2] = [-3, 1]$$

3. MULTIPLICATION

$$A * B = [C^-, C^+], \text{ where}$$

$$C^- = \min\{\bar{a}\,\bar{b}, \ \bar{a}\,b^+, \ a^-\,\bar{b}, \ a^+\,b^+\}$$

$$C^+ = \max\{\bar{a}\,\bar{b}, \ \bar{a}\,b^+, \ a^+\,\bar{b}, \ a^+\,b^+\}$$

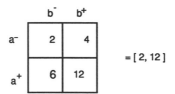

$$= [2, 12]$$

4. DIVISION

$$A/B = [C^-, C^+], \text{ where } 0 \notin B$$

$$C^- = \min\{\bar{a}/\bar{b}, \ \bar{a}/b^+, \ a^+/\bar{b}, \ a^+/b^+\}$$

$$C^+ = \max\{\bar{a}/\bar{b}, \ \bar{a}/b^+, \ a^+/\bar{b}, \ a^+/b^+\}$$

	b^-	b^+
a^-	.500	.250
a^+	1.50	0.75

$$= [.250, 1.50,]$$

Figure 6. Fundamental rules of interval arithmetic.

parameters of the dynamical model. Mathematically, we assume that the bounds over which the parameters range are bounded and connected. This gives us intervals that are defined as representation (3). For example, suppose we have a discrete model like the feeling of wellness model described by equation (2), which states that $\text{WELL}_{k+1} = A(\text{WELL})_k$. More generally the mood model will have the form $X_{k+1} = AX_k$. Let us modify the form of this equation somewhat by introducing a constant exogenous source of change, Parameter B that will add something to the new value of the variable X, even if WELL equals 0.0. Then the new model becomes

$$X_{k+1} = \begin{cases} AX_k + B & \text{for } 0 \le k \le K \\ 1.0 & \text{for } K < k \text{ or if } AX_k + B > 1.0 \end{cases} \tag{9}$$

where $0 \le A \le 1.0$, $0 \le B \le 1$ are real numbers. Equation set (9) might also be a simple way to represent how individuals learn a new skill on the assembly line of an automobile plant. In this case X_k might be the percentage of total aspects of the new skill the individual has learned by time period k. Thus X ranges from 0.0 to 1.0. The constant A might represent the baseline amount of knowledge from previous experiences that could be transferred to the present new task. Moreover, we would need to know the initial level of the skill at time k = 0. Of course, if we were to actually try to develop this into a model, A, B, k and X_0 would have to be obtained and we would need to have a reasonable level of confidence that the model does approximately describe how people learn the new skill. Because we want to illustrate sensitivity analysis using interval arithmetic, we set the model validity issue aside and assume that the model fits reality for the moment. The time series behavior of the model with A = .5, B = .5, k = 10, and $X_0 = 0.0$ is given in Figure (7).

The Interval Version of the Model. There are four places in this model where we might want to know how inaccuracies in estimation or measurement might affect the succeeding values of the amount the individual has learned. These are the values of A, B, k, and X_0. From a practical and/or computational point of view, and especially if equation set (9) also involved nonlinearities, transcendental functions, integral, and/or differentials, numerical truncation, and roundoff errors due to the use of a floating point representation of numbers on a computer also come into play. All these issues are of interest in a complete sensitivity analysis. However, we concentrate on the parameter perturbations and set aside the issues of numerical truncation and roundoff (which are handled in a similar manner), because they are beyond the

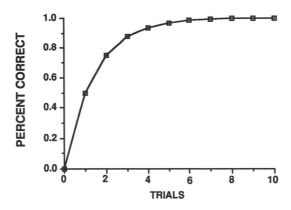

Figure 7. Baseline time-series behavior of a model that did not use the interval approach.

scope of this chapter. The interested reader is directed to Alefeld and Herzberger (1984), Moore (1979), or Kulisch and Miranker (1981).

The model as expressed as equation set (9) can be rewritten as

$$X_{k+1} = A^{k+1}X_0 + B \sum_{i=0}^{k} A^i \qquad (10)$$

and because the parameters A and B as well as the initial value X_0 are bounded below by zero, when we apply the rules for addition and multiplication to equation 10, we obtain:

$$X_{k+1} = [X_{k+1}^-, X_{k+1}^+], \qquad (11)$$

where

$$X_{k+1}^- = (A^-)^{k+1}X_0^- + B^- \sum_{i=0}^{k} (A^-)^i \qquad (12)$$

and

$$X_{k+1}^+ = (A^+)^{k+1}X_0^+ + B^+ \sum_{i=0}^{k} (A^+)^i \qquad (13)$$

Figure 7 was generated using standard arithmetic in which $A = 5$, $B = .5$, and the initial value of the state variable, X_0, was set equal to 0.0. Let us now vary those parameters by 10% and use Equations (10)–

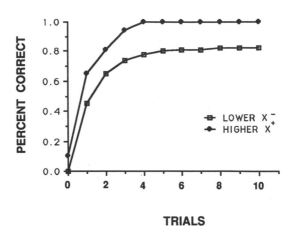

TRIALS

Figure 8. Projected time series of model that was formulated using the interval arithmetic approach.

(13) to analyze the system through an interval approach to a sensitivity analysis. For example, if $A = [.45, .55]$, $B = [.45, .5]$, and $X_0 = [0, .1]$, then we obtain the following band of output over time as found in Figure 8.

Note that with a plus or minus 10% perturbation on the initial value x_0, the left end point indicates that the individual always has at the minimum a capacity of learning 82% of the new process that is reached essentially in eight time periods, whereas the right end point indicates that the individual learns the whole process in four time periods. By contrast, the individual represented in Figure 7 learns the whole process in 10 time periods. Of course, the totally perturbed model illustrated by Figure 8 cannot indicate which one of the parameters or initial value causes the highest degree of change. There is a way, however, to prioritize these parameters with only a few additional simulation runs. This is accomplished by applying the same perturbation, say 10% to parameters A, B, and X_0 one at a time; e.g., $a = [.475, .525]$, $b = .5$, $x_0 = 0.0$.

Figure 9 shows a comparison of the efficacy of utilizing the interval approach on single parameters, keeping everything else constant, as seen in the previous example. Comparing the three graphs, the first impression is that the general patterns are the same in this linear model. It can be seen that when the 10% perturbation took place in the initial value of the state variable, the Figure 9 indicated that the upper and lower curves were very close to each other. Moreover systematically varying the initial value gave rise to a faster developing curve,

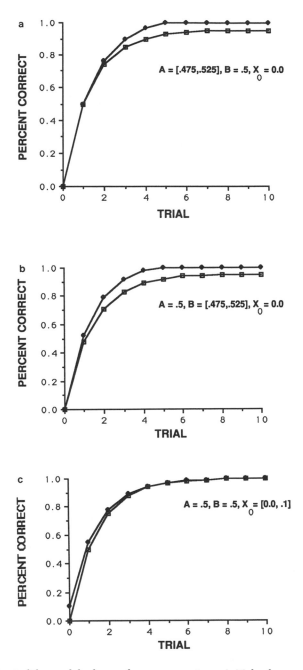

Figure 9. Output of the model when only one parameter or initial value was specified in interval terms.

although in all three cases, the upper curve in each graph reached 1.0. Comparing Figure 9 with Figure 8, which shows the output of the system when A, B, and X_0 were simultaneously defined as intervals, one can see that the *global* output of the system was more extreme. First, in terms of the upper curve seen in Figure 8, X^+ rises faster than the other upper curves and reaches 1.0 sooner than the other curves. Also, X^- increases at first faster than comparable lower curves in Figure 9 but reaches the lowest asymptote, namely .85, whereas all the other lower curves leveled off at least at .95.

At this point it might be interesting to compare the results of this interval analysis with the traditional approach described earlier. There are two parameters and one initial value. Each parameter represents a point number instead of an interval. Nevertheless we can systematically vary each parameter from its baseline value, that is, varying each simulation run from the $A = .5$, $B = .5$, and $X_0 = 0.0$. systematically. Thus the parameter values of a typical run might be $A = .525$, $B = .5$, and $X_0 = 0.0$. This would generate a total of six runs to perform a sensitivity analysis. This is compared to only three runs to obtain sensitivity information using an interval approach. When the number of parameters and initial values are larger, the savings in computer runs is even greater when using the interval approach. In this example, the six simulation runs using the traditional sensitivity method yield approximately the same kind of information as the interval arithmetic method. However, there are some situations where this is not the case, so the interval approach can become an important source of knowledge concerning the potential behavior modes of the system, besides being more efficient in terms of reducing the number of simulation runs needed to perform a complete analysis.

Some Inherent Mathematical Problems and Their Solutions. The care that must be taken in computing the resulting intervals is easily illustrated if in the model the parameter A were allowed to be negative; as, for example, if A varied between -1.0 to 1.0. In this case, any even power of A has a lower bound of 0.0, whereas if one applies Equation 7, then negative values always appear as lower bounds when the left end of the interval is negative. The reason for this is that in the definition of operations, Equation 4, both intervals are considered as independent. However, when the same interval is multiplied by itself, it is clearly *dependent*, and an exception needs to be made. Interval arithmetic as it is currently formulated is not equipped to deal with dependencies of this nature. Moreover, interval arithmetic is not *distributive*, precisely because of this lack of dependence. Interval arithmetic is *sub-*

distributive that can be best seen by an example. Let numbers $A = [0,1]$, $B = [-1,2]$ and $C = [2,3]$. Then subdistributivity says that,

$$A(B + C) \subseteq AB + AC. \tag{14}$$

For our example, we have the following interval arithmetic

$$A(B + C) = [0,1]([-1,2] + [2,3]) = [0,1][1,5] = [0,5] \tag{15}$$

$$
\begin{aligned}
AB + AC &= [0,1][-1,2] + [0,1][2,3] \\
&= [-1,2] + [0,3] = [-1,5]
\end{aligned}
\tag{16}
$$

The problem is that, given Equation (4), the two instances of the interval A in the right-hand side of the distribution shown in (14) are considered as independent. Therefore, the most "factored-out" the form of the operations are, in general the narrower the resulting intervals. Clearly what is needed is a dependent type of arithmetic to be developed. For the application of interval arithmetic to dynamical models and sensitivity analysis, the user must take care in setting up the way the arithmetic is performed. Although interval arithmetic will always produce bounds guaranteed to include all the perturbations, these inclusions may be too wide to be useful unless care is taken in performing the interval arithmetic.

Summary. Sensitivity analysis using interval arithmetic for the general linear dynamical systems model consists of replacing the variables of interest by interval variables and the algebraic operations by Equations (4)–(8), taking care that the excessive inherent widening of the intervals due to the way interval arithmetic considers each instance of a variable as independent (whether or not it is) is minimized. Non-algebraic relations and functions must be approximated by algebraic ones, and the associated approximation errors can be incorporated (added) into the bounds of the induced perturbation(s) in a sensitivity analysis.

There are at least two types of sensitivity analysis that benefit from interval methods. The first is an overall sensitivity that we call *global sensitivity analysis*, and the second is a sensitivity that uncovers the parameters that are responsible for the greatest change in the solution outputted time series called *local sensitivity analysis*. Global sensitivity is depicted by Figure 8 where the maximum variations of all the parameters are analyzed once. Local sensitivity is depicted by Figure 9, where each parameter is subjected to the same fixed percentage variation around its assumed value and analyzed one at a time. Some examples of the application of interval analysis to sensitivity of linear systems theory can be found in Galperin (1975), Skelboe (1979), and Deif (1981).

Interval Sensitivity Analysis and Nonlinear Dynamical Models

The first approach to sensitivity analysis of nonlinear models would be to linearize the model around equilibrium points and then apply the techniques described in the previous section. On the other hand, sensitivity analysis for nonlinear problems can be dealt with directly. We will point the reader to the literature for more advanced issues while touching upon the preliminaries. Toward this end, we take as the generic discrete nonlinear dynamical model, the set of equations,

$$X_j(k + 1) = f_j[X(k), X(0), k], j = 1, \ldots, J, \tag{17}$$

where
$X(k)$ = a vector of state variables at time period k
$X_j(k)$ = the value of the jth state variable at time period k
$X(0)$ = the initial values of the state variables
f_j = a function that is generally nonlinear.
The continuous time dynamical model can be represented as a system of differential equations thusly:

$$X_j' = f_j[X(t), X(0), t] , j = 1, \ldots, J, \tag{18}$$

where
$X(t)$ = a vector of state variables at time t
$X_j'(t)$ = the rate of change of the jth state variable at time t
$X(0)$ = a vector of the initial values of the state variables
f_j = the jth function associated with $X_j'(t)$ that is generally nonlinear.
In these two equations, the j's represent each of the different J state variables, the X represents a vector of state variables, and all the f_j's are functions of either discrete or continuous time, as represented by k or t, respectively. Finally, the new value of X_j or rate of change (the derivative on the left-hand side of Equation 18, in each of the J state variables is also an explicit function of time, at least in some models.

Interval Analysis and Nonlinear Discrete Models. Given that perturbations are represented as intervals, these intervals are input into f_j

in Equation 17, in place of the real numbered components (point intervals) of the vectors $X(k)$ and $X(0)$ and the parameters of the function f. The output of this equation system is an interval that is the representation for $X_j(n + 1)$. Thus, to use the interval approach for sensitivity analysis, the definition of an interval-valued function whose input or domain is interval numbers must be given. This has been worked out by Moore (1966, 1979) and is a specialization of set-valued functions. If the nonlinear functions, f_j, are continuous, then in principle the problem of obtaining the interval for the succeeding X_j's has a solution in theory because a continuous function over a special set known as a *compact set* has a maximum and a minimum, and these are the precise bounds for the interval representing $X_j(n + 1)$. Thus we assume that the f_j's are continuous, and the problem boils down to obtaining a global maximum and global minimum of the right-hand side of Equation 17. In general this is a very hard problem. However, we are interested in finding "quick-and-dirty" ways of obtaining the inclusions because what is often necessary is not the precise maximum and minimum but an inclusion of the maximum and minimum that is close, because in most models, the model representation and supporting data do not attain such precision.

There is an extensive literature dealing with interval methods for optimizations that have been developed and applied. In particular, quick-and-dirty as well as quite exhaustive techniques and procedures to obtain extrema using interval analysis can be found in Hansen (1988) and Ratschek and Rokne (1984).

Interval Analysis and Nonlinear Continuous Models. Solutions to differential equations as represented by Equation (18), using interval domains have also been extensively studied, and software exists that explicitly solves such problems. In fact, differential arithmetic software has been developed (see Rall, 1981; Corliss & Chang, 1982; and Corliss, 1988). However, it may be desirable to transform the continuous model into its discretized version in which case, global estrema have to be found.

An example of a sensitivity analysis given a nonlinear dynamical model is found in Lodwick and Levine (1985). These authors present an interval version of a predator–prey model, the Kaibab Plateau Model, analyzed by Goodman (1980). The model has 14 major variables, 4 table functions, 3 sectors, and a section that increments the populations and time-dependent variables to the next period. All parameters of the model were subjected to change and allowed to run over 20 time periods. The results were presented as interval values of the state variables that represented the total variation in the time series over the entire time horizon. This was a sensitivity analysis of a discrete nonlinear model.

A continuous nonlinear model was studied by Ames and Adams (1975). They develop methods for finding upper and lower bounds that they call "two-sided bounds," for a differential birth and death process. The question of how to incorporate uncertainties in the probability distribution or density function in a nonlinear socioeconomic or biological model has been developed by Ahmad (1975). These arise in dynamical models as table functions or are used directly in the nonlinear models themselves.

A Statistical Approach

The interval arithmetic method described generated bands of output that could be useful in finding behavior modes, given the same underlying structure. Another approach to simultaneous sensitivity analysis was developed by Ford, Amlin, and Bakus (1983). Their method relies on more traditional statistical methods that are perhaps more familiar to psychologists than are interval arithmetic methods. Both methods assume the existence of uncertainties in one's knowledge of the exact values of parameters, principally because all parameters will vary somewhat from one situation to the next. The statistical approach requires the decision maker/modeler to specify an interval of uncertainty for each parameter of the model. In addition, one has to specify a probability distribution over the interval of uncertainty. This distribution often is either a random or normal distribution, although in some cases, discrete distributions can be used if, for example, there are separate regions of parameter values.

The statistical approach uses an efficient parameter sampling scheme so that generalizations can be made concerning the behavioral characteristics of the system under study, especially for forecasting future behaviors. The simultaneous sensitivity methods emphasize looking at combinations of parameter values, rather than the effects of each parameter separately. One wants therefore to find good representative samples of parameter values to base conclusions on. One example is to use a systematic sampling scheme, such as sampling at the two extremes of each interval and perhaps at the midpoint. Another approach is to randomly sample each interval using Monte Carlo techniques. However, these sampling methods are less precise and efficient than a sampling plan principally developed by McKay, Conover, and Beckman (1979) called *latin hypercube sampling*, which is a variation of the latin square approach to experimental design (Winer, 1971). Two pieces of information about each parameter are required, namely (1) the interval of uncertainty and (2) the distribution of relative frequency of occurrence around the interval. Briefly, the interval describing the uncertainty of each parameter of the model is subdivided into N subintervals of equal probabilities. For all N subintervals, values are taken at random, giving N values per parameter. At that point, on a random basis one can set up N computer runs, assigning at random one value per parameter for that run. This sampling scheme insures representing combinations of values over all ranges of the parameter space, so that the resulting N simulations give an excellent picture of the diverse behaviors inherent in the structure of the system.

To illustrate, suppose one were working with a model that had three parameters of interest. This would generate a parameter space of three dimensions. Experts would be required to specify a modal value of each parameter and an interval of uncertainty around the modal value. Next they would specify the nature of the distribution of each parameter. For example, if they felt that there was a normal distribution over the range of uncertainty of a parameter, the experts would specify the mean and standard deviation of this normal distribution. If they felt that a uniform distribution were appropriate, they would so specify.

The choice of the number of simulation runs also dictates the number of subintervals one would use in the sampling scheme. Because one is going to use statistical methods in analyzing the sensitivity of the system, usually one chooses to use an N of at least 40 runs (subintervals). For example, suppose one chose an N of 45. Each parameter interval would be divided into 45 subintervals of equal probability; then values would be next chosen, and finally 45 simulation runs would be constructed by randomly assigning one parameter value per interval to each simulation run. This scheme would generate a good representative sample of combinations of the three parameters. The modeler would record the outcome of all 45 simulation runs and analyze the results in terms of a variety of indexes that will be described shortly.

Let us take another simple example used by Amlin (1982) to illustrate the sampling process to insure adequate representation of the parameter values in simulating the system over time. In this illustrative example, we assume that there are two parameters of interest, and it is decided to use only four simulation runs with combinations of the values of the two parameters. Figure 10 shows the stages one goes through to generate the four simulation runs. For each parameter, four subintervals are defined. In this example, one assumes a uniform dis-

1. **DIVIDE THE RANGE OF EACH DISTRIBUTION INTO N EQUAL-PROBABILITY INTERVALS, WHERE N IS THE NUMBER OF RUNS**

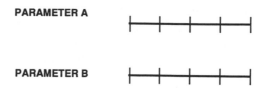

2. **SELECT A VALUE FROM EACH INTERVAL ACCORDING TO THE CONDITIONAL DISTRIBUTION**

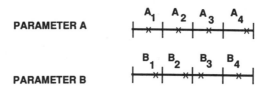

3. **ASSIGN THESE VALUES AT RANDOM TO THE N MODEL RUNS**

	PARAMETER	
	A	B
RUN 1	A3	B4
RUN 2	A1	B2
RUN 3	A2	B3
RUN 4	A4	B1

Figure 10. The steps used in formulating simulation runs using the latin hypercube sampling scheme.

tribution of parameter values over each subinterval. In this example, one would choose a point somewhere within the subinterval, thus making four values per parameter. The last step is to randomly assign, without replacement, the four values of one parameter with their counterparts associated with the second parameter to form four simulation runs.

The HYPERSENS Package

Ford, Amlin, and Backus (1983) describe a sensitivity analysis program called *HYPERSEN*, which utilizes a latin hypercube sampling scheme (see Amlin, 1982). If we return to our example of attempting to understand the sensitivity of a three-parameter dynamic model, HYPERSENS can first be used to aid the decision maker in specifying the size of the three intervals, their respective distributions, and the number of subintervals desired for the analysis. Once this information is given to the program, the latin hypercube sampling scheme would be used to specify the 45 combinations of parameter values that would be the basis of the next series of simulation runs undertaken by the HYPERSENS program. After the system had been simulated 45 times at varying combinations of parameter values, the results of several sets of analyses can be displayed.

Maximum and Minimum Values. For each state variable, the first major set of indices would be a table or time plot of the maximum and minimum output values of the 45 simulation runs, along with the mean time curve (see Figure 11). It should be clear that the maximum and minimum values of the dependent variable were found at each time period, so that, in some models, the maximum (minimum) value of time k may not be taken from the same simulation run as the maximum (minimum) value found at time $k + 1$. If the experts initially specify the modal values of the three parameters, the "nominal" values, the results of a run using these parameter values can be presented on the same plot to compare with the mean time series curve, which should be fairly close. In addition, the standard deviations are plotted over time to give the modeler a picture of the amount of variability in the 45 simulations as a function of time. Figure 11 shows the maximum and minimum values, the average curve, the standard deviations, as well as the curve one would obtain when parameters are set under "nominal" conditions. The figure might represent an analysis of a model of the acceptance of a new technology that, over time, might be displaced by new technology that directly competes with it. One can see the wide range of potential responses to the sampling of parameter values.

This information provides us with a picture of potential variability in the behavior of the system over time. HYPERSENS also will be helpful in finding qualitatively different behavior modes by allowing the modeler to look at individual and grouped time series curves. Thus one might perform what is analogous to a profile analysis on the 45 curves, clustering curves into different patterns, and then perhaps look-

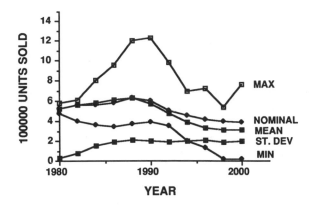

Figure 11. Output of the initial HYPERSENS analysis showing the maximum, minimum, and nominal responses as well as the range of behavior one would expect given an optimal sample of parameter values.

ing at the particular constellation of parameter values that generated those classes of qualitative behavior modes.

Relative Importance of the Parameters. The next major analysis performed by HYPERSENS is to obtain an index of relative importance of each of the parameters over time. In this case, the partial correlation coefficient is used to provide this index of importance. To see how this works, suppose the parameters of the model were time constants associated with the growth and acceptance of some new technology. One of the dependent variables might be the number of units sold. For any particular year, HYPERSENS evaluates the impact of a parameter on the number of units sold. One could fill in a table in which the rows would represent the 45 simulation runs, and the columns would represent the value of the parameters and the number of units of the new technology telephones sold for a particular simulation run and time (see Figure 12a). This table only represents one time period, period 14, let us say. It should be noted that there is a corresponding table for all the other time periods. Figure 12b shows the scatter plot between Parameter A and the state variable, the number of units sold, as well as the scatter plot between the two parameters for one particular time period. For the same time period and dependent variable, one could also obtain correlations between the number of units sold and the other two parameters. The partial correlation between the first parameter and the dependent variable, with influence of parameter B partialled out, can be used as an index of relative importance.

How does one interpret these partial correlations? Although one is a   DATA TAKEN FROM TIME PERIOD 14

SIMULATION RUN	UNITS SOLD (100,000)	VALUE OF PARAMETER A	VALUE OF PARAMETER B
1	12.1	70.0	3.2
2	8.0	31.4	2.9
3	11.6	72.6	6.5
4	9.6	35.7	4.9
5	10.4	67.3	4.1
6	4.8	55.7	2.8
.	.	.	.
.	.	.	.
.	.	.	.
45	11.9	66.7	6.1

b

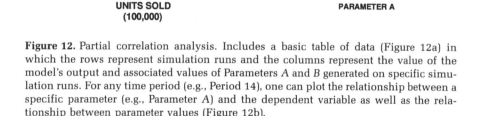

Figure 12. Partial correlation analysis. Includes a basic table of data (Figure 12a) in which the rows represent simulation runs and the columns represent the value of the model's output and associated values of Parameters A and B generated on specific simulation runs. For any time period (e.g., Period 14), one can plot the relationship between a specific parameter (e.g., Parameter A) and the dependent variable as well as the relationship between parameter values (Figure 12b).

only looking at the system at one time point, which is a static picture of the system, high positive or negative partial correlations indicate that this parameter can take a dominant role in "determining" the behavior of the system, if the parameter can be controlled and changed in the right direction. Thus this information might be used later to control the system problem under study. These partial correlations provide a way to discover the policy levers and sensitive places in the system where change might occur.

Thus far what we have described is the generation of partial correlations at a single point in time. However it is the long-term trends in the partial correlation coefficients that are important here. HYPER-SENS calculates these partial correlations at every time period and puts the results on a single plot, so one can determine the trend in the degree and the polarity of influences of each parameter over time. Ford, Amlin, and Backus (1983) provide a excellent example of the use of partial correlational analysis to find and interpret the degree of sensitivity of a set of parameters. They analyzed a complex system dynamics model of electrical usage that was part of a larger model of national supply and demand for energy in general. The model maps out the competition between oil usage in electrical generation, (OUEG) versus coal and nuclear usage in electrical generation, (CNEG), over time. One of the major parameters in the model is called IDGC, or Indicated Demand_Growthrate_Constant, which helps determine the rate of increase in electricity demand, (ED). Figure 13a shows a causal diagram of a portion of the model. The parameter IDGC plays the role of an exogenous input into the rate of change of electrical demand. Large values of IDGC would lead to faster electrical demand and vice versa. The growth of electrical demand has two major impacts. First it increases the use of oil for electrical energy (OGEG). Second, growth of electrical demand spurs on the use of coal and nuclear electrical generation. Finally, the reliance on coal and nuclear power plants, according to the model, will inhibit the usage of oil-fired generation plants, which eventually may be phased out in favor of coal and nuclear generation.

Ford, Amlin, and Backus (1983) used the HYPERSENS program to generate 45 simulation runs that provided the simulation data for the analysis of the sensitivity of each parameter of the model. The model was used to forecast oil usage from 1980 to the year 2000. Figure 13b shows the partial correlations between IDGC and the state variable Oil_Usage_for_Electrical_Generation (OUEG) over time. From 1980 to 1990, the partial correlations are all positive, some of which are quite large in value. The figure also shows some relatively large negative partials from 1990 to 2000. Our first comment is that not all of the plots of the partial correlation between OUEG and the other parameters were this dramatic. This makes this sensitivity analysis of parameters relatively useful because one can eliminate minor insensitive parameters from parameters that may provide us with leverage for policy interventions. It certainly gives us a handle on potentially what parameters could be controlled to manage a given system.

We note that over time the partial correlations first rapidly grow in a positive direction and then change polarity. How does one interpret this trend? The answer to that question can be found by examining Figure 10a again. During the 1980s, according to the model, the growth of electrical demand increased the use of oil-fired generation plants. The speed in which electrical demand will increase depends on the value of IDGC, and the larger the value of IDGC the faster demand

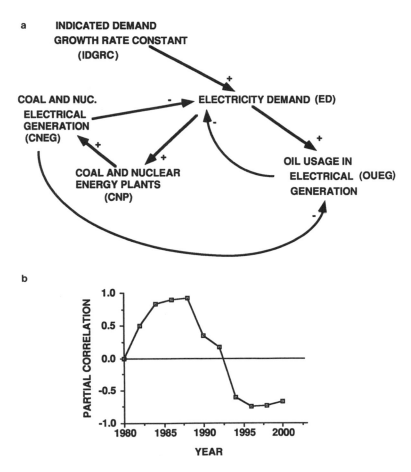

Figure 13. Basic dynamic structure of an electrical power usage model (Figure 13a) and the graph of the partial correlation of a particular parameter with the dependent variable as a function of time.

would grow. We see from the diagram that growth in electrical demand would stimulate oil usage for electrical generation OGEG, and large values of parameter IDGC would be associated with more oil generation than small values of IDGC, thus indicating at any point in time between 1980 and 1990 a positive correlation. However, something else is happening during this period. A rise in the growth of electrical demand stimulates the creation and startup of coal and nuclear technology plants, CNP, that takes time to come on line. When these plants increase in number and usage, oil-fired plants are used less and less. There is a rapid growth in the size of the correlation between the years

1990 and 2000 in a negative direction. Simulation runs associated with large values of IDGC tended to generate the growth of coal- and nuclear-fired plants faster than small values of IDGC and therefore also had a greater impact on inhibiting the use of traditional oil-fired plants for generating electricity. This would account for positive relationships from 1980 to 1990 and negative partial correlations during the last forecasting period from 1990 to the year 2000.

Confidence Bands. The final set of output generated by HYPER-SENS deals with understanding the variability of projections and forecasts as a result of the uncertainty in the parameter values. Companies, for example, would like to know the demand for their products in order to make reasonable decisions concerning investments in labor, supplies, raw material, and manufacturing capacity. The analysis performed by HYPERSENS will give the decision maker a fairly precise range of outcomes generated by the uncertainties in parameter values. HYPERSENS provides the decision maker with information concerning the tolerance intervals of the simulation forecasts. These values indicate the range of behavior one could expect in, let us say 90% of the simulation runs of the model to certain confidence level. Figure 14 shows the time pattern expected for 90% of the forecasts generated by the energy demand model. It shows that the upper bound of the 90% interval under different values of the parameters diverges at first and then decreases in size over time.

Interdependencies Among Parameters. Statistically, in order to interpret these tolerance intervals correctly, one must assume that all parameters are independent of each other. Sometimes on the first pass, however, this may not be true. In the model-building process, one builds a model, does some preliminary testing, and then carefully goes through a procedure of finding potential loops that need closing, by following the flow of information throughout the system. This may take the form of determining what aspects of the situation decision makers pay attention to and act upon. In some cases, one can discover actual structural dependencies among the parameters themselves, which violate the assumption of parameter independence. Consider a parameter as a potential variable that might change over time either by some unknown set of circumstances or through direct management. If there is feedback involved, a decline in Parameter A might make a company or individual attempt to change Parameter B to compensate for the original decrease in A. These parameters then take the role of variables that must conceptually be coupled together as part of the model. The use of HYPERSENS encourages reevaluating the potential interdepen-

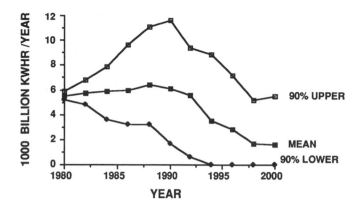

Figure 14. The output of the electrical power model in terms of 90% confidence bands projected over time.

dencies among parameters, and the subsequent modifications and fine tuning of the model generates more confidence in the model's forecasting power. In most cases there will only be a few interdependencies among parameters. Once these modifications have been performed, one runs the HYPERSENS program again to maximum values, partial correlations, and tolerance intervals that usually are not only more precise numerically but somewhat tighter because the colinearities among parameters have been eliminated.

In summary, the HYPERSENS approach to parameter sensitivity provides a powerful way to sample the potential range of values and uncertainties among parameters. New qualitative behavior modes can be discovered by routinely using the program. Moreover, through the use of partial correlation analysis, the relative importance of different parameter inputs can be ascertained over time. This provides insight into possible management of the system through a knowledge of which policy levers are dominant at particular time points. Finally, HYPER-SENS can provide tolerance intervals for future projections of the system. In performing this task, it was also necessary to attempt to verify the independence of parameter values, an important exercise in deriving confidence in one's model. Ford, Amlin, and Backus (1983) also suggest using the HYPERSENS procedure to discover those time plots that lead to a preselected target for one's set of state variables or perhaps lead to a certain trajectory (time series). This would indicate the combinations of parameter values that would get one to the target state, and obviously, if one can control some of these parameters, it might lead to assessing the feasibility of reaching one's goals. We feel that the HYPER-SENS procedure is very powerful, especially in working with very nonlinear complex systems where one might observe bifurcations and chaotic behavior patterns. HYPERSENS could provide a great deal of information concerning the effect of control parameters in projecting future behavior.

Sensitivity of Loop Structures

Normally, most sensitivity analyses concentrate on the reactions of the system to small changes in the initial values, table functions, and the parameters of the model. Another very fruitful endeavor is to also assess the efficacy of the loops themselves in understanding the role of feedback on model behavior. In model formulation, system dynamicists hypothesize certain dynamic loop structures to account for problem behavior. In isolation, each loop usually has relatively known behavior modes. When two or more loop structures are then coupled together, new behavior patterns may emerge. We end this chapter by discussing the procedures used to assess the sensitivity of the feedback structures that comprise the network of interactions among the variables in the model.

To test the sensitivity of a given loop structure, one can frequently modify the model slightly to partial out the action of the loop by changing the shape of a table function so that the output of the table function is constant for all values of its input. Another way to assess the role of a given loop is by means of the *partial model* approach. The action of each loop in isolation is first assessed, and then the loops are carefully combined to discover the synergy effect of coupling them together to understand the nature of the problem. A study by Morecroft (1983 a,b) provides a good example of how one can study the sensitivity of the loop structure. This model was briefly discussed in Chapter 4 of Volume 1. Morecroft addressed a problem found in many organizations where salespeople attempt to meet a periodic deadline of sales. If either the population of customers changes or there is a downturn in the economy, it is hard to make a sale. If either or both of these occurred, the sales force would work harder, but after awhile stress and physical fatigue might set in making sales effectiveness even lower, and therefore sales might slump even further. On the supervisory side, it is not always clear whether the decrease in sales is caused by exogenous "forces" or by a decrease in worker motivation, especially at later stages, where fatigue has set in. Typically in such organizations the supervisor may decide to ease up on the sales force, but it seems rational, because the source of the sales slump is not known, to lower the sales goals very cautiously. As a result, the morale and performance of the sales force become quite low over time.

Morecroft (1983b) hypothesized that each salesperson attempted to meet a specific target of sales per month, and when off target, the natural response was to work harder by going into overtime to make up the sales. This compensatory process can be represented as a negative loop that can be seen in Figure 15a. If for any reason MONTHLY_SALES decrease, RELATIVE_PERFORMANCE would go down also. A decrease in the perceived RELATIVE_PERFORMANCE increases OVERTIME, which will increase SALES_EFFECTIVENESS. Finally, to complete Loop 1, an increase in SALES_EFFECTIVENESS would increase

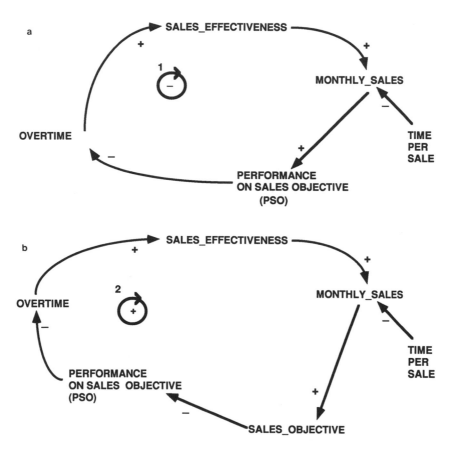

Figure 15. Two major loops of Morecroft model of stress among salespersons (Morecroft, 1983b). Loop 1 describes a mechanism of working harder when performance goes below a target level. Loop 2 describes how supervisors might set those targets.

MONTHLY_SALES. Generically this loop represents the *working harder* mechanism.

Morecroft's (1983b) model also includes a second loop that is also sensitive to a decrease in MONTHLY_SALES. He hypothesizes that the supervisors set the new monthly SALES_OBJECTIVE from information about past trends in sales performance. Thus a decrease in MONTHLY_SALES would eventually lower the SALES_OBJECTIVE. Because RELATIVE_PERFORMANCE can be represented as the ratio of MONTHLY_SALES TO SALES_OBJECTIVE, decreasing the denominator of that ratio would therefore increase RELATIVE_PERFORMANCE. Following the loop shown in Figure (15b), one can see that a increase in RELATIVE_PERFORMANCE decreases the use of overtime, leading eventually to a further decrease in MONTHLY_SALES. The purpose of this positive loop is to ease pressure on the sales force, when conditions that helped to lower MONTHLY_SALES were not under the control of the salespeople in the first place. For example, if the economic environment had worsened, the TIME_PER_SALE might have increased, making it more difficult to maintain sales.

Note that Loops 1 and 2 form a generic structure as discussed previously. Recall Chapter 5 in Volume 1, which described qualitative approaches to the dynamics of social processes, and where this generic structure was described behaviorally as a *drift to low performance*. For example, in classroom dynamics, if a teacher's standards of performance are influenced by the student's current lack of achievement, eventually the teacher's standards may drift lower and lower over time (see Clauset and Gaynor, Chapter 2, this volume). That is a case where lowering objectives leads to problems because the teacher will work less and less to improve the performance of the student as standards are lower. On the other hand, in the situation addressed by Morecroft (1983b), the opposite is true, for problems may occur if the supervisor *delays* in lowering the monthly SALES_OBJECTIVE when the source of the decrease in MONTHLY_SALES comes from a worsening of the economic environment. It is not the fault of the sales force, and lowering the monthly SALES_OBJECTIVE would be a realistic policy.

To test the sensitivity of Loop 1, Morecroft (1983b) first modified the model slightly to "turn off" the influence of Loop 2 (and other loops) by, in effect, keeping the supervisor unresponsive to changes in MONTHLY_SALES. Thus in this initial simulation run, the SALES_OBJECTIVE was held constant over time while Loop 1 responded to a decrease in MONTHLY_SALES by the mechanism of working OVERTIME to increase the sales again. Under this condition, Figure 16a shows the response of the system to a drop in MONTHLY_SALES caused by doubling the time it takes to make a single sale. Loop 1 helps

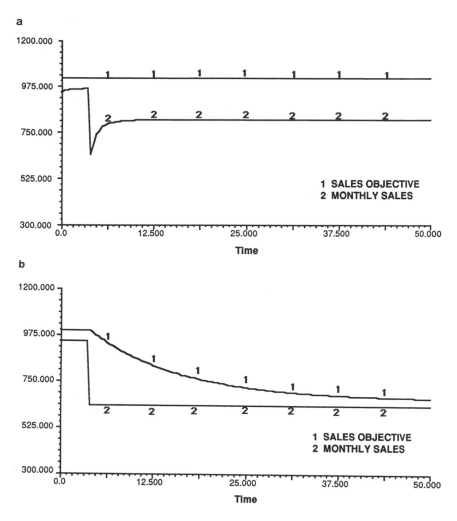

Figure 16. The behavior of Loops 1 (Figure 16a) and 2 (Figure 16b) when completely isolated from each other and from the rest of the loops associated with the system.

to get the sales force back to its monthly targets through overtime. In isolation, the loop responds to the initial drop in monthly sales by working its way back up. In this simulation run, SALES_OBJECTIVE was held constant, implying that Loop 2 was inoperative. It can be seen that there is a steady-state error between the two curves. Nevertheless, Loop 1 serves as a major mechanism for the sales force to bring up its MONTHLY_SALES in response to an external source of change in the TIME_PER_SALE.

Using the partial model approach, one can also assess the action of Loop 2 in isolation. In the next simulation run, the first loop has been made inert, meaning that the only action portrayed by the model is the response of the supervisory staff to decreasing MONTHLY_SALES. In this run, Loop 2 will be the only loop operating. Figure 16b shows the response to the system. It can be seen that over a 50-unit time period managers lower the SALES_OBJECTIVE in a smooth fashion attempting to match the lowered level of MONTHLY_SALES. Again Loop 2 is an appropriate response for management if the source of change in the situation came from the external environment, which in this case it did.

Loops 1 and 2 as a unit operate under normal conditions to solve the problem of poor morale and poor performance levels. These mechanisms do not generate the problem; they help to solve it. Morecroft (1983b) hypothesized other sets of loop structures that "kick in" when Loops 1 and 2 operate too slowly. Figure 17 illustrates the additional loops that may play an important role in generating the problem behavior. Loop 3, a positive loop, goes into action as a natural consequence of prolonged overtime. A lowering of MONTHLY_SALES initiates OVERTIME, which if high enough will decrease the worker's MOTIVATION level. Lowered MOTIVATION decreases SALES_EFFECTIVENESS, which lowers MONTHLY_SALES even more.

Loop 4, another positive loop, deals with the direct lowering of MOTIVATION by a decrease in perceived RELATIVE_PERFORMANCE. The action is the same as Loop 3. In the first case, MOTIVATION decreases because of the effect of fatigue generated by OVERTIME, whereas Loop 4 captures the psychological effect of poor RELATIVE_PERFORMANCE. These two loops represent two ways of discouraging the maintenance of SALES_EFFECTIVENESS.

Consider the salesperson's point of view. He or she is working very hard and yet shows poor RELATIVE_PERFORMANCE. That is quite disheartening. We also note that these two mechanisms come into play when the system goes into extreme conditions as a natural consequence of OVERTIME and periods of sustained poor performance. The model assumes that the two loops are highly nonlinear in their action.

Now consider the supervisor's point of view. Morecroft (1983b) hypothesized a relationship between the setting of the SALES_OBJECTIVE and MOTIVATION. Loops 5 and 6 (Figure 17b) come into play when the manager decreases the monthly SALES_OBJECTIVE in response to a decrease in MONTHLY_SALES. Loop 5 deals with lowering the amount of OVERTIME, which in turn would raise the salesperson's MOTIVATION and consequently MONTHLY_SALES. Loop 6 indicates how the manager, by lowering the monthly SALES_OBJECTIVE, can directly impact MOTIVATION. Loop 5 is the manager's way of counteracting the action of Loop 3, and Loop 6 counteracts the action of Loop 4.

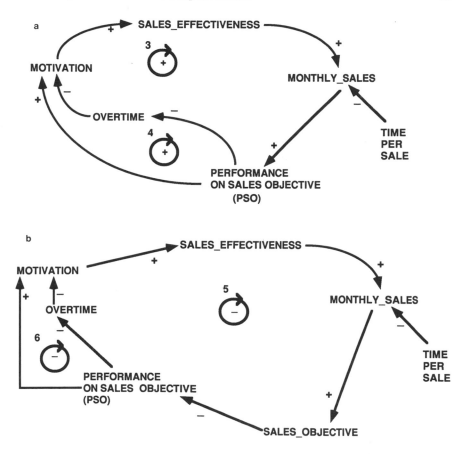

Figure 17. Representation of two motivational loops that come into play as a result of prolonged use of overtime. Loop 3 deals with the direct motivational effects of overtime and Loop 4 represents the negative effect of prolonged poor performance.

These major loops can help to understand the pattern of poor MONTHLY_SALES, low MOTIVATION, and poor SALES_EFFEC-TIVENESS. From the sales force perspective, if Loop 1 fails to make up for the change in MONTHLY_SALES, Loops 3 and 4 kick into operation to drag the system down. Obviously, these two loops contribute greatly to the creation of problem behavior. An individual salesperson may perceive that he or she is spinning wheels and not performing well.

From a management perspective, Loops 2, 5, and 6 contribute to realistically ease the pressure to perform at a level that cannot be maintained by the sales force. However, the major problem in this situation is that these loops respond too slowly to counteract the effects of Loops 3 and 4. Why does management take too much time to lower the monthly SALES_OBJECTIVE? First, from a very limited rational point of view, at first the managers may not know whether the decrease in MONTHLY_SALES is a just a temporary setback or a more permanent trend to pay attention to. In setting monthly SALES_OBJECTIVES, managers would have to wait somewhat to tell how temporary the slump in MONTHLY_SALES actually was, which means that if they lowered the targets at all, it would be done cautiously.

Supposing the managers determined that the downward slump in MONTHLY_SALES was an actual trend. The next decision would be to assess the source of the change in MONTHLY_SALES. It may not be clear to the supervisory staff that either the population of customers or the economic environment had changed enough to effect MONTHLY_SALES. The other alternative is that there is something wrong with the sales force *per se*, such as a lack of MOTIVATION and/or SALES_EFFECTIVENESS. If it takes time to determine the cause of the sales slump, then the "rational" strategy would be to decrease the SALES_OBJECTIVE rather slowly waiting for more conclusive evidence to come in. Loops 3 and 4 represent the impacts of too much OVER-TIME and sustained poor RELATIVE_PERFORMANCE. They impact on MOTIVATION and SALES_EFFECTIVENESS. As a result, the manager obtained a very mixed picture because the loss of MOTIVATION and SALES_EFFECTIVENESS might indicate a defect in the sales force, yet there may also be evidence that the population of customers has changed or that the sales slump was due to a poor external economic climate.

Figures 18a and 18b show the behavior of the full model for several major variables. Note that at first the salespeople increase their SALES-_EFFECTIVENESS through OVERTIME. However, with sustained levels of OVERTIME, MOTIVATION drops decreasing SALES_EFFEC-TIVENESS and MONTHLY_SALES. Figure 18b indicates that MONTHLY_SALES continue to fall until the managers finally lower the monthly SALES_OBJECTIVE to a more realistic level. In terms of SALES_EFFECTIVENESS, one can compare this falling index with NORMAL_SALES_EFFECTIVENESS, which is a baseline figure of how the salespeople would react if uninfluenced by overtime or motivational factors. In Figure 18a, the lost effectiveness is represented by area between Line 2 and Curve 1. The drop in MOTIVATION due to OVERTIME and the psychological effects of poor performance took its toll on the sales force. Note also that at the end of 50 time units, SALES_EFFECTIVENESS only returns to the baseline level of NOR-MAL_SALES_EFFECTIVENESS, which is a bit lower than its initial entry level.

As part of the sensitivity analysis of loop structure, we now can compare the full model, with all loops intact, with the behavior of the partial model in which Loop 2 is isolated. In terms of setting SALES-

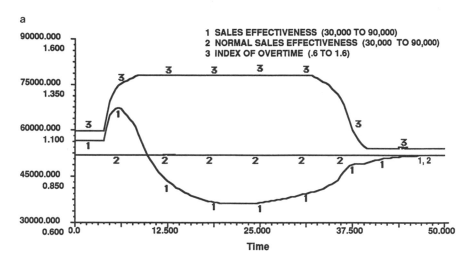

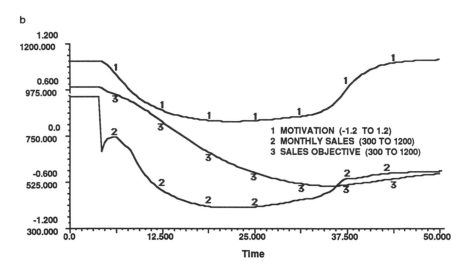

Figure 18. The problem behavior of the full model in terms of the SALES_EFFORT, OVERTIME, MOTIVATION, MONTHLY_SALES, and SALES_OBJECTIVE.

_OBJECTIVE, Figure 19 indicates that the managers hesitated at first to lower the SALES_OBJECTIVE due to the gallant effort of the salespeople in bringing up the MONTHLY_SALES. Figure 19 also shows just how much and how fast the supervisors eventually had to lower their SALES_OBJECTIVE to handle the situation after MONTHLY_SALES fell precipitously.

In performing a sensitivity analysis of the loop structure, one

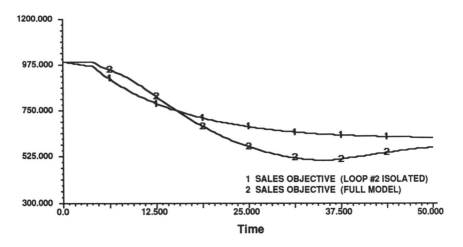

Figure 19. The behavior of the SALES_OBJECTIVE variable when Loop 2 was isolated and when the same loop was embedded in the full model.

would carefully tease out the importance of each loop in isolation as well as in combination with other loops. In addition, as in the case of the Morecroft model, some of these loops do not come into domination at first because of the nonlinearities inherent in these complex relationships. In order to help prevent this problem from occurring again, one strategy is to search for new sources of information flows, such as monitoring morale, motivational levels, and overtime in the work force to be able to decide sooner whether there was a change in the environment or in the motivation of the salespeople. This would lead to hypothesizing one or more additional informational feedback loops that would be integrated into Morecroft's model.

Summary

We covered three approaches toward performing a sensitivity analysis of a model to discover leverage points. The first approach simply systematically changes parameters one at a time, holding everything constant. This classical approach can lead to a deep understanding of the behavior modes inherent in the system when extreme values of the parameters are sampled and the behavioral output is systematically recorded. This is essentially the methods used to discover bifurcations and chaotic behavior in nonlinear systems.

A second approach to sensitivity analysis deals with defining each parameter in terms of an interval or range of values. Two ways of accomplishing this task were described. Interval arithmetic can be applied to dynamic models to obtain much information concerning the behavioral output of the system in a minimum of simulation runs. The statistical approach to interval analysis involves (1) the use of a sampling scheme to obtain a representative sample of parameters and (2) application of statistical methodology to assess the dominance of parameters as a function of time. This latter approach also generates confidence limits for the dependent variables over the time horizon. In general, the HYPERSENS approach appears to be promising to assess the independence of parameters, their efficacy, and the confidence one would have concerning the future behavior of the system.

Finally, one can perform a sensitivity analysis of the loop structure of the model. The concept of a partial model was introduced, in which parts or components of the model were first viewed and tested in isolation and then systematically combined to see the points of synergism emerge. This method then leads to asking questions concerning the possibility of new policies and new sources of information that might be put into play to correct the problem of interest.

References

Ahmad, R. (1975). A distribution-free interval mathematical analysis of probability density functions. In Karl Nickel (Ed.), *Interval arithmetic Lecture notes in computer science*, No. 29 (pp. 127–134). New York: Springer-Verlag.

Alfeld, L. E., & Graham, A. K. (1976). *Introduction to urban dynamics*. Cambridge, MA: Wright-Allen Press.

Alefeld, G., & Herzberger, J. (1983). *Introduction to interval computations*. New York: Academic Press.

Ames, W. F., & Adams, E. (1975). Monotonically convergent numerical two-sided bounds for a differential birth and death process. In Karl Nickel (Ed.), *Interval arithmetic: Lecture notes in computer science*, No. 29 (pp. 135–140). New York: Springer-Verlag.

Amlin, J. S. (1982). *Users manual for HYPERSENS: Sensitivity analysis using latin hypercube sampling*. Advanced Modeling/Simulation Group, Control Data Corporation, Dayton, Ohio.

Amlin, J. S. (1982). *Hypersens: Sensitivity analysis using latin hypercube sampling*. Technical report prepared by A. Ford, Los Alamos National Laboratory.

Corliss, G. (1988). Applications of differential arithmetic. In R. E. Moore (Ed.), *Reliability in computing the role of interval methods in scientific computing* (pp. 127–148). New York: Academic Press.

Corliss, G., & Chang, Y. F. (1982). Solving ordinary differential equations using Taylor series. *A.C.M. Transactions in Mathematical Software, 8,* 114–144.

Deif, A. (1986). *Sensitivity analysis in linear systems*. Berlin: Springer-Verlag.

Dinkel, J., & Tretter, M. (1987). An interval arithmetic approach to sensitivity analysis in geometric programming. *Operations Research, 35*(6), 859–866.

Ford, A., Amlin, J., & Backus, G. (1983). A practical approach to sensitivity testing of system dynamics models. *International System Dynamics Conference* (Plenary Session). Chestnut Hill, MA, 261–280.

Forrester, J. W. (1969). *Urban dynamics*. Cambridge, MA: M.I.T. Press.

Galperin, E. A. (1975). The condition problem in solution of linear multistage systems. In Karl Nickel (Ed.), *Interval arithmetic: Lecture notes in computer science.* New York: Springer-Verlag.

Goodman, M. R. (1980). *Study notes on system dynamics.* Cambridge, MA: M.I.T. Press.

Hansen, E. (1988). An overview of global optimization using interval analysis. In R. E. Moore (Ed.), *Reliability in computing, the role of interval methods in scientific computing* (pp. 289–308). New York: Academic Press.

Kulisch, U., & Miranker, W. (1981). *Computer arithmetic in theory and practice.* New York: Academic Press.

Lodwick, W., & Levine, R. L. (1985). Finding qualitative behavior modes: The use of interval analysis to perform sensitivity, stability, and error analysis on dynamical models. *Proceedings of the 1985 International Conference of the System Dynamics Society.* (pp. 502–512) Keystone, CO.

Mass, N. J. (Ed.). (1974). *Readings in urban dynamics: 1.* Cambridge, MA: Wright-Allen Press.

McCain, L. J., & McCleary, R. (1979). The statistical analysis of the simple interrupted time-series Quasi-Experiments. In T. D. Cook & D. T. Campbell (Ed.), *Quasi-Experimentation: Design and analysis issues for field settings* (pp. 233–293). Chicago: Rand McNally.

Mckay, M. D., Conover, W. J., & Beckman, R. J. (1979). A comparison of three methods for selecting values of input variables in the analysis of output from a computer code, *Technometrics, 21,* 239–245.

Moore, R. E. (1966). *Interval analysis.* Englewood Cliffs, NJ: Prentice-Hall.

Moore, R. E. (1979). *Methods and applications of interval analysis.* Philadelphia: SIAM.

Moore, R. E. (1988). *Reliability in computing: The role of interval methods in scientific computing.* Academic Press, New York.

Morecroft, J. D. W. (1983a). System dynamics: Portraying bounded rationality. *Omega,* 11(2), 131–142.

Morecroft, J. D. W. (1983b). Rationality and structure in behavioral models of business systems. *Proceedings of the 1983 International Conference of the System Dynamics Society,* (pp. 1–51). Chestnut Hills, MA.

Mosekilde, E., Aracil, J., & Allen, P. M. (1988). Instabilities and chaos in nonlinear systems. *System Dynamics Review,* 4(1–2), 14–55.

Parker, T. S., & Chua, L. O. (1987). INSITE—A software toolkit for the analysis of nonlinear dynamical systems. *Proceedings of IEEE, 75*(8), 1081–1089.

Rall, R. E. (1981). Automatic differentiation: Techniques and applications. *Lecture notes in computer science,* No. 120. Berlin: Springer-Verlag.

Ratschek, H., & Rokne, J. (1984). *Computer methods for the range of functions.* New York: John Wiley.

Skelboe, S. (1979). True worse-case analysis of linear electrical circuits by interval arithmetic. *IEEE Transactions on Circuits and Systems, CAS-26,* 874–879.

Schroeder, W. W., Sweeney, R. E., & Alfeld, E. (1975). *Readings in urban dynamics II.* Cambridge, MA: Wright-Allen Press.

Tank-Nielsen, C. (1980). Sensitivity analysis in system dynamics. In J. Randers (Ed.), *Elements of the system dynamics method.* Cambridge, MA: M.I.T. Press.

Winer, B. J. (1971). *Statistical principles in experimental design* (2nd ed.). New York: McGraw-Hill.

Zadeh, M. (1973). Outline of a new approach to the analysis of complex systems and decisional processes. *IEEE Transactions of Systems, Man, and Cybernetics, 3,* 28–44.

4

Psychological Scaling and Filtering of Errors in Empirical Systems

Ralph L. Levine and Weldon Lodwick

Introduction

This chapter will address many of the issues that confront research psychologists who conduct empirical studies based on the system modeling methods described in this volume. In this chapter, we shall cover the problems of the design of proper empirical studies, emphasizing the use of longitudinal and time series studies to describe the dynamic behavior of the system. The chapter deals with the problem of measurement error in time series associated with system behavior and possible solutions to measurement error through the application of filtering techniques borrowed from the engineering literature. Finally, the topic of developing psychological instruments from the perspective of scaling will be covered in some detail. A way of developing theoretical curves for system variables that will be congruent with interval scales will be suggested.

Most previous chapters in this volume considered the theoretical side of system modeling. Indeed, one can learn a lot from analyzing the loop structure of the model and going through a formal sensitivity analysis of the model's parameters (see Chapter 3, this volume). Al-

Ralph L. Levine • Department of Psychology, Michigan State University, East Lansing, Michigan 48824.   **Weldon Lodwick** • Department of Mathematics, University of Colorado, Denver, Denver, Colorado 80217-3364.

Analysis of Dynamic Psychological Systems, Volume 2: Methods and Applications, edited by Ralph L. Levine and Hiram E. Fitzgerald. Plenum Press, New York, 1992.

though the model was based on the empirical literature and one's experience with the system under study, many times it is obvious that there is a need for further empirical studies to be made. From the data side, there are many activities associated with predata collection. These include such activities as picking the appropriate level of aggregation to fit the model, that is, determine whether to predict individual time series or an average of many times series. At this point, one might also assess the available indexes of system behavior suggested by the model for their scale properties, reliability, and the like. Finally one has to understand how fast variables change to obtain samples of behavior, so that one can determine appropriate sampling frequencies and the overall duration of the study.

Once the data are obtained, the next stage may be to attempt to filter the raw time series data to smooth out the inherent measurement errors. This chapter will describe several ways to filter out measurement errors from time series data. The next chapter will describe methods for feeding the smoothed data points into an estimation program, obtaining estimates of the parameter values, and evaluating the fit of the model to the empirical data set.

Considerations Prior to Data Collection

Assume that one has developed a model, which was perhaps based on the literature and/or on a preliminary set of data. In addition to leading to a better understanding of the underlying dynamics, a model can provide a framework and direction for the data collection process. It can be the basis of an information-gathering system to be used by the modeler and perhaps others doing research on the same topic. The model should tell the researcher the major variables to measure over time, and, if the study includes one or more interventions, such as various approaches to therapy, the model will point out where and when the intervention should be introduced in the study.

Handling the Time Dimension

Perhaps the first problem in obtaining one's data in the empirical phase of system research is to decide whether to use a cross-sectional or longitudinal (time series) design. Because these models principally involve dynamic processes, the obvious answer appears to be that one has to generate data using a longitudinal (repeated measures) design to obtain time series data. However, up to the present moment, in the field of path analysis and causal modeling, the majority of data are collected from cross-sectional designs. This puts the path modeler at a particular disadvantage because of the restrictive assumptions of equilibrium (see Tetrick, Chapter 2, this volume). Any social problems that lead to, say, explosive growth or perhaps oscillatory behavior would not fall into the modes of behavior that can be captured by a cross-sectional approach to the data collection process.

Although it is most advisable to study behavior over time, occasionally cross-sectional studies can provide a great deal of information concerning the *variety* of behaviors found in the population of individuals or companies that one might want to study over time. Doing a preliminary study of, let us say, 200 companies in a cross-sectional design might provide information concerning the size and distribution of the intervals associated with many of the parameters of one's model. For example, suppose one developed a model of organizational entry and problems of adopting company values. In order to study the adoption of a company's values upon organizational entrance, one would like to survey the range of organizational attitudes and values in the industrial population under study as well as the range of original worker attitudes towards those issues. This would be helpful in performing an interval sensitivity analysis or using HYPERSENS (see Chapter 3), which, as you recall, requires specifying intervals of uncertainty around all model parameters.

Picking the Level of Aggregation

Accepting the assumption that longitudinal studies are most appropriate for fitting dynamic models, another problem is to match the level of aggregation associated with the model to the level of aggregation of the data. For example, typically system dynamicists are called upon by companies to model problems associated with that particular organization. These may include such problems of growing too fast, low levels of worker satisfaction, being topheavy, not coordinating goals across departments, and so forth. The model might have been developed at the company level and would be most appropriately fit to data from individual companies. However, it would be less desirable to fit the model to mean data taken over companies. Although one might design a study that sampled 50 organizations, the dynamic model would be fit to all 50 companies individually, not to the mean data curves. The main reason for this suggestion is that companies might start out at different initial values and be in different phases of their development, so that by combining data without taking different developmental stages into consideration, the mean curve may not correspond to the behavior of individual cases.

Certainly there are some times where one would want to fit a model to average data. It fundamentally depends on the level of aggregation of the model. For macrolevel models, such as models of regional problems, average data might be most appropriate. In the majority of cases, however, one should fit the model to individual cases. If one attempted to formulate the model as a set of generic structures capturing the essential dynamics of the original problem, then most of the differences among cases would be variation in parameter values, assuming no interaction between cases and models, that is one group of individuals would be fit by one class of dynamic models, and another group might be fit by a different class of model possessing divergent causal structure. Although this appears to be rare, in some cases this has happened. For example, in the Gaynor and Clauset (1983) model of classroom dynamics, teachers in effective schools, when finding drops in academic performance, put more time and effort into teaching than before to bring their students back up to an external standard. In contrast, teachers in ineffective schools tended to tie their teaching standards to the performance level of their students, rather than using externally generated standards (see Chapter 3 on qualitative dynamics, this volume). In a sense, one has an additional loop structure that is not there in the effective schools, so that there appears to be an interaction between school types and structures.

Filtering Measurement Errors

In elementary physics and chemistry labs, students learn that every measurement has some degree of error in it. This error could be a systematic error due to bias in the measurement instrument itself, or the error could be just a process due to the fuzziness or uncertainties of the measurements per se. Another example of measurement errors in physical data is in the processing of information from satellite signals. Engineers and space scientists find varying degrees of noise in the signal. This noise causes the signal to be distorted somewhat so that measurements of the amplitude of the signal are slightly in error.

Measurement Error in Social Science Data

Another area where measurement errors appear is in social science research. For example, in attempting to assess the severity of drug involvement among teenagers, researchers may find that their instruments systematically underestimate teenage drug abuse due to fact that many teenagers do not want to divulge information about their illegal activities to parents and other authorities. This underestimation of drug abuse is an example of a systematic error of measurement. Unsystematic errors due to the unreliability of the data also occur in social science research. To deal with unreliability, psychologists have developed sophisticated techniques for the formulation of reliable measures of constructs to keep the standard error of measurement down. Even so, the size of the measurement errors can nevertheless become a problem in fitting models to time series data. The problems become particularly acute when attempting to use least squares estimation techniques when fitting one's model to a fallible data set. In the context of a single time series, for example, suppose Y is a behavioral index of interest, and the reliability of the instrument, r_{yy}, were known. Then the standard error of measurement is defined as

$$S_e = S_Y\sqrt{1 - r_{YY'}}, \tag{1}$$

where S_e is the standard error of measurement, S_Y is the standard deviation of Y, and $r_{YY'}$ is the reliability of the Y instrument. In a dynamic situation, represented by time series data, even with high reliability, the standard error might be relatively large if the time series covers a wide range of Y values over time. In order to compensate for measurement error inherent in time series, we suggest adopting one or more filtering techniques, to be described shortly. These filters will remove some of the measurement error before estimating parameters.

According to Equation (1), the greater the variability of the variable over time, the greater the standard error of measurement, even with substantially high reliability coefficients. A manifestation of this problem is that ordinary least squares techniques for estimation parameters are particularly sensitive to measurement errors. In most cases the problem of measurement errors is compounded by the fact that it is very difficult in many research areas to obtain large numbers of time points, so that one can do something about those measurement errors by means of filtering techniques. This section will therefore cover two situations separately: (1) when one can only obtain at most five to seven time points, and (2) when many time points can be obtained, such as in the study of biorhythms and psychophysiological processes, measured continuously.

Short Time Series Studies

By far the most frequent repeated measures study is one that follows the behavior of the same subject over a small finite number of points.

Typically a number of measures are taken periodically to provide a small number of time points per variable. For example, one might study potential changes in attitudes toward an issue by following the same subjects over a year's duration, obtaining attitude measures every other month. Studies like this take large amounts of time, money, energy, and other resources, and yet only six time points are obtained. In this case, one may not be able to use sophisticated filtering techniques because these methods frequently require more data points to adjust their parameters, so that they can skim off the noise due to measurement error. However, we can suggest a general procedure for using a "dumb" filter or smoothing technique that will eliminate much of the raggedness of one's time series data due to measurement error.

Representation Functions. Here is a concrete situation where a smoothing approach can be used. Suppose one were measuring attitudes toward an issue in a laboratory situation in which people were given highly favorable messages toward one side of the issue over a time period of six consecutive weekly sessions. Assume that the issue is not one that is discussed outside of the laboratory between sessions. One might measure those attitudes each session by instructing the subjects to describe their endorsement of statements concerning the issue on a scale ranging from 0 to 100, 0 meaning no endorsement at all and 100 meaning maximum endorsement of the statement. Suppose further that the attitude change process could be adequately represented by a dynamic model in which change in one's attitude was a function of the discrepancy between one's position and the message, as well as a *polarity effect* that states that subjects with extreme positions, that is, close to 0.0 or 100, do not respond quickly to extreme messages. This polarity effect assumes that if you are in the moderate range of the scale, then it is easier to change. Suppose subjects were given a message highly favorable to one side of the issue. This would imply that individuals who were initially extremely low on the attitude scale, would at first move off of their initial position slowly, but as they gradually move into "neutral" territory, their attitudes change more rapidly, and finally, as they begin to take an extreme position moving toward 100, attitude change would begin to slow down again.

If the polarity effect holds, one would predict an S-shaped curve over time for subjects starting at low positions on the attitude scale. However, that would be in theory, without the presence of measurement errors, for usually empirical curves are jagged, reflecting noise around the smooth curve due to measurement errors. If there are only five or six time points, the theoretical S-shaped curve may not be manifested in the empirical (fallible) time series. Assuming now that one has developed a dynamic model of the process and one also has a set of fallible time series data, we suggest the following procedure when working with this small time series situation:

1. *Model exploration and sensitivity analysis.* Learn as much about possible behavior modes generated by the system model (see Chapters 4 and 6, this book). In the case of the attitude change model example, one should expect S-shaped time curves for persons initially holding extremely low positions compared to others whose initial attitudes were neutral or extremely high (close to 100).

2. *Use of representation function.* Pick a *representation* function that captures the theoretical behavior of the system. For example, S-shaped theoretical curves may be approximated by a third-degree polynomial.

3. *Estimate parameters of representation function.* Fit the empirical time series data to the representation function.

4. *Use fitted representation function as empirical data.* For each time point, evaluate the representation function to obtain an estimate of the smoothed behavior of the system. This will generate a smooth curve and help to skim off outliers generated by measurement errors.

5. *Fit the system model to smoothed data set.* Using these points as one's empirical data, fit the original *theoretical* model to the smoothed time series.

Some explanation of this suggested procedure is in order. First let us differentiate between the *theoretical* system model, the nature of which was described in detail in previous chapters. Mathematically these system models are composed of one or more differential or difference equations. Usually system models are not explicit functions of time *per se*. They represent the mechanisms of change. One uses a computer to find a *solution* to these differential equations that is the time series behavior of the system. Thus, when working with the actual systems model, one goes from a set of laws of change, in the form of differential equations, to their solutions, which is the behavior of the system in the form of a time series. On the other hand, in using *representation* equations, one never addresses specific change mechanisms but only sticks to describing time series behavior.

Representation equations are always explicit functions of time, which describe the general trends in the original time series. An example of this approach is the use of the ANOVA for trend analysis (see Keppel, 1982). One systematically partitions the main effect, usually time, into linear, quadratic, and perhaps higher order effects to describe polynomial trends in the time series data. This would be a specific application of the representation equation approach. Usually in the context of trend analysis one only tests for the significance of each orthogonal effect without actually obtaining the coefficients of the polynomial being fit to the time series. In the context of modeling the mechanisms that explain the behavior of the system, representation functions are approximations of the solutions to the original set of differential equations.

This suggested procedure is somewhat similar to the methods currently used in understanding physical systems. Acton and Squire (1985) describe a method for approximating the time behavior of systems represented by very complex differential equations. Differential equations are mathematical equations describing how the set of state variables is changing over time. One proceeds from the equations describing the change in the state variables to a solution of the differential equation (if one exists), either by mathematical analysis, numerical approximation, or simulation techniques. The solution is the state variable as a function of time. The method of approximation described by Acton and Square (1985) requires the researcher to develop a rough qualitative sketch of the solution to see its shape. At that point, they suggest picking what they call a *trial function*, which describes the state variable as an explicit function of time to approximate the solution to the original differential equation. They describe the use of this approximation function from the point of making the original differential equation even more understandable than before, because the parameters are being interpreted from the point of view of a structurally simpler model. Frequently the characteristics of these simpler models are well known, understood, and perhaps more intuitive.

Acton and Squire (1985) suggest several general classes of functions that can serve as trial functional approximations to physical processes. They are (1) sinusoidal curves, (2) exponential draining and growth curves, and (3) parabolic curves. To this we might add (4) logistic curves and other curves derived from polynomials besides from the parabolic functions. It should be made clear that Acton and Squire's approach is to approximate the time solutions to *theoretical* differential equations and essentially stop there. What we suggest is that the approximating function, whatever it is, should be fitted to actual *fallible data* to smooth out the measurement errors. Once the function, acting as a filter, has lowered the size of the measurement errors, one can then fit these newly obtained smoothed data, which hopefully have less measurement error at this point, to the original set of differential equations to estimate the parameters of the model.

There are some philosophical issues to be resolved here that may immediately come to mind. In terms of the sequence of events, it is suggested that it is best to develop a dynamic model, perform a sensitivity analysis, and spend time thinking about the potential impacts of change in one or more policies to solve the original social problem. All of this can be performed prior to extensive data collecting. However, once having collected empirical data, is it appropriate to use the information about the shape of the curve generated by the theoretical model to filter the empirical data set? Is that not cheating somehow, for after all, it assumes that the original model is correct? To be clear, using the model as a set of change rules, the computer will generate a solution to the original differential equations in the form of a time series that has a specific shape. The researcher, using the shape of the outputted time curve as a guide, picks a trial function to approximate the empirical time series data set. Then, after that specific curve has been fit to the data set, the original system model is fit to the smoothed data instead of the original data set, that had greater measurement error. The appropriate answer is that this two-stage data-fitting process is essentially analogous to a "smart" filter that (1) has a knowledge of the dynamics of the process, that is, has an underlying process in mind, and (2) attempts empirically to go through the time series data to learn about the size of the error term to filter out measurement error and leave the basic signal generated by the deterministic process alone. This is opposed to so-called "dumb" filters that have no underlying mathematical model concerning what deterministic process is generating the time data.

A Numerical Example. To illustrate this approximation method, let us return to the polarity model of attitude change. We ran two sets of 50 simulation runs of this model over a time horizon of seven time periods. For each run and each of the six time points, we added or subtracted a normally distributed error term to the latent (theoretical) variable generated by the model. The mean of the normal probability distribution was set equal to zero with a standard deviation (standard error of measurement) varying over the two sets of simulation runs. In the first set of 50 simulation runs, the standard deviation was set to 5.0. Setting this value equal to the standard error of measurement and calculating the variance of the dependent variable, one can solve Equation 1 for the reliability coefficient, which in the first set of simulation runs, was approximately .92. For the second set of 50 runs, we set the standard error of measurement equal to 10 units, which is comparable to having a reliability of only approximately .6. This resulted in a very insightful exercise. In the runs where reliabilities were in the 90s, the standard S-shaped curves, that is, the smooth curves one expects from the pure theory, held up reasonably well. These curves that contained only a moderate amount of noise generally looked like jagged S-shaped curves. On the other hand, when reliabilities were approximately .6, the points were quite wild! Frequently these simulated empirical curves were not monotonically increasing functions of time.

In the two series of simulation runs, a third-order polynomial function was fit to the simulated fallible time series to smooth out each set of time curves as much as possible. Figure 1a shows an example of fitting the polynomial filter to fallible data which roughly has an overall reliability of .9. The theoretical curve and the filtered scores overlap almost completely except for the last time points. Figure 1b shows a comparison among the original theoretical point, the filtered scores and the fallible scores that were the input into the polynomial filter. In this particular run, the filtered data were much closer to the theoretical prediction of the model than the fallible scores except for time point 1, where filter did a bit worse.

If the fallible data have relatively low reliability, the filter has a much more difficult job eliminating measurement error. Figure 2 shows the results of filtering a specific time series, run 28, in which the reliability of the time series was approximately .6. In this particular run, the polynomial filter did a reasonable job, yet not as well as in the previous example where the reliability of the time series was .9. Figure 2b shows how well the filter reduced the fallible data, which is used as input. Except for the last point, the filter reduced measurement error. For example, for the fourth point, the value of the attitude model was 68.38. The comparable fallible point was 72.6, having a measurement error of 4.22 attitude units above the theoretical value. The value of the polynomial filter was 67.09, generating a smaller measurement error of -1.29.

As a rough indicator of fit we chose the absolute deviation of the smoothed series from the original time series taken without error. For each time point, one calculates the difference between the smoothed value taken from the fitted polynomial equation with the original value of the curve that was generated by the model. Thus,

$$\text{Abs\_error} = \Sigma |X_{\text{SMOOTH}} - X_{\text{ORIGINAL}}|, \tag{2}$$

where

X_{SMOOTH} equals the value coming out of the polynomial filter and

X_{ORIGINAL} is the value generated originally from the theoretical dynamic model.

For any one simulation run, the Abs_error is the sum of the discrepancies between the smoothed polynomial output and original the-

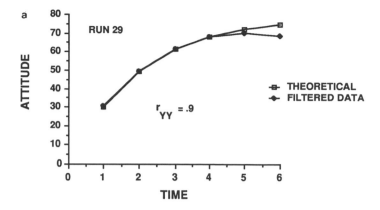

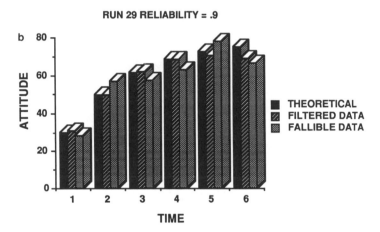

Figure 1. A comparison between theoretical (model) and filtered attitude measures for simulation run 29 when the reliability was set to .9. The lower figure compares the original theoretical curve with the fallible and filtered curves for all six time points.

oretical curve. This was compared to the absolute deviation of the original fallible time series to the series of six points without measurement errors. Figure 3a shows the percentage reduction of the smoothing operation from the original fallible absolute errors when the reliability of the attitude scale was set at approximately .92. Thus for example, if the original average absolute error per point was 4.42 and the fitting of a third-degree polynomial reduced it down to 3.09, then the percentage change was −.30, or 30% reduction. For the case where reliabilities averaged .92, the average percentage reduction was approximately

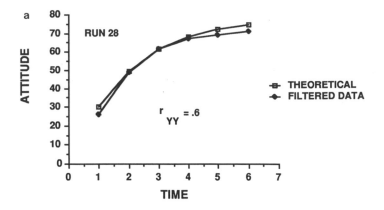

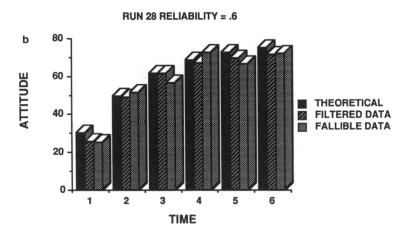

Figure 2. A comparison between theoretical (model) and filtered attitude measures for simulation run 28 when the reliability was set to .6. The lower figure compares the original theoretical curve with the fallible and filtered curves for all six time points.

22%. Figure 3a also indicates a relatively large number of cases at or where the reduction of absolute error was greater than 20%. The figure also indicates about eight cases where the smoothing process actual increased the error, but one notes that, when this happens, the average absolute error per point is relatively low, having a *maximum* of 11% increase.

For the results of the analysis on data that were generated by assuming a reliability around .6, Figure 3b indicates a similar pattern, although the effect of the smoothing operation was not as dramatic as in

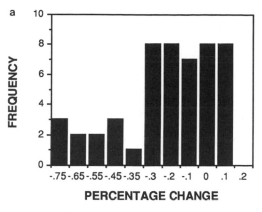

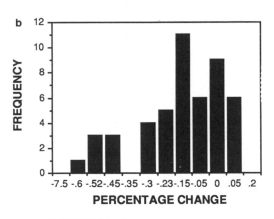

Figure 3. The distribution of the percentage change from the original fallible error rate by using a polynomial filter for data whose reliabilities were (a) .9 and (b) .6.

the case of starting with highly reliable data. In this case, the average change in percentage of absolute error was about −.15, 15%. The distribution again was skewed to the left, showing several cases where the smoothing operation lowered the average error per point by a substantial amount. In terms, however, of the absolute size of error, the smoothing operation suggested here is less impressive. For example, for the case of the fallible curves generated to have a reliability of .93, the mean absolute deviation of the theoretical values from the fallible scores was 3.93 on an attitude scale of 100 points. The polynomial filter reduces that mean figure down to 2.97, on the average. For the case where the fallible time series depicted data with reliabilities approximately at .6, the mean absolute error between the theoretical point from the fallible point was much higher than in the other case, namely 8.41. Smoothing of these very jagged time series reduced the error per point down to 7.16 on the average.

We conclude from the results of this pilot simulation study that using polynomial filters on fallible data of this nature should be recommended but only guardedly. Both Figures 3a and 3b indicate substantial variability in percentage changes. Luckily, in the few cases where smoothing actually led to making the fit worse, the percentage change was close to zero.

Alternative procedures to least squares estimation procedures do exist and may be of use here. For example, there are some regression techniques that may do a better job on outliers that show up so often in unreliable data. Although there may not be much commercial software available for this type analysis yet, there is a growing literature on techniques that get around the lack of robustness of least square estimation processes (see Beaton & Tukey, 1974; Bloomfield & Steiger, 1983; Rousseeuw & Leroy, 1987). These techniques use other criteria for fitting data, such as the *least median of squares* criterion, and are less influenced by extreme outliers, which might occur when measurement error is large. Estimating the parameters of the same third-degree polynomial by using this criterion, for example, may reduce measurement errors even more.

Filtering Longer Empirical Time Series with an Alpha–Beta Tracker

In many areas of psychology and related fields, it is difficult to obtain many time points. However, in some areas, such as neuropsychophysiology, one can obtain longer time series under a variety of circumstances. Under these conditions, the filtering techniques used by engineers can become quite useful for smoothing out measurement errors. One such filtering technique is called the *alpha–beta tracker*. This is what we would call a *smart tracker*, which modifies itself as more and more of the time points are fed to the tracker. The device is really a data-fitting model; unlike the previous approach to simultaneously fitting a set of points to a polynomial, in a global manner, the alpha–beta tracker fits the data locally by starting from the first point, and sequentially modify itself, as each point is included as input into the model.

The idea of calling this filter a *tracker* comes from its historical roots in the tracking of moving objects by means of radar technology.

The alpha–beta tracker was used to measure the distance from moving objects, such as airplanes, by means of reflecting signals that were unfortunately unreliable due to the presence of noise in the data. Consider the simplest situation. The radar station might send out a single pulse moving at the speed of light that would hit the object, be partially absorbed, and partially reflected back to the origin of the radar signal. The time it took for that process to occur provides information concerning the distance between the radar source and the object. In this context, the alpha–beta tracker is a device that not only filters noisy return signals but has to track the target or object as it moves in time. The dual roles of reducing noise in the signal and tracking the movements are sometimes incompatible. The filtering process uses information concerning the present location of the object as well as how fast the object is traveling to predict the next position. In order to work with noisy data that are changing, a series of pulses must be sent out and processed. The tracker helps to trace the signal and ignores noise in the system.

The alpha–beta tracker is particularly successful in the radar case. The same filtering technique can be used on any time series data set if one considers movement in terms of changes in the underlying process away from a constant value. Assuming that the trend in the latent or true scores are not constant over time, the alpha–beta tracker attempts to reduce the error term and at the same time follow trends in the data by estimating the rate of change of the variable as a function of time. Thus if one is studying the short- and long-term behavioral impacts of a change in the criminal code, the number of arrests may rapidly increase when the new code goes into effect. The alpha–beta tracker must sense this rapid change in the number of arrests, even though the data may only be moderately reliable.

Here is how the alpha–beta tracker works. First let us differentiate between the terms *prediction* and *estimation* used in the filter literature. The term *prediction* will be used here to forecast the value of a *future*, fallible score, whereas we shall use the term *estimation* to represent the process of using some data to guess the value of an unknown set of underlying true scores and/or the rates of change of a set of true scores over time.

Suppose we look at how the alpha–beta tracker begins its dual function of noise reduction and tracking the trend in the measured data. In keeping with the J, K, and L notation used by system dynamicists for differentiating time points, let us start at the initial time point, J, and understand how the tracker filters and tracks the next two points at times, K and L. Given an empirical (fallible) time series of a key variable, the alpha–beta tracker has to have an initial estimate of the true score at time J, which we shall call EST(TRUE.J), and an estimate of the rate of change in the true score, EST(RATE.JK), as it moves from time J to K. Given these two initial estimates, the alpha–beta tracker will predict the next fallible score in the time series by the following prediction equation:

$$\text{PRED(FALL.K)} = \text{EST(TRUE.J)} + \text{DT*EST(RATE.JK)}, \qquad (3)$$

where
PRED(FALL.K) = the prediction of the next measured, fallible point in the series
EST(TRUE.J) = the estimate of the value of the true score at the Jth period
EST(RATE.JK) = the estimate of how fast the trend in the true scores is changing, and
DT = the time interval associated with this analysis that will be set to a value of 1.0 in this case.

This equation will generate a prediction of the next fallible point. At the beginning of the time series, this prediction will differ from the actual measurement of the next fallible score, so that we can define a *prediction error* as the difference between the actual fallible measurement at Time K and its prediction that was made in the previous time period. Although a prediction of the fallible point is of interest, in reality, the alpha–beta tracker strips off the measurement error by estimating the true or latent score at Time K. The alpha–beta tracker then goes through a process of updating itself, by comparing the actual point at Time K with its prediction and using this information to make a new, updated estimate of the true score located at the second time period, K. The calculations used for estimating the true score is seen in the following equation:

$$\begin{aligned} \text{EST(TRUE.K)} = {} &\text{PRED(FALL.K)} + \text{ALPHA*} \\ &\text{[FALL.K} - \text{PRED(FALL.K)]}, \end{aligned} \qquad (4)$$

where
EST(TRUE.K) = the estimate of the value of the true score at Time K,
PRED(FALL.K) = the prediction of the fallible score at Time K calculated at Time J,
FALL.K = the value of the variable actually measured at Time K, and
ALPHA = a positive parameter of the alpha beta tracker.

A second piece of information needed to make a prediction of the next fallible score in the time series deals with the rate of change in the true trend in the data. This can be accomplished by using the following equation:

$$\text{EST(RATE.KL)} = \text{EST(RATE.JK)} + \text{(BETA/DT)}*$$
$$[\text{FALL.K} - \text{PRED(FALL.K)}] \qquad (5)$$

where
EST(RATE.KL) = the new estimate of the rate of change from periods K to L,
EST(RATE.JK) = the old estimate of the rate of change from periods J to K, and
BETA = a positive parameter of the alpha–beta tracker.

Once estimates of the value and rate of change of the true score have been obtained, one can make a new prediction about the next fallible score at Time L. In order to use the three previous equations to move the system over time, a computer program implementing the alpha–beta tracker would then update the indexes of time by reassigning K to J, L back to K, and then designate the fourth point in the series as point L. When this is accomplished, the estimates of the true score and the rate of change in the trend of underlying true scores calculated in the previous step can be fed into Equation (3) for a prediction of the next empirical time point.

One can consider this set of equations as describing a linear dynamic system. Technically the alpha–beta tracker is characterized as a second-order dynamic system that has as its state variables, the predicted value of the next empirical time point, and the estimated rate of change of the true score during any period of time. Equation (4) is an algebraic equation that attempts to locate the value of the true score at Time K. If this estimate is closer to the true score than the comparable fallible score, FALL.K, then the filtering role of the alpha–beta tracker will be useful. Likewise, if the second state variable, which is the estimate of the rate of change in the true scores is accurate, then the alpha–beta tracker can be sensitive to rapid changes in the trend of true scores.

An excellent description of the alpha–beta tracker, along with a mathematical treatment of its properties can be found in Cadzow (1973). In general the output of the tracker displays oscillatory behavior. In order to *critically* dampen the system so that the tracker will converge toward the trend in the true scores without oscillating, the ALPHA parameter must relate to the BETA parameter in the following manner:

$$\text{ALPHA} = 2\sqrt{\text{BETA}} - \text{BETA} \qquad (6)$$

This equation implies that once the value of BETA has been specified, ALPHA is determined as well. Cadzow (1973) discusses the filtering power of the alpha–beta tracker in some detail. Consider a time series of fallible measurements that is generated by a deterministic social process and a set of Gaussian errors, the variance of which can be designated as S^2_{error}. This would correspond to the square of the standard error of measurement. Although the alpha–beta tracker attempts to predict the true scores, the *estimates* of the true scores will not necessarily correspond to the actual true scores. However, under suitable values of BETA, the variance of the error term associated with the output of the alpha–beta tracker will be smaller than the variance of the Gaussian noise associated with the fallible scores. These two variances are related to each other such that

$$S^2_{fallible\_error} = Q^*S^2_{tracker\_error}, \tag{7}$$

where
$S^2_{fallible\_error}$ is the square of the standard error of measurement,
$S^2_{tracker\_error}$ is the variance of the error generated by the tracker, and
Q is the parameter called the *variance reduction ratio*.

When Q is less than 1, then the alpha–beta tracker acts as a filter to reduce the noise inherent in the empirical data. As Q approaches zero, measurement error becomes more suppressed. It can be shown that the variance ratio is functionally related to BETA. Figure 4 shows Q as a function of Beta. It can be seen that after a BETA of 1.8, the functional values rapidly go beyond a value of 1.0, generating a larger error instead of reducing the noise component. Cadzow (1973) suggest that BETA be kept strictly between zero and one for best results. In general, given the same time horizon, but differing in the number of time points spanning over that horizon, for few number of time points, we suggest that BETA range between .25 to .5, and for many time points, BETA should optimally be set in the range of .05 to .1.

Here is an example of the use of the alpha–beta tracker on simulated data from a relatively long time series. The underlying dynamic model is of attitude change. The model was simulated for a total of 24 periods of time. At each time period, corresponding to a measurement of the attitude, a small Gaussian measurement error was superimposed on the smooth curve. To generate the fallible time series, the size of the standard deviation for the normal probability generator was set at .75, with a mean of zero, which might represent a moderately small amount of measurement error.

The original smooth attitude curve, the curve obtained from simulating the model over time, and the fallible time series is shown in

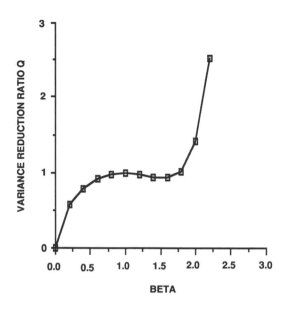

Figure 4. The nonlinear relationship between the Variance ratio, Q, and the BETA parameter.

Figure 5. The jagged curve in Figure 5a roughly approximates the smooth theoretical curve. Next, we attempted to use the alpha–beta tracker, with BETA set to .1 and the rate of change set to 1.0, to filter out the measurement error. The results of this filtering process are found in Figure 5b. Note that at the first three points or so the tracker was systematically off the mark but improved with more exposure to the series. The tracker has to adjust to the poor initial estimates of the location and rate of change in the curve. By the tenth time period or so the tracker was starting to attenuate the measurement error and to better approximate the smooth theoretical curve.

Short Time Series Revisited

It would be of interest to see how well the alpha–beta tracker can do when the number of time points is small. Previously we had attempted to smooth out a fallible time series by means of approximating an S-shape time curve by a third-order polynomial. How well would the alpha–beta tracker do against the polynomial filtering technique for the same data set? We next ran the alpha–beta tracker on the same 50 fallible simulation runs under conditions where the reliability of the

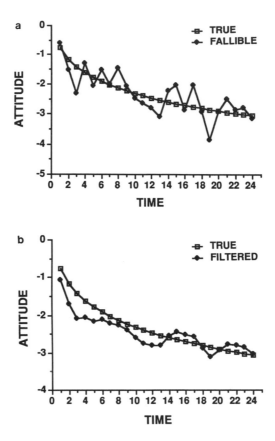

Figure 5. A comparison between the theoretical attitude curve and the fallible curve (a) before and (b) after the filtering by means of the alpha–beta tracker.

data was approximately only .6. Because we were working with only a few points, we set BETA equal to .25. One of the big problems was having an *initial* estimate of the true rate of change in the curve. We applied the following rule: If the second fallible point in the series was greater than the first fallible point, then set the rate of change in the curve equal to 15 attitude units per time. On the other hand, if the fallible curve went down, then set the rate equal to −15 units per time.

In reality, with this set of parameters, the theoretical curve goes up, so when we set the initial estimate of the rate of change in the curve to be negative, then we are taking a conservative strategy because the tracker has to reverse itself in future time periods to follow the theoretical curve that was going up, even though initially the fallible curve with measurement error was going down.

The results of this analysis when the reliability of the fallible data was approximately .6 showed that the average absolute error was reduced by the alpha–beta tracker (for $N = 50$) from 8.34 to 7.21, a reduction of approximately 13%. In this situation, the alpha–beta tracker was approximately equivalent to the polynomial filter approach previously described.

In summary, filtering techniques have not been applied widely to psychological time series. It is felt that filtering techniques, such as the alpha–beta tracker and the substitution method, as represented by the polynomial filter, may be useful in future studies to reduce measurement errors. Such reductions may be needed in later stages where one wants to obtain estimates of parameters and assess the fit of the model to one's data. The filtered data may be closer to true scores than the original fallible data set, and therefore the estimates of parameters may be more exact. More extensive studies should be performed to assess the utility of these filtering techniques on behavioral data. At this point, they show some promise. It also should be stressed that even if these techniques can filter measurement errors effectively, there is no substitution for developing highly reliable psychological scales in the first place, using sound psychometric principles. If one's scales are reliable, then filters have less to do to eliminate measurement errors.

Interval vs. Ratio Scaling of Instruments

In Chapter 5 of Volume 1, which deals with qualitative dynamics, a point was made that there is a difference between quantification and measurement. Quantification deals with assigning a range of potential values to a key concept in one's models. Often this was considered necessary to formulate the model, for if a variable is considered important in understanding the dynamics of a problem, then it should be included in the model, even if one cannot measure it directly. Going beyond a qualitative analysis, however, it is time to consider measurement problems inherent in applying system modeling to psychological data in detail and to relate those problems to statistical and psychometric methods. The first problem one runs into is that in system dynamics and in dynamical systems theory (introduced in Chapters 3 and 4 of Volume 1), complex processes are theoretically represented by nonlinear differential equations. The data associated with these nonlinear equations should be measured on a ratio scale, if the equations are to make sense mathematically. Also, from a scaling point of view, if the theoretical concept underlying a state variable has a natural zero, then the concept should be measured on a ratio scale, if at all possible.

Indeed, many psychological variables may fall into this category. For example, any motivational variable, such as hostility, interest, enthusiasm, energy stores, appetite, or coping mechanism might fall into this category. Along the same line, aptitude variables that might change in a situation, such as skill, knowledge, strength, and other personal resources may have a natural zero level. Conceptually one might have no appetite, no enthusiasm, no knowledge about a subject, and so forth.

When to Use Multiplicative Expressions

All of the variables mentioned might be measured by means of developing psychological scales formed by either averaging or summing over a set of items designed to tap off the dimension under study. Psychologists have developed an elegant methodology for generating instruments, evaluating scale properties, such as internal consistency reliability, and relating one's scale to other similar scales found in the literature. Indeed every graduate program in psychology and related fields have one or more courses that deal with basic psychometrics, factor and cluster analysis, and scale reliability theory. Although most psychological scales produced this way are at best interval scales, correlational techniques can easily handle interval scaled data. We can even use regression models to predict behaviors such as suicide, executive performance, or whether a given organization will adopt the latest computer technology.

Necessary Variables and Multiplicative Relationships

In applying system modeling to psychological problems (see Chapters 4 and 5, Volume 1), one must capture nonlinear dynamic processes as accurately as possible. There are many variables that play the role of being *necessary* for generating action. More specifically, action is associated with rates of change in the state variables. If a variable is necessary, it means that if it is missing, no action occurs. This suggests that multiplicative relationships, not additive or linear relations are operating here, so that if one really wants to accurately describe how the system functions, those descriptions must be in the form of multiplicative equations. Let us take one or more of the motivational variables mentioned. If people are not interested in supporting the local symphony orchestra, very few if any people go to concerts. If a person has no appetite, he or she usually does not eat. Likewise, when a necessary resource is missing, action stops. Thus, for example, when energy stores are empty around the twentieth mile on a marathon run, the runner gives up.

Ratio Scales. All of these variables are necessary conditions for action and therefore can best be represented in terms of ratio scales. Another common technique associated with system dynamic models is the use of relative ratios that affect other variables in the model. The usual purpose here is to represent a perceptual process whereby the individual is affected by a variable *relative* to a baseline or expected level of that variable. Thus, for example, when building a model of crowding in prisons, people are assumed to view the crowding situation in terms of the actual prison density to some maximum tolerable level of density. This relative figure might be expressed as the ratio of the actual prison density to the maximum tolerable density. Likewise, in the case of a psychological example, one might want to view achievement in light of a person's desired achievement. If, in a study, achievement is operationally defined as, let us say, the number of problems solved in a experimental session, the subject's evaluation of achievement might relate to the number of problems solved relative to the person's level of aspiration, that is, the number of problems the subject thinks he or she could or would want to solve. Ratios are convenient vehicles for representing relative quantities. In order to use ratios of this nature, one should work with ratio scales, if at all possible.

Interval Scales in Time Series Studies

If many psychological concepts bear a motivational role or act as resources for the initiation and preservation of action, then ratio scaling techniques are most theoretically compatible with fitting of system models. We could attempt to persuade the reader to apply magnitude estimation techniques to the development of psychological scales. Magnitude estimation techniques were first developed in the twentieth century (S. S. Stevens, 1936, 1955, 1957, 1961, 1968, 1975) and can be used in a variety of settings, ranging from classical psychophysical experiments (see Gescheider, 1988, for a modern review of this area) to the study of attitudes (Lodge, 1981). Moreover, Woefel and Fink (1980) have successfully developed a multidimensional scaling technique, called the *Galileo method*, which generates ratio scale attitude measures. Although ratio scale techniques are currently available, the interval technology associated with the development of psychological scales, through the analysis of item responses, has served us well over the years. Interval scales formed from combining intercorrelated items have been part of the psychologist's toolbox that has enabled social scientists to learn much about psychological processes. We feel that, rather than encouraging a radical shift toward ratio scaling techniques, it would be best to first indicate the type of errors one might make using interval time data and then suggest ways to compensate for the lack of ratio properties inherent in one's data through a transformation procedure. This might allow one to use the "bread-and-butter" methods one is taught in graduate school when working with system models, just as one would use this type of scale in performing statistical analysis such as factor and regression, and ANOVA, and the like. All of those tools can be retained in applying systems thinking to a psychological problem. In addition, we will suggest a way of transforming one's time series data to a set of numbers that have ratio scale properties.

An Example

Although we have made an obvious argument for stating that ratio-type scales are theoretically better than interval scaled measures when multiplicative relationships predominate, we have discovered another important reason for wanting to change interval measures into ratio measures. Going back to our example of a nonlinear attitude change model, suppose a person's initial attitude were quite extreme, and he or she were given a message in the opposite direction. The S-shaped behavioral response made some sense, because, if the person's initial attitude was quite extreme, and a polarity effect operated, then change in the opposite direction would start off very slowly. The polarity effect states that change is slow at the extremes. On the other hand, if initially the person's attitude is in moderate territory, then change would be faster. Thus, for changing extreme attitudes in the opposite direction, attitudes change ever so slowly at first; then, as attitude values go into intermediate levels, attitudes change quite rapidly and then slow down again as one's attitudes become extreme at the opposite end of the scale. This would generate a S-shaped curve over time.

The S-shaped curve is, of course, a theoretical curve. As we have seen, empirical curves, which contain measurement errors give approximations to smooth S-shaped curves in varying degrees according the size of the error variance. Let us assume that the theoretical model's variables (without error) can be considered as latent variables that are tied together theoretically by the model's equations. The first question one might want to ask is whether or not these latent variables theoretically have a true zero point? If so, the optimal strategy is to obtain measures of these latent traits by means of magnitude estimation techniques. However, as we have indicated, it is more likely that the researcher developed a set of interval scales to measure the set of latent traits. The next question asks whether the information obtained by using interval scales essentially matches the shape of the curves one would obtain from ratio scaling techniques? In other words, would one come to different conclusions concerning the fit of the model if one used an interval-scaled curve instead of a ratio-scaled curve? Indeed are they equivalent for purposes of predicting future behavior?

The Relationship between Ratio and Interval Scales

Again consider the S-shaped behavior found in our polarity model of attitude change or changing motivational variables such as interests. The S-shape is a major characteristic of this dynamic model. If this characteristic curve is not observed in using interval attitude data, assuming the model to be valid, but ratio scaled measured variables do yield the S-shape, then something is wrong with using interval data for time series. In order to see this potential artifact of using the wrong scaling technique on time series data, we have to go to a bit of psychophysical scaling theory. Stevens (1975) discussed at length the functional relationship between interval (categorical) and magnitude estimations of the same latent trait. If interval measurements are plotted on the Y axis and the ratio counterparts are plotted on the X axis, one would observe a concave downward curve that is obviously nonlinear. Based on theory, Baird and Noma (1978) suggest the following equation to describe the relationship between a latent trait that has a natural zero and an interval measure of that trait:

$$X_{int} = m\log(L_{ratio}) + k, \tag{8}$$

where
X_{int} is a measurement of the trait on an interval scale and
L_{ratio} is the value of the latent trait that has ratio properties.

Equation (8) indicates that the interval measurement is a linear function of the $\log(L_{ratio})$. Thus in theory if the values of parameters m and k were known, and one were given a specific value of the latent trait, say 100, then one could plug that information into Equation (8) to obtain the interval measurement of that value. In general, the only way one can obtain estimates of the parameters m and k is to have highly reliable measurements of both L_{ratio} and X_{int} so one could use standard regression techniques. Normally one might take time to obtain two sets of measures, one interval and one ratio scaled set of measurements to estimate m and k. Instead, to investigate what the curves would look like, we suggest combining Equation (8) and simulation techniques to ascertain differences in the shape of the curves if any. Then we shall suggest a way to only obtain one set of measures with interval scales and to map back to the latent counterpart using the inverse of Equation (8).

Suppose the polarity model of attitude change was a good representation of how interest in listening to hard rock music changes over time. Let the variable INTEREST be the latent attitude, L_{ratio}. In formulating the model, we might want to quantify the ratio version of INTEREST to range from a true 0.0 to a maximum of 100. Now usually people who encourage working with magnitude estimation techniques like not to have a ceiling on the maximum value of the continuum. But in this case, think of the attitude as being something like a percentage of maximum intensity, with a standard of 10 and a maximum of 100. That is the latent trait side of Equation (8). Now let us examine the categorical or interval measurement scale on the left side of the equation. Typically, a subject might rate a series of bands, some of which are hard rock groups, according to the subject's interest in listening to each band's music. Thus the subject who will be asked to indicate the degree of interest in the topic, ranging from little interest (coded 1), through moderate (coded 3) to strong interest (coded 5), so, for each item in the attitude scale, one could obtain a number from 1 to 5. Usually scale scores are either summed or averaged to obtain a total hard rock interest scale score. Assume that the responses are averaged, so that the interval measurement X_{int} ranges from 1 to a maximum of 5. That will be the way we shall represent X_{int} variable that is found on the left-hand side of Equation (8).

At this point we have a specific representation of both X_{int} and L_{ratio}. Equation (8) indicates that X_{int} is a linear function of the log of L_{ratio}. Figure (6) shows the relationship between the two variables on semilog axes. The log of 100, the maximum value of L_{ratio}, is 2. On the other end of the scale, let us pick a very small value of L_{ratio}, say 1.0, which is very close to a zero value of L_{ratio} 0.0. Taking the log of 1.0 gives a value of 0.0. Figure 6 shows the linear relationship between the log ratio scale and the interval measure. This transformation is complete except for a relatively small portion of the ratio scale, namely the interval between 0.0 and 1.0. Here information is lost, the cost of converting to an interval scale.

What we want is to find the slope and intercept of this straight line, which turn out to be 2.0 and 1.0, respectively. With this specific scaling scheme, we now have specific values of m and k, so that now Equation (8) becomes

$$X_{int} = 2 \log(L_{ratio}) + 1 \tag{9}$$

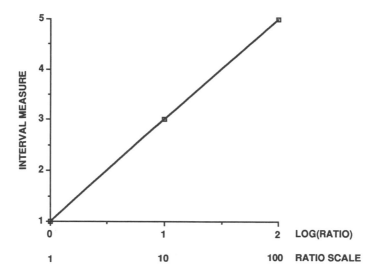

Figure 6. The relationship between an interval measurement and its log ratio counterpart when the ratio measure goes from zero to 100 and the interval measurements range from 1 to 5.

The General Transformation Equation. In order to tie the two variables together, so that a very low value of L_{ratio} equals 1.0 on the interval scale (little interest) and the maximum possible value of L_{ratio} equals the maximum value of X_{int} (high interest), whatever that happens to be, the general equation is

$$X_{int} = .5(X_{maxint} - 1)\log(L_{ratio}) + 1, \tag{10}$$

where X_{maxint} is the maximum value of the interval scale.

Thus, if one is using a categorical scale that goes from 1 to 7, instead of 1 to 5, then the slope will become 3.0. Once the two parameters have been specified, the next step is to compare the results of a simulation run that maps out the time course of both L_{ratio} and X_{int}. Figure 7a indicates the nonlinear relationship between the two types of scales. This is a phase space representation of the behavior of the model, looking at the relationship between the two methods of measuring INTEREST. Figure 7b shows the more traditional view of the behavior of the two different indexes of INTEREST over time. At every time point, the computer first obtained L_{ratio}, the ratio version of INTEREST. Then it substituted that value into Equation (9) to obtain the interval measurement, X_{int}. One can see that the two theoretical time curves do not correspond to each other in shape! The predictions of obtaining an

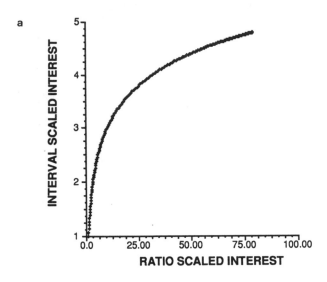

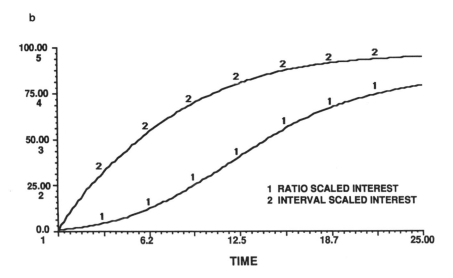

Figure 7. The dynamic representation of changes in attitudes when attitudes are measured on either an interval or ratio scale. The upper figure is a phase portrait, and the lower figure shows the same information in terms of two time series of the attitude variable.

S-shaped curve, which is the dominant behavioral characteristic of the polarity model, would be missing when using interval scaled variables. Over time the interval curve rises rapidly at low values (around 1.0) and then does begin to rise slowly, as would be expected as the values of the variable increase toward 5.0, the top of the scale.

Does the discrepancy between the two curves indicate that interval measures somehow are artifactual. It seems to us that this is not the case. In this figure, the interval curve is above all ratio points generated by the dynamic model. As long as the researcher who is testing the model realizes that one would not expect S-shaped behavior when measuring interest on a Likert-type interval scale, there is no problem sticking with this popular scaling technique. The problem would occur if the researcher rejected the model on the basis of not seeing an S-shaped curve when examining empirical time series based on interval data.

Staying with What Is Familiar. When formulating a dynamic model of a problem, it was suggested previously that it is best to quantify all important variables. Further, at that stage, one should decide whether all variables that have been quantified have a natural absolute zero. If so, one should then take time asking questions about the operationalization of each variable of interest. Many of the psychological variables would be measured by means of interval scaled variables. If that is the case, it would be trival to use Equation 10 to generate predictions of the major psychological variables if and when they might be measured on an interval scale. Now suppose one collected empirical measures of the psychological variables at several time points. The model's interval predictions such as what one saw in Figure 7b could be fitted to the empirical interval data, so that estimates of the time parameters of the model could be obtained. In other words, we might assume that most of the data of interest will be interval data. Use the interval predictions to fit these interval data curves.

Hidden Potential Costs of Using Interval Data. Although at this point one might feel comfortable with making interval predictions, note that in deciding to develop instruments that measure the intensity of a psychological variable in terms of interval measures, some information may be lost. For example, the true zero on the latent ratio variable cannot be mapped into the interval scale, although the magnitude value of 1.0, which is close to zero can. The other potential problem with this approach is that the ratio scale is bounded above at 100. This is probably not too restrictive for our purposes, looking at the summary of magnitude estimation studies reviewed by Stevens (1975). The problem is at the lower end of the magnitude scale, between a value of 1.0 and zero. Anything below 1.0 cannot be mapped onto the interval scale because the scale does not go below 1.0. This presents something like a floor effect, which may or may not be operating in a given situation. In general, as long as the ratio scaled variables do not take on exactly a value of zero, but stay low, interval scale values will operate fine as a substitute for ratio scaled instruments.

Summary

This chapter addressed some crucial methodological issues often left out when discussing the application of systems thinking to psychological problems. Characteristically, system dynamicists have not spelled out in detail the relationship between the development of system models and the world of time series analysis. This chapter provides a brief introduction to the issues involved in designing empirical studies and handling the data once it has been collected.

The first issue dealt with the need to obtain time series data instead of cross-sectional results, especially in attempting to study how the system changes. Although there are advantages of attempting to study average time series, it was argued that the system model and data set should be derived and studied at the individual or organizational level, rather than attempting to simultaneously study many companies changing over time, if their initial states are different at the time of study.

A second topic covered in this chapter dealt with the devastating effects of measurement error on time series analysis. The notion of filtering out these measurement errors through finding a representative function or through a formal filtering technique, such as the alpha–beta tracker, was introduced. The small simulation studies reported here seem somewhat promising. It would be of interest to perform more complex studies to validate these initial results of filtering errors from social science time series data.

The final topic mentioned in this chapter dealt with the scale characteristics of the major state variable measured in an empirical systems study. Before working on this problem of translating scale types, the author had been upset with the trend in relying more and more on interval measures without looking at the consequences for representing processes. Theoretically, few people were asking whether or not the latent trait underlying one's measures should have an absolute zero. The suggestion of using the log function associated with Equation 10 goes a long way to making one's systems model compatible with Likert-type scaling procedures. Going back to Figure 7b, as long as one knows that both curves are legitimate theoretical predictions and which curve is which, there will be little problem in working with interval data. In terms of using filtering techniques described previously, from Figure 7b, one would expect that a second-degree polynomial would do a reasonable job smoothing noisy data based on Curve 2. The results of using this filter then could be put into the alpha–beta tracker to skim down the measurement errors even more. The empirical time series would then be ready to be used in assessing the model's fit, which is the topic of the next chapter.

References

Acton, J. R., & Squire, P. T. (1985). *Solving equations with physical understanding.* Boston: Adam Hilger, Ltd..

Baird, J. C., & Noma, E. (1978). *Fundamental of scaling and psychophysics.* New York: John Wiley & Son.

Beaton, A. E., & Tukey, J. W. (1974). The fitting of power series, meaning polynomials, illustrated on band-spectroscopic data. *Technometrics, 16,* 147–185.

Bloomfield, P., & Steiger, W. (1983). *Least absolute deviation.* Boston: Birkhauser.

Cadzow, J. A. (1973). *Discrete-time systems.* Englewood Cliffs, NJ: Prentice-Hall.

Gaynor, A. K., & Clauset, K. H. (1983). Implementing effective school policies: A system dynamics policy analysis. *Proceedings of the 1983 International Systems Dynamics Conference, 1,* 307–314.

Gescheider, G. A. (1988). Psychophysical scaling. In M. R. Rosensweig & L. Porter (Eds.), *Annual Review of Psychology, 39,* pp. 169–200. Palo Alto: Annual Reviews, Inc.

Keppel, G. (1982). *Design and analysis a researcher's handbook.* Englewood Cliffs, NJ: Prentice-Hall.

Lodge, M. (1981). *Magnitude scaling: Quantitative measurements of opinions.* Beverly Hills: Sage Publications.

Rousseeuw, P. J., & Leroy, A. M. (1987). *Robust regression and outlier detection.* New York: John Wiley and Sons.

Stevens, S. S. (1936). A scale for the measurement of a psychological magnitude: Loudness. *Psychological Review, 43,* 405–416.

Stevens, S. S. (1955). The measurement of loudness. *Journal of the Acoustical Society of America, 27,* 815–820.

Stevens, S. S. (1957). On the psychophysical law. *Psychological Review, 64,* 151–181.

Stevens, S. S. (1961). To honor Fechner and repeal his law. *Science, 133,* 80–86.

Stevens, S. S. (1968). Ratio scales of opinion. In D. K. Whitla (Ed.), *Handbook of measurement and assessment in behavioral sciences* (pp. 171–199). Reading, MA: Addison-Wesley.

Stevens, S. S. (1975). *Psychophysics: Introduction to its perceptual, neural, and social prospects.* New York: John Wiley & Son.

Woefel, J., & Fink, E. L. (1980). *The measurement of communication processes: Galileo theory and method.* New York: Academic Press.

5

Parameter Estimation and Assessing the Fit of Dynamic Models

Ralph L. Levine and Weldon Lodwick

Introduction

This chapter will cover those aspects of systems analysis that are more numerical in scope, dealing with ways to assess the relationship between dynamical models and empirical data. We shall describe techniques that have been found useful in estimating parameters of dynamic models. The chapter will also cover ways in which the modeler can assess how close the model predicts various quantitative and qualitative aspects of the real system under study. Finally we shall describe a method for pinpointing the exact nature of the model's specification errors in terms of various parts of the model.

Fitting of Old Data vs. Forecasting

Within this context, let us differentiate between two different tasks. The first is to fit the model to a set of data, such as a time series, that represents the past behavior of the system. The second task is to use the model as a forecasting tool to predict future behavior of the system

Ralph L. Levine • Department of Psychology, Michigan State University, East Lansing, Michigan 48824.   **Weldon Lodwick** • Department of Mathematics, University of Colorado, Denver, Denver, Colorado 80217-3364.

Analysis of Dynamic Psychological Systems, Volume 2: Methods and Applications, edited by Ralph L. Levine and Hiram E. Fitzgerald. Plenum Press, New York, 1992.

under perhaps unknown or historically rare conditions. The two tasks can be relatively independent of each other. What may tie them together is the insight generated by developing the model. In general, if the model is not insightful, one might be able to fit *past data* extremely well but have a difficult time predicting the future of the system.

Consider a team of recreation economists and/or geographers who work for a state commerce or natural resource department. Their administrator wants them to forecast the next 5 years' recreation demand for a state park. Suppose during the last few years, recreational use has been rising rapidly and the economists are aware of these increases in use. Figure 1 shows the potential curves the group is evaluating. The team thought it best to extrapolate from past performance. The problem is that they do not even know what the shape of the curve will be in the next few years. The first member of the team wants to use the last two data points to fit a straight line through the end of the curve. Another team member feels that recreation demand will level off in the future. She forecasts that the curve will flatten out for the next several years, which would generate Curve 2. On the other hand, the third team member feels that what goes up must eventually go down. As a result this person forecasted Curve 3.

The point is that past behavior, *per se*, cannot always give enough

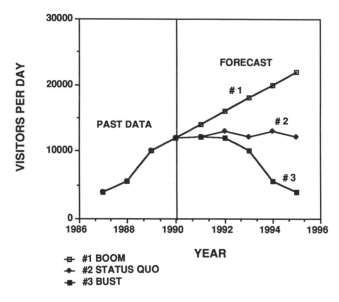

Figure 1. Recreational forecasts of three specialists making different assumptions about the future, based on the past.

insight to predict future behavior. Unless one has a specific dynamic model concerning recreation demand, the task of extrapolation would be problematic. If the model under consideration is dynamic, then one can identify and study the loop structure associated with the model. Qualitative analysis may reveal the potential behavior modes associated with that loop structure (see Chapter 5 in Volume 1). It will not tell the team what numerical values to expect. However, once the qualitative aspects have been taken care of, then one might want to make more refined numerical predictions by fitting this qualitatively insightful model to the existing data and then perhaps extrapolating a single curve or a confidence band of future values of the system under study.

Parameter Estimation

In the behavioral sciences, the process of estimating parameters of one's model plays a central role in validating the modeling process. On the other hand, as indicated before, when taking a system dynamic approach to time processes, less stress is placed on sophisticated estimation techniques. There are a number of reasons for this. First, in order to develop the model, the researcher has already looked over what time series data existed at that moment and used the qualitative information contained in the series to generate dynamic hypotheses concerning the processes underlying the problem under study. Although there is no guarantee that the hypothesized feedback processes will eventually generate curves that correspond to the empirical time series, in a sense, the data have already been used once before being used again to estimate the parameters of the model. The second reason for not emphasizing the parameter estimation stage is that in most cases a preliminary sensitivity analysis has shown that most parameters are insensitive to major changes, so that changes in parameter values usually do not change the qualitative aspects of the time curves, only how fast or slow the processes move over time.

Sometimes, however, a sensitivity analysis will point out policy levers, that is, parameters that can be controlled somehow to guide the system in a favorable direction. In this case, the researcher must be very careful to estimate the values of these parameters as precisely as possible. Sensitivity analysis can permit the researcher to spend time and other resources in an efficient manner to focus mainly on accurately estimating the most important parameters, rather than appropriating equal time to all parameters. Finally, in most applications of system dynamics, one wants to know what would happen to the system under nonhistorical conditions, that is, under conditions that were not in effect when the data were collected. Some of the estimation methods described in this chapter appear to be more robust than traditional statistical estimation methods that use historical time series data as the basis for finding the value of the time parameters.

Using the Results of Prior Analyses

Three major prerequisites for fitting a dynamic model to behavioral data deal with (1) justifying the level of aggregation, such as an individual, organization, or a group of individuals, (2) using the proper measurement scale to match the dynamics of the latent trait being modeled, and (3) attempting to filter out as much measurement error as possible (see Chapter 4, dealing with the scaling and filtering of empirical data). The next stage is to estimate the parameters of the model by a variety of methods to be described shortly. In dynamic models, a parameter consists of time constants, scaling transformations, initial values, and table functions that by definition are approximately constant during the period of interest. In the behavioral sciences, we have observed that estimation of parameters and the assessment of fit plays the most important role in later stages of the modeling process, especially in regression modeling. The value of the parameters becomes a fundamental dependent variable of interest. For example, in multiple regression analysis, one may want to predict suicidal behavior with a set of predictor variables, such as whether the adolescents were using drugs, told their friends and/or parents of a plan to commit suicide, had given away their prize possessions, and the like. All of these variables might be put into a regression equation. Obviously one would want to know the value of the multiple correlation coefficient, R, which tells one the correlation between the observed suicidal behavior and the behavior predicted by the multiple regression model. Moreover, the size and statistical significance of the beta weights would be of interest to the researcher, especially if the input variables were relatively independent of each other.

The relative size of the coefficients take on even more importance in and interpreting path analytic models than in analyzing coefficients in traditional regression models. For example if predictor variable, X_1, was directly related to variable X_3, and also related indirectly through a path going first to X_2 and then to X_3, it would be of interest to learn of the relative strength of the two pathways.

There is a qualitative difference between a regression (or the usual way one performs a path analysis) approach and the use of parameters in system dynamic models. In regression modeling, the size and direction of each parameter is purely empirical. One *waits* for the estimation process to be performed before saying what the results will be. On the other hand, with a systems approach to modeling, one has devoted much time to prior sensitivity analyses and thinking about the process from a theoretical point of view so that the approximate values of the parameters are frequently known even before all of the data may be in hand. Simulation methods assume that one can supply reasonably good ball park figures for all initial values of the state variables and time constants. Without these, the computer could not generate theoretical curves in the first place!

To illustrate the differences between the two approaches, the "wait-and-see" versus the "ball park figure" approach, the first author remembers serving on a graduate master's committee where a student hypothesized reciprocal relationships between two variables in a structural equation context. After the model was formulated, he performed a path analysis on the data in which all variables were standardized. The analysis showed that the loop being tested was positive because the two path coefficients associated with the loop were significantly greater than zero. As an afterthought, he decided to do an analysis of the same model using the original raw scores, that is, unstandardized variables. In this case, the results showed that indeed the loop under question was significant, but this time it turned out to be a negative loop! Empirically, these results are very baffling. However, after scrutinizing the statistical properties of the structural equations, the members of the committee assured themselves that it is possible to have a significant positive loop when using standardized scores and a negative loop when using raw scores. The real problem comes when attempting to make theoretical sense out of this empirical analysis. Using the systems approach, one usually knows enough about the structure and function of the loop to say for sure whether a feedback process should generate growth (collapse) or should be counteracting. That is the nature of the model building process.

Unaggregated Parameter Estimation Approaches

One of the principles of dynamic model building is that all parameters should be meaningful to both the model builder and/or the decision maker who will be using the model to solve the problem under study. In most circumstances, the parameter should not be an arbitrary number. In many cases the parameters are related directly to time constants, which deal with how fast the system adjusts to changes in the variables.

The time constants themselves are functions of many subprocesses and events that determine how long the system takes to react to internal and external influences. Thus model parameters are aggregations of many small processes that take time. In estimating those parameter values, one can estimate them with the time series data available, fitting the model to the series, or estimates can be obtained by dropping down a level of analysis and coming up with estimates that are developed from unaggregated data. It has been shown by Mass and Senge (1980) that unaggregated estimates are as accurate or more accurate than parameter estimates obtained by traditional statistical techniques, such as generalized least squares methods, which use the aggregated data.

To illustrate the difference between aggregated versus nonaggregated approach to estimation of parameters, consider the task of an electrical engineering student who has to measure or estimate the value of a parameter, such as the resistance, R, in a circuit equation. Suppose the circuit's capacitor dynamics were described by this differential equation:

$$\dot{C} = V * 1/R * C, \tag{1}$$

where
$\dot{C}$ is the rate of change in capacitance,
V is the voltage,
1 is the amperage,
R is the resistance of the circuit, and
C is the capacitance.

In this equation, the resistance, R, is the constant to be estimated. How would the engineering student estimate or measure R in a laboratory situation? One way is to generate some data, namely the value of C, V, and 1 over time and then use a statistical procedure to estimate R from the preceding equation. This would roughly correspond to the statistical procedures used by most social scientists. The other more common way is not to use the equation per se to obtain the value of R but to use an ohm meter to measure the resistance. This is a more disaggregated approach to the estimation problem.

A second example closer to home might be in estimating the value of parameters in a model describing the dynamics of support groups or perhaps neighborhood organizations as studied by Florin, Chavis, Wandersman, and Rich (Chapter 9, this volume). Suppose one were to model the dynamics of a support group over time. One of the major state variables would be the number of people in the. group at any moment in time. In describing the growth or decline of the group over time, the decline in group size might simply be represented by the following equation:

$$DEC\_MEMBERS = MEMBERS/LENGTH\_ACTIVE, \tag{2}$$

where
DEC_MEMBERS is the rate of decline of members of the support group,
MEMBERS equals the number of members at any given time, and
LENGTH_ACTIVE is the average length of time being an active member.

The equation's unknown parameter is the constant, LENGTH_ACTIVE, which typically is operationalized as the average amount of time a person stays in the group before dropping out. Again there are two major approaches to estimating a parameter like this. One is to obtain the value of this time constant using the disaggregated approach. Suppose the support group met every week. If someone were taking attendance, then there would be data at the individual level concerning who attended the meetings. This information could be analyzed to obtain the number of weeks or months a person stayed in the organization that could then be averaged to obtain a value of the LENGTH_ACTIVE constant. Another way to estimate the value of this constant is to ask members who have been in the group for a long time to provide a guess of the upper and lower ranges of active group membership. The estimate would be somewhere in the midrange of the interval and averaged.

These methods of estimating the value of this parameter do not use aggregate information provided by the variable MEMBERS over time. One went down a level of analysis to individual members to obtain one's estimate and then worked back up to the aggregate level. Of course there are pitfalls to this approach, but in general, once these estimates are obtained this way, the researcher then has the freedom to fit the model that now has all of the parameters specified independently of the time series, *per se*. In other words, the fit can be assessed in a more honest way, with more understanding of the meaning of the parameters.

Aggregated Parameter Estimation Approaches

A second approach to estimation is through the application of traditional statistical methods. For the sake of argument, let us assume that

one has made the proper scale transformations and has filtered most of
the measurement error out of the empirical time series data prior to the
estimation process. Chapter 4, this volume, introduces these topics.
Once proper transformations and filtering have been performed, there
are generally two classes of statistical methods used to estimate parameters; a single equation approach and a full informational approach,
which utilizes all of the equations simultaneously. The first aggregated
approach would to use only the data that pertained to each specific
equation. For example, returning to equation (2), one would first solve
for the unknown parameter, LENGTH_ACTIVE. Thus

$$\text{LENGTH\_ACTIVE} = \text{MEMBERS}/ \text{DEC\_MEMBERS} \tag{3}$$

One simple way to obtain an aggregate estimate of LENGTH_
ACTIVE is to pick a given month from the time series, count the
number of members who attended meetings and the number of members who quit the organization that month. By dividing the second
number into the first number, one can obtain the estimate of the
LENGTH_ACTIVE parameter. This is an example of estimating the value of this parameter at the aggregate level.

The second method for estimating parameters of the model deals
with using all or most of the data simultaneously. The reader may be
familiar with this approach to parameter estimation, for these methods
are used in a computer package called LISREL (Joreskog, 1979) by means
of a full information maximum likelihood technique that uses all of the
data simultaneously. Two major estimation techniques *have* emerged to
estimate the parameters of nonlinear models that also use all of the data
simultaneously. The first technique that is a "quick-and-dirty" method
is the *polytope* algorithm (see Gill, Murray, & Wright, 1981, pp. 94–98 or
Press, Flannery, Teukolsky, & Vetterling, 1986, pp. 289–93). These references are reworkings of the original development of the polytope algorithm given by Spendley, Hext, and Himsworth (1962) and modified
by Nelder and Mead (1965). Press, Flannery, Teukolsky, and Vetterlling
have companion software disks containing PASCAL, FORTRAN, and C
implementations. In addition, an article on this method appeared in
Byte Magazine that described the method in detail and demonstrated
how easily the algorithm could be programmed for a microcomputer
(see Caceci & Cacheris, 1984). It should be stressed that the polytope
algorithm works quite well for models containing less than 500 parameters. For larger models, care must be taken in how this algorithm is
implemented. The solution to this problem goes beyond the scope of
this chapter. The interested reader can consult Gill, Murray, and Wright
(1981), who describe limitations and modifications of the method.

The second algorithm is the *complex* method, which was developed by M. J. Box (1965). These quite similar methods estimate the total number of parameters in a model. In the terminology associated with the complex procedure, the number of parameters in a model equals the *number of search variables* that can be called NS. Thus, if one wants to statistically estimate 25 parameters, NS would be equal to 25.

Basic Definitions

A *complex* is also an object in a space of NS + 1 dimensions. In our work here, a complex may have more points in it than the NS parameters. It could be a simplex with NS points or more than NS points. One sets up a so-called *objective function, U,* which can be minimized or maximized to find the best set of parameter values. In both the polytope and complex methods, the objective function used is the sum of the squared deviations of each data point from the corresponding point predicted from the model. One can think of a space where all but one axis represents the values of the model's parameters, with an axis left over for the value of the sum of squares, *U.* Each point in this NS + 1 space represents a vector of parameter values, the height of which is the sum of squares. The closer the point to the origin, the better the fit.

Complex

Although the polytope algorithm has been implemented on computers, we shall spend the remaining time describing how the complex algorithm works, because we have direct experience with software implementing the complex algorithm. To use the complex routine for estimating the parameters of the model properly, it is necessary to understand the meaning of each parameter. If each parameter is well understood, then one can provide a accurate and feasible range of values for each parameter. For example, suppose one of the parameters was the length of time, on the average, that housing lasts before being demolished. We might consult with a few builders and come up with a range of estimates, going from 35 to 75 years. One might then take the center of the interval as our best initial guess concerning that parameter and use the two ends of the range as constraints in defining the feasible region in parameter space.

Data Sources

Assume that one has time series data from a single individual, organization or region, whose behavior is represented by many variables over time. The variables usually are in raw score form, rather than standard-ized form, especially if one is only studying one unit over time. The units of each dependent variable may therefore differ. These data usu-ally contain measurement errors, unless it has been filtered prior to the use of the computer program that implements the complex algorithm. We recommend assessing whether in the quantification stage each vari-able was theorized as a ratio scale. If empirically these variables were measured on an interval scale, then each latent variable in the model must be transformed slightly to give predictions of the shape of the theoretical curve that one would predict for variables measured on interval scales. As we have seen in Chapter 4, in some cases these curves would be different. Also, in this chapter we strongly recom-mended filtering out as much random measurement error from the time series as possible before using the complex algorithm.

Software Availability

We would like briefly to describe a computer package, called M-OPSIM, developed in Michigan State University's Department of Elec-trical Engineering and Systems Science by Thomas Manetsch and his colleagues. This package contains programs for simulating systems and using the complex method for optimization of objective functions. Thus the package can be used for estimating parameters of one's model. M-OPSIM was written in BASIC for the IBM PC and compatible micro-computers and is available for a nominal price from the developers. In what follows, when describing the complex procedure, we will refer to the subroutine COMPLEX, which is part of the M-OPSIM package.

The Objective Function, U

The program, COMPLEX, part of M-OPSIM, finds the best-fitting pa-rameter values by means of an optimization process that will be de-scribed later. Again best fitting, in the sense we are using it, implies that the value of some function, called the *objective function*, is maximized or minimized, depending on the situation. In our current uses of COM-PLEX, we minimize a function by means of the weighted least squares criterion. The objective function is the *weighted* sum of squared devia-tions of the model's predicted time series from historical empirical time series. Because the various dependent variables are measured on poten-tially different scales, we have adopted a *weighted* least squares ap-proach. We weight the squared residuals inversely proportional to the mean value of each dependent variable. Dividing one set of units by the same unit insures a more equal influence of all dependent variables in minimizing the objective function. Thus our objective function takes the form of

$$U = \sum_{j=1}^{J} \sum_{k=1}^{N} \left(\frac{(X_{jk} - O_{jk})^2}{M_j} \right), \tag{4}$$

where

U is the value of the objective function, low values of U associated with better fit of the model to data,

X_{jk} is the theoretical prediction of the jth dependent variable at the kth time period,

O_{jk} is the observed point for the jth dependent variable at the kth time period, and

M_j is the mean observed response for the jth dependent variable over time.

Customarily, COMPLEX minimizes this objective function. In theory, given a point in the parameter space, one can substitute the particular parameter values into the model equations, run the simulation program to generate the model's predictions, and then compare these predictions with the actual behavior of the system, the real time series, by summing the weighted squared residuals, which is what this formula says to do. COMPLEX finds the configuration of parameters that gives the smallest value of U, subject to certain constraints.

Constraints on the Parameter Space

In specifying the ranges of the parameter space, we are actually limiting somewhat the potential value of certain parameters. For example, if the parameter of interest is a time constant, one would not expect the parameter to be negative. Likewise, in most cases, one would not expect a time constant to be infinitely long. Efficient use of COMPLEX requires a good initial estimate of the approximate location of the point in parameter space. The specification of the intervals generates a feasible region from which COMPLEX can have a good chance of converging on the best-fitting parameters. Utilizing some of the material described in Chapter 3 (this book), one could perform a HYPERSENS or an interval arithmetic sensitivity analysis to better understand where the sensitive parameters lie in the first place and then use those intervals for constraints in estimating the parameters of the model in the context of a set of empirical curves.

Type of Constraints. In the context of using the COMPLEX program, one can distinguish between two kinds of constraints, explicit and implicit constraints. Explicit constraints are constraints on each of the parameters separately, while implicit constraints deal with constraints on the combination of parameters. As an example, suppose one had a model which had three probabilistic parameters associated with it, X_1, X_2, and X_3, such that the explicit constraints are

$$0 < X_1 < 1$$
$$0 < X_2 < 1 \tag{5}$$
$$0 < X_3 < 1,$$

and the implicit constraint is

$$X_1 + X_2 + X_3 = 1 \tag{6}$$

Given this information, the COMPLEX algorithm can be described as the following set of stages:

Stage 1: Generating the Complex of Points

The first task is to generate the complex of vertices in parameter space. This is accomplished by first telling COMPLEX how many vertices one wants in the configuration of points, for example, 5. The number of points or vertices in the parameter space is denoted by NV, and this must be greater than the number of unknown parameters (search variables). Again we denote the number of search variables as NS.

Once this is accomplished, COMPLEX then asks for the best guess concerning the sought-after parameter values. Supplying your initial guess allows COMPLEX to place the first point in the parameter space. Then on a random basis, subject to the constraints also specified by the researcher for each parameter, COMPLEX will then randomly generate the rest of the NV vertices around the initially specified point. The program will also check each new point to see that it is not out of bounds. Suppose we want to estimate three parameters and have chosen to do this with a complex of seven points or vertices in the parameter space. When asked for "guestimates" of the coordinates of the initial set of parameters, perhaps three time constants, one might type in 5, 7, and 10 years for the values of the first, second, and third parameters, respectively. The computer will use the coordinates 5, 7, and 10, as the first vertex of the complex and then randomly generate the other six vertices around the starting point.

Stage 2: Evaluating the Objective Function

Once the initial complex is generated, the program evaluates the objective function, U, for each point in the complex, rank orders those values, and finds the value of the worst-fitting point in the COMPLEX, which, in this context gives the largest value of U. This step is done by taking the first point in the space, placing those parameter values into the system model's computer equations, running the model over time to as many time points as one has in one's time series data, and calculating the sum of the weighted squared residuals, our objective function, U. This value is saved, and the program then goes to the second point in the complex, the parameter values are substituted into the model, and the theoretical time series is matched with the data as before to obtain a new value of U. The process is repeated for each of the remaining points, so that now one has a series of values of U, one for each of the seven vertices in the complex.

Stage 3: Replacing the Worst Point

The values of U are now rank ordered. Remember that large values of U mean poor fit, whereas values of U close to zero means that those parameter values provide a good fit to the historical data. However, COMPLEX was designed to provide an iterative procedure where over iterations, the value of U decreases toward zero. To accomplish this goal, the program has to shift the location of the points in the complex in a systematic way. Shifting points will change to values of the objective function, and COMPLEX is designed to keep the same number of vertices, but each iteration is oriented toward lowering the value of U. This is accomplished by eliminating the worst point (the one with the highest value of U) and replacing it by a new point whose value of the objective function, U, is less than the value of U associated with the old point.

The next point is chosen somewhere near the center of the remaining swarm of points, in the direction of the center of gravity of the complex. The program calculates the *centroid* of the remaining vertices. If each of these points can be represented as a vector of parameter values, one for each time constant in the model, then the centroid is the average vector, composed of the mean value of the first parameter, second parameter, and finally the mean of the third parameter. It is the center of gravity of the remaining points of the complex.

COMPLEX passes a line between the worst point and the centroid of the *remaining vertices* and then locates the replacement point some distance *beyond* the centroid. This distance is in proportion to the distance between the worst point and the centroid. The constant of proportionality is called *Alpha*, and typically Alpha is set at 1.3. The hope is that the new point, being far away from the bad point and nearer the center of gravity of the better points in space, will have a value of U that is less than the value of the objective function for the old

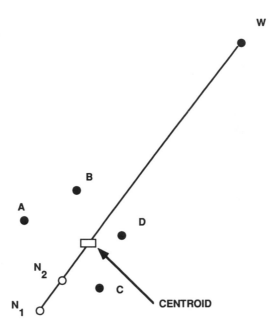

Figure 2. An illustration of the COMPLEX procedure showing a line from a poor point, W, through the centroid of the remaining points in the parameter space. COMPLEX will evaluate a new point, N_1, to see whether SS_n is smaller than point W.

discarded point. It should be made clear that the centroid itself is not the optimal place to terminate the estimation procedure. Putting the new point beyond the centroid, and away from the old worst point, helps to get the system out of temporary hangups and local depressions in the parameter surface, which, in effect, would lead to incorrect estimates of the unknown parameters.

Figure 2 shows a two-dimensional view of the worst point, W, the remaining points, A to D, the centroid of those remaining points, as well as a tentative new point, N_1, which is α distance away from W. N_1 may or may not be better than W. Therefore let us therefore consider what can happen in each case:

Case 1: New Point Violates Constraints. Three things can happen after COMPLEX generates a replacement vertex. First the replacement point can be placed in a region of the space where one or more of the parameters may be violating an explicit constraint. For example, if the parameters are probabilities, the new point may have been placed where the value of P_1, the first parameter, was 1.15, obviously over 1.0. If this is the case, COMPLEX will move the first parameter down to a value of 1.0 and then down again by a small number, DELTA, so that it is not on the boundary.

Case 2: New Point Is as Bad or Worse Than Old Point. A second thing can occur, however. The new point may be within the appropriate boundaries, as defined by the constraints, yet the new vertex has a value of U that is actually *larger* than the value of the discarded point. The original attempt was to find some territory in the space where the value of U was small by placing the new point away from the old point and on the other side of the centroid. When this fails, the strategy is to pull the point back toward the center of gravity of the good points. Therefore COMPLEX moves the replacement point from its original assignment to one-half the distance from the centroid of the remaining points. Figure 2 shows the rejected point, W, the original location of the replacement point, N_1, and the centroid of the remaining "good" points. If the value of U for the replacement point was greater or equal to the discarded point's value of U, then COMPLEX would move the replacement point half way in along the line going through the centroid to point N_2, which would in turn be evaluated to see if this new combination of parameters at point N_2 help the model to fit better than at W.

Case 3: New Point Is Better Than the Old Point. Indeed the value of the objective function, using the new point, N_1, could be smaller than the discarded point. If that is the case, then COMPLEX would accept the location of the new point in the parameter space, and if a stopping rule does not indicate that the estimation procedure should be terminated, the program will begin the next iteration by inserting the value of U for the new point and not considering the value of U for the discarded point, W, because it is not needed any more. The value associated with the replacement point will be somewhere in that rank order of U values. In the second iteration, COMPLEX would start all over again locating the new worst point, and attempt to find another replacement point that does better than the discarded vertex. The rules for the eventual placement of the new point are the same as in the first iteration. After this has been accomplished, the next iteration would occur in the same manner, namely finding the new worst point, replacing it with a better set of parameter values, and so forth until the iterations are terminated by some predetermined procedure.

Stage 4: Stopping the Process

When does this procedure stop? COMPLEX has a rather elaborate termination mechanism that determines whether or not the parameter estimates converge toward a final value. At the end of every iteration, when the values of U for each point in the complex are known, the program calculates the difference among all of these values and then compares these differences to a cutoff point BETA, a small number. If all of the differences are extremely small, that is, less than the set value of BETA, then before going to the next iteration, a counter is set to 1. On the next iteration, differences among the new set of U values are calculated and compared to BETA, and if they are again less than BETA, the counter is set to 2, and so on until this scheme generates GAMMA consecutive iterations of being less than the BETA criterion. The value of GAMMA usually is set at around seven consecutive iterations. This gives the system ample room to move out of local minimum values of the objective function and to move toward the true global minimum value of U. Suppose the BETA criterion was met for five iterations, and then on the next iteration the new point generated some differences in U values that were greater than BETA. COMPLEX would reset the counter to zero and start all over again. The idea behind the BETA test is that as time goes on the complex begins to converge on the optimal solution to the parameter estimation problem. The swarm of points contracts in space so that the distances between points becomes quite small. When the interpoint distances become small enough, the values of the objective function associated with this tightly packed swarm of points also become so small that they are less than the value of BETA, which is set to a number close to zero, like .00002.

An Example

There are few examples of applying the complex algorithm to obtain parameter estimates of dynamic models associated with social system data. We take the attitude change model, given in Chapter 4 in Volume 1, to illustrate how COMPLEX can be applied to estimate the parameters of a social process model. Recall that this model dealt with the influence of higher order attitudes and beliefs on more concrete and specific attitudes and beliefs. The model had two parameters to look for, namely the *time constant* associated with the impact of a message given at the higher level, X_1, and the time to respond to a message at the lower level of the belief hierarchy, X_2. The basic idea to test COMPLEX was to first specify what the "true" parameters should be, namely we set the values of X_1 and X_2 to be 8 and 3 weeks, respectively, to represent the differential response to messages at both levels of the hierarchy. Once we specified the values of the parameters, we then generated a time series to represent the data and submitted this time series to

COMPLEX to attempt to recover the parameters, eight and three weeks respectively.

Running COMPLEX over 40 to 50 iterations, the program attempted to minimize the value of the weighted least squares objective function. After some wandering around, the value of the objective function dropped down steadily, and by the time it met the Beta criterion, .00005, seven times, the value of U, the objective function to be minimized, associated with the estimate of X_1 and X_2 is extremely small, $2.11968e^{-7}$.

Figure 3 gives a slightly different perspective to how COMPLEX finds the minimum point. This scatter plot, which represents the parameter space, shows the location of the centroids of the three vertices with respect to the 50 iterations. Originally, the first centroid, Point 1, was down in the lower middle of the figure. The location of this centroid was at approximately 7.38 and 2.17. The next additional iterations moved the centroid away from the original point, shifting toward the upper right-hand corner. The program had nearly found the value of X_2 by eight iterations but still had a long way to go to accurately estimate the first parameter, which originally was set at 8 weeks. One can see, however, that with additional iterations, by the thirty-second itera-

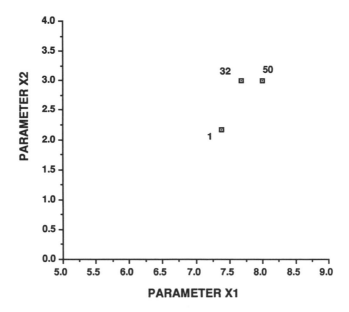

Figure 3. The numerical values of the model's parameters as a function of iterations followed by the COMPLEX program.

tion the program's estimate of P_1 was 7.672. Finally, after seven consecutive iterations where the BETA criterion was met, the program terminated with an estimate of X_1 equal to 8.00058 and X_2 equal to 2.999979, with a value of the objective function equal to $2.11968e^{-7}$!

In this illustration, COMPLEX appears to have sought out the true parameter values. However, we warn those who want to use COMPLEX or a similar estimation program that one must be aware of the role of measurement error in processing social science data. We highly recommend assessing the reliability of one's scales and filtering the time series. In some of the other simulation runs, where we did not filter the time series, the estimated value of X_1 was very far away from 8.0, although the estimates of X_2, which had a true value of 3.0, was 2.96. In general, unfiltered time series data having low reliability will lead to obtaining poor estimates of the model's parameters. Filtering the same data series may result in more accurate estimates. Filtering techniques were discussed in Chapter 4.

Other Full Information Approaches

We have found that the COMPLEX approach to estimating parameters works relatively well if the original data series is filtered first. However, there is another source of errors that also should be considered when dealing with parameter estimation. These errors are called *specification (model)* errors and can also be a source of potential problems as well. Specification errors are the cumulative effects of either missing variables or important relationships among the variables, especially in specifying inherent feedback loops representing decision making and information usage. In simulation, even a good model that captures the most important variables and relationships may be off from what the *true model* should predict, because it may have left out a few minor loops. Because the next iteration of the model is based on the previous prediction, which was slightly off, the next prediction of the model will be off as well, causing some cumulative drifting.

Gpsie

An approach to filtering out the effects of both measurement errors and specification errors is to use a Kalman filter (see Kalman, 1960; Schweppe, 1973; Peterson, 1975). Gregson (1983) provides a fine introduction to the topic of Kalman filters when applied to time series in psychology. The Kalman filter approach attempts to iteratively estimate parameters of the model over time points so that the filter can, as it is exposed to more and more data, keep estimating the size of the two error terms, that is, the size of (1) the measurement and (2) cumulative model error. At each time point, the filter estimates the next starting place for the coming simulation period. This starting point is adjusted to partial out or compensate for the past effects of measurements and specification errors. Peterson (1980) describes a program called GPSIE that use the Kalman filter technology to estimate parameters. The program is written in PL1 computer language, and as far as we know, GPSIE has not been translated into other more common languages, such as FORTRAN or Pascal. It is unfortunate that GPSIE is not available, currently. However, there are other pieces of software available to the modeler that uses the Kalman filtering technique. For example, the Tau Corporation, 485 Alberto Way, Los Gatos, California, offers several Kalman filter software packages written in FORTRAN that might be integrated with one's dynamic model to estimate parameter values. The package runs on a variety of mainframe and microcomputers.

Summary

There are at least four approaches to the estimation of parameters: disaggregate methods, single equation techniques, full information methods, and partial-modeling estimation techniques. Of all methods discussed, disaggregative methods have an appeal because one attempts to find a value of a given parameter by independent means from a knowledge of the subprocesses involved in the formulation of the parameter. It is like taking an independent meter reading of a resistor using an ohm meter. Frequently the purpose of estimating parameters is to use the model for forecasting how the system will respond to a new set of circumstances. If the disaggregation cannot be used, then the partial-modeling approach has much appeal. Actually the single equation estimation techniques appear to be a subset of the more general partial-modeling approach. Partial modeling comes from a desire to understand and to keep track of what is going on in a complex model. It appears to be a very insightful approach to both sensitivity analysis of feedback loops and estimation of parameters.

Finally, there are the full information approaches to estimation of parameters. These are by far the most sophisticated techniques, especially those methods that rely on filtering out measurement and specification errors as the filter "sees" more and more data. We have found the COMPLEX program to be quite useful in estimating parameters from filtered data sets. In general, research on the estimation of parameters for nonlinear dynamic models is just developing. It is a field that will expand rapidly.

All the methods mentioned here have their advantages and disadvantages. In general it depends on what one wants to know. Disaggregate methods generate parameters that lead to better fits in newer situations and poorer fits for historical data sets. On the other hand, remaining techniques, all of which fit the model to the historical time series, do better on past data than on forecasting future behavior, especially under new conditions.

Assessing the Fit of Dynamic System Models

Once the parameters have been estimated, one would like to assess the degree of fit to a set of historical data. As we have indicated, fitting historical data relatively well is a necessary condition to gain confidence in one's model, at least from a behavioral point of view (as opposed to structural and policy perspectives). Unfortunately one can fit historical data well and yet the model, as a forecasting tool, can fall down. In this spirit, we now review the current methods used to verify the fit of the model to the time series data. When most social scientists use regression analysis and path analyses, it is customary to assess the overall fit of the model and be satisfied the assessment of one general index of fit. There does not seem to be an interest in assessing parts of the model for different degrees of fit by partitioning the error variance into several sources of error. Again in fitting one's model to a set of historical time series, the residuals, the discrepancy between the points simulated by the model and those associated with the actual data, can come from unsystematic sources, such as noise, or systematic sources, such as the lack of a very important feedback mechanism that was omitted from the model. We start our review of the assessment of fit by describing a set of appropriate statistics suggested to give the modeler an overview of the relative contribution of each class of error to the total error variance. For each state variable, this global analysis can provide the modeler information concerning where the model went wrong as well as assure him/her that most of the error in fitting the model is associated with random, unsystematic error variance.

Appropriate Statistics for Fitting System Dynamic Models

Suppose one has a model of a social process that has been fit to a set of one or more time series so that all of the values of model's parameters have been estimated. Let the actual data series be denoted as A and the simulated counterpart to A as S. Assessing the fit of S to A is a bit like performing a profile analysis of two personality patterns taken from MMPI protocols. The first pattern would be the theoretical time curve, S, and the second pattern corresponds to the time series pattern A, the empirical data series. One way of understanding a profile is to answer two major questions: Is the shape of the two profiles the same, and are there differences in the overall intensity or magnitude of the two profiles? So given two time curves, our profiles, one can focus on differences in the shape of the curves and in the general level of the two time profiles. For example, both the simulated curve and the actual data time series may have the same upward trend, but one curve could be higher than the other, indicating a systematic bias in the model.

Much of the statistics to follow will be a bit easier to understand if one thinks of questions asked in a profile analysis. Sterman (1983) suggests using Thiel's inequality method of partitioning the errors (see Thiel, 1966; Pindyk & Rubinfeld, 1981) to gain a more thorough understanding of the source of biasness in one's dynamic model. Although there may be many potential indexes of errors in making predictions, the *mean-squared-error* (*MSE*) is perhaps the most commonly used statistic. We define the mean-squared-error as

$$MSE = \frac{1}{n} \sum_{t=1}^{n} (S_t - A_t)^2, \tag{7}$$

where
n = Number of observations ($t = 1, 2, \ldots, n$),
S_t = Simulated value at Time t, and
A_t = Actual value at Time t.

In what is to follow, we shall also want to have a normalized, relative index of the size of the error associated with fitting the model to a given time series of a state variable. We shall use the so-called root-mean-squared percent error (*RMSPE*), which indicates the percentage deviation of the predicted points from the actual time points. We define this relative measure as

$$RMSPE = \sqrt{\frac{1}{n} \sum_{t=1}^{n} \left[\frac{(S_t - A_t)}{A_t} \right]^2} \tag{8}$$

Generally this index can be considered as a percentage value. Sterman (1983) illustrated the use of this index when fitting an energy-economy model to a set of actual economic time series data that ranged from 1950 to 1980, a 30-year period. The data set included a number of economic variables, such as REAL_GNP, REAL_CONSUMPTION, REAL_PRIVATE_INVESTMENT, PRIMARY_ENERGY_CONSUMPTION,

ENERGY_IMPORT_FRACTION, and so on. Sterman's dynamic model was fit to the set of actual time series, A, and in every case the residuals were analyzed first in terms of the absolute size of the mean-squared-errors (MSE) and in terms of the root mean-squared percentage error (RMSPE). The units of each MSE is in terms of squared units and in general are fairly hard to understand and compare with other MSE's. For example, the MSE associated with fitting the Sterman's model to REAL_PRIVATE_INVESTMENT was 3.5×10^2 billion 1972 dollars per year squared and the MSE associated with fitting his model to time series data on PRIMARY_ENERGY_PRODUCTION was 9.1×10^0 quads per year squared. These two numbers per se are hard to understand and to relate to each other. However, the RMSPE was 11.7% and 7.6% for REAL_PRIVATE_INVESTMENT and PRIMARY_ENERGY_PRODUCTION, respectively. These two numbers are more comparable. Indeed, Sterman (1983) found that in fitting his model to 11 major dependent variables (time series), the RMSPE's ranged from a low of .025 to .14, which on a relative basis alone indicates a small amount of error in predicting these economic time series.

Decomposing the Sources of Error Associated with the RMS

Thiel (1966) partitioned the RMS in the following way:

$$MSE = \frac{1}{n} \sum_{t=1}^{n} (S_t - A_t)^2 = [(\bar{S} - \bar{A})^2] + [(\sigma_S - \sigma_A)^2] + [2(1 - r)\sigma_S - \sigma_A]$$

(9)

where
$\bar{S}$ and $\bar{A}$ = the means of S and A, respectively,
σ_S and σ_A = the standard deviations of S and A, respectively, and
r = The zero lag correlation between simulated data, S, and actual data, A.
If we now divide both sides by the MSE, we obtain separate terms on the right side of the equation:

$$U^M = \frac{(\bar{S} - \bar{A})}{\frac{1}{n} \Sigma(S_t - A_t)^2},$$

(10)

$$U^V = \frac{(\sigma_S - \sigma_A)^2}{\frac{1}{n} \Sigma(S_t - A_t)^2},$$

(11)

$$U^C = \frac{2(1 - r)s_S s_A}{\frac{1}{n} \Sigma(S_t - A_t)^2},$$ (12)

and $U^M + U^V + U^C = 1.0$.

The logic and derivation of these proportions of the total MSE is very similar to the procedures used to partition the sum of squares in the analysis of variance. It should be emphasized that one calculates each of those three proportions for every measured variable in one's model. This provides the opportunity to evaluate the model locally to see where the strong and weak parts of the model lie. In terms of computer programs for computing these model evaluation statistics, Sterman (1983) illustrated how it is easy to do this with system dynamic models by augmenting a DYNAMO™ implementation of an economic model with a set of macros, that is, subroutines, which calculate the MSE, $RMSPE$, U^M, U^V, and U^C for all of the major dependent, time series variables in the model. From Chapter 4 in Volume 1, one might recall that DYNAMO™ is a major simulation language used for simulating nonlinear (and linear) continuous dynamic processes. The latest version of STELLA™, another simulation language, allows long historical data series to be input into the model. It therefore seems feasible to develop procedures for generating these statistical indexes of fit in the STELLA™ programming environment and to save them in a file, so that they can routinely be used for evaluation of a model.

Systematic Bias

The first component, U^M, which is called the *bias proportion*, deals with the differences between the mean or average height of the theoretical and actual time series. Large values of this proportion mean that the two curves are systematically off, usually due to errors in *model assumptions*. Pindyck and Rubinfeld (1981) suggest that, for any model variable, values of U^M above .1 or .2 would be quite problematic, indicating the necessity for revising one's model, at least for the portion of the model dealing with that specific variable. Figure 4 shows an example where U^M might be close to 1.0. One can see that the trends are the same, but the average level is quite different. This systematic displacement of the simulated curve, S, needs to be addressed.

Differences in Standard Deviations

U^V is the component of MSE that deals with the variability of the two time series. It is directly related to the difference between the standard deviations of S and A, the actual time series. When U^V is large, then

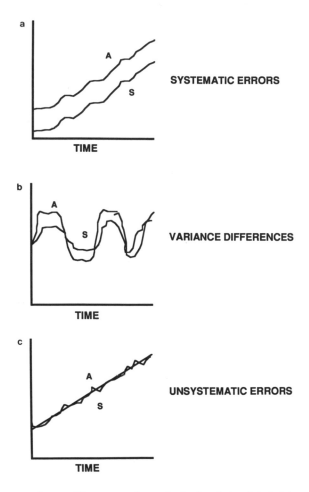

Figure 4. Sources of error: The comparison of actual and simulated data with respect to systematic errors, variance-difference errors, and unsystematic errors (after Sterman, 1983).

either S or A has more fluctuation than the other. An example where a large difference between the variances of the S and A can be found in Figure 4b showing oscillatory behavior. The timing and length of periods are the same, but the actual data fluctuates more than the simulated data. This most likely would be due to systematic error, such as misspecifying the order of the delay. For example, Homer (1983) used what he called a "partial modeling" approach to estimating parameters associated with a variable called REPORTING RATE. One of the parameters was the order of the delay. The order of a delay process is associated with whether or not, in response to a step input, the major portion of the output of the delay occurs immediately or occurs later in the delay period. Third-order delays, or higher, typically generate a S-shaped time output curve, starting slowly and then increasing rapidly over time. At first, when adjusting the model to fit another parameter, he used a first-order delay. Had he then calculated U^V, he would have found that this component would have contributed the major proportion of the total mean-squared-error. Changing the order amounts to rethinking about the delay process with respect to the submechanisms involved in writing, submitting, revising, and so forth that go on in research. The value of U^V would have decreased after adopting a higher order delay in the model.

Unsystematic Errors

The third component, U^C, called the *covariance* proportion, deals mainly with unsystematic error. It is in essence what is left over when the systematic errors associated with difference in the means and variability have been accounted for. It is related to the zero lag correlation between the two time series, A and S. Most of the jaggedness, that is, little bumps and jumps in the actual time series, A (e.g., see Figure 4c), frequently are due to measurement errors and unsystematic external disturbances. Normally, system dynamicists are concerned about systematic errors picked up by statistics such as U^M and U^V. These two indexes deal with predicting patterns of behavior rather than making good point-for-point predictions, especially if many of the points are filled with measurement error. If one does not have perfect prediction, that is, if $MSE > 0.0$, then the best error profile to have would be $U^M = 0$, $U^V = 0$, and $U^C = 1.0$. This would mean that the model addressed all the major substantive processes that generated the data and did not, in a sense, overfit the data by attempting to predict random error.

The Purpose of the Model and the Thiel Statistics

Sterman (1983) also points out that sometimes one can tolerate higher values of U^M and U^V if (1) the cause of the systematic errors are known and (2) the purpose of the model is not compromised by these types of errors. Sterman (1983) fitted an economic model to a set of energy data. In particular, he compared the actual change in REAL_PRIVATE_ INVESTMENT over a 30-year period with the model's predictions. For this dependent variable, the root-mean-square percent error (RMSPE) was relatively high, 11.7%, which is the first signal to ask what went wrong with the model. For this dependent variable, the profile of Thiel statistics was .02, .10, and .88 for U^M, U^V, and U^C, respectively. Although only about 20% of the errors due to bias was small, systematic differences in the variability of the curves accounted for 10% of the errors. Sterman (1983) first points out that the 88% accounted for by U^C indicates that the model predicts the upward trend quite well but does not fit it point-by-point because of random errors. He then examined the residual plot of the difference between actual REAL PRIVATE INVESTMENT curve with the model's predictions. According to Sterman, the actual data series fluctuates with the business cycle around the simulated REAL PRIVATE INVESTMENT. However, when developing the model, Sterman explicitly excluded those structures that would generate short-term business cycles because those mechanisms had little to do with the long-term *purposes* of the model. Because that is the case, the large value of U^V has little meaning since the purposes of the model were not compromised. There is no need to tune the model to get the simulated trend to display those short-term cycles because, from a policy point of view, nothing is gained. However, if the short-term business cycles were theoretically important, a large value of U^V would point out the inadequacies of the model.

Additional Statistical Tests

There are several other quantitative and statistical techniques for assessing the fit of one's model that augment the Thiel proportion analysis. Barlas (1985) considered the problem of assessing the fit of system dynamic models again assuming that one wants to assess a model in terms of its purpose and its use. Moreover, Barlas (1985) emphasized the need to be *pattern oriented*, rather than assessing a point-for-point fit. He describes a software program, written in FORTRAN, that can be used as an external subroutine in a DYNAMO™ implementation of the model to test the fit of a model along many dimensions. All of the indexes used in this analysis can be considered to be absolute measures as opposed to relative measures. It is like the difference between calculating the size of the correlation coefficient as opposed to r^2, the proportion of variation due to the linear relationship between the two variables. The present analysis tests differences between trends in the data as opposed to trends in the theoretical curves, the average level of the two curves, and the periodicity of the curves, and so forth.

   In this section, we describe the various tests of these dynamic models used by Barlas (1985). It is important to know that Barlas's testing procedure forms a sequence, so that some tests later in the sequence are not valid if earlier stages indicated that there were differences in the two types of curves being compared. The first analysis attempts to compare the trends in the data with the trends in the output of the fitted model. For example, suppose an oscillatory empirical time series drifted upward over time. An example of such a phenomenon might be the gradual increase in an animal population with the introduction of a new food or water resource. Barlas suggests the upward trend in the data and in the predictions of the model, which are a function of time, might be approximated by two simple regression equations, one for the data series and one for the predictions over time. Differences between the two regression slopes would lead to indicating a bias in the model. Failure not to reject the null hypothesis would give further credence to the model to capture the trend in the time series. If the upward trends in A and S were exponential, one might fit two exponential curves to the A and S by taking logs of both sides of the exponential equations. One could then make a statistical test of the difference between slopes as before.

Comparing Theoretical and Observed Periods

Barlas (1985) designed a comparative procedure based on the autocorrelation function, $r(t)$, where t is the length of the time lag. The autocorrelation (see Gregson, 1983) is the correlation of a time series with itself, as the name implies. One can think of a time series of T numbers in a row. An autocorrelation of Lag 1 would be correlation between this time series and the same time series, represented in a second row, but slid over by one time period. Let $r_S(t)$ and $r_A(t)$ be the autocorrelations of S and A, respectively, for a lag of t time points. Given $t = 1, 2, \ldots T$, the autocorrelation function gives information concerning the periodicity of the behavior patterns observed in the data and in the predictions from the model. As the series cycles again, the autocorrelation $r(t)$ approaches 1.0. Barlas's program generates an autocorrelation function from the empirical time series, A, and from the predictions of the fitted model, S. This will give us two patterns to compare. Although the distribution of $r(t)$ is generally unknown, Barlas (1985) generates an approximate test of significant differences in the two autocorrelation patterns from the computed standard errors. The program can generate a confidence band as a function of the lags in the system. This is a relatively elegant test of the pattern of autocorrelation coefficients.

Comparing the Means

Figure 4a was used as an example where U^M would be relatively high, indicating a bias. Although one might be able to find that the trends in the data match the trends in the predictions of the model, one can separately test the difference between the mean empirical response and the mean predicted response. It might very well be that the difference in the means of the two time series is due to this type of systematic error. We would suggest that one assess the discrepancy between the two means statistically by finding the confidence interval of the differences between means. In addition, Barlas (1985) suggests the following percentage change statistic when comparing the mean response of A and S.

$$E1 = \frac{|\bar{X}_S - \bar{X}_A|}{|\bar{X}_A|}, \tag{13}$$

where
$\bar{X}_S$ and $\bar{X}_A$ are the mean responses in series A and S, respectively.

Amplitude Variations

Barlas's program also calculates the standard deviations of the two time series. Recall that U^S is the proportion of the MSE due to variance differences. We suggest performing a test of significance between two dependent variances. From a descriptive point of view, Barlas defines a percentage change measure of the discrepancy between S and A as

$$E2 = \frac{|S_S - S_A|}{S_A}, \tag{14}$$

where
S_S and S_A are the positive standard deviations of time series A and S.

Comparing Phase Lags

The autocorrelation function deals with the self-correlation of a given time series with itself for various time lags. On the other hand, the correlation between series A and S at various time lags tells us whether the actual time behavior is out of phase with the output of the model. Thus Barlas (1985) suggests using this cross-correlation index at various lags. Let us interpret the zero lag cross-correlation. This correlation is obtained by lining up the two time series so that the time periods match, that is, there is no lag difference. Recall also that the zero lag cross-correlation is used in calculating U^C to find the proportion of

MSE accounted for by random errors. In general, if the zero lag cross-correlation is lower than the cross-correlations associated with other lags, then the two time series are out of phase. However, Barlas warns that if there are period differences as well, the cross-correlational analysis will be greatly affected by period differences so that there may not be actual phase differences. One has to test period differences before performing phase tests by means of the cross-correlation coefficients.

A Discrepancy Coefficient: Summarizing Fit

All of these tests of fit give the modeler a good idea about the nature of the errors. Recall that initially, for each key variable one can assess the amount of error involved in fitting one's model by calculating the MSE and the RMSPE suggested by Sterman (1983). Barlas (1985) suggests also looking at an index of fit that is a modification of a discrepancy index developed originally by Thiel (1958), called U. Thiel defined U_O as

$$U_O = \frac{\sqrt{\Sigma(S_i - A_i)^2}}{\sqrt{\Sigma S_i^2} + \sqrt{\Sigma A_i^2}}. \tag{15}$$

where
U_O = Thiel's original discrepancy index,
A_i = the value of the actual time series at the ith time period, and
S_i = the value of the simulated time series at the ith time period.

U_O has been scaled in the denominator of the preceding equation to range from 0 to 1.0. If $U_O = 0$, the two curves will be on top of each other, that is, the model will exactly fit the empirical time series associated with that particular key variable. On the other hand, if $U_O = 1$, the model fails maximally. Griner et al. (1978) consider a technical inadequacy of U_O, namely that for two different situations the sum of the squared errors, the numerator of Equation 15 can be the same, yet the model with the larger S would have a smaller U_O. Moreover Barlas (1985) points out that it is even possible for a model to have a larger MSE than another model and yet obtain a lower value of U_O. Griner et al. (1978) modify U_O by centering the data at the mean values of S and A. Thus, let U_{Mod} be a modification of Thiel's basic discrepancy index such that

$$U_{\text{Mod}} = \frac{\sqrt{\Sigma((S_i - A_i) - (\bar{S} + \bar{A}))^2}}{\sqrt{\Sigma(S_i - \bar{S})^2} + \sqrt{\Sigma(A_i - \bar{A})^2}} \tag{16}$$

Let the error $(S_i - A_i)$ equal E_i and $(\bar{S} - \bar{A})$ equal $\bar{E}$. Then

$$U_{\text{Mod}} = \frac{\sqrt{\Sigma(E_i - \bar{E})^2}}{\sqrt{\Sigma(S_i - \bar{S})^2} + \sqrt{\Sigma(A_i - \bar{A})^2}} \tag{17}$$

Finally, if we multiply each of the three expressions *within each radical sign* by $1/T$, those sums of squares become the standard deviation of the E, S, and A scores. Thus, we can express U_{Mod} in terms of its standard deviation form as

$$U_{\text{Mod}} = \frac{\sigma_E}{\sigma_S + \sigma_A}, \tag{18}$$

where
σ_E = the standard deviation of E,
σ_A = the standard deviation of A, and
σ_S = the standard deviation of S.

We note that (1) U_{Mod} also ranges from 0.0 to 1.0 and (2) it does not represent errors in the means anymore. In terms of interpreting the size of this index, Barlas (1985) suggests that one would normally find that for most "good"-fitting models, U_{Mod} might be as high as .6, according to the results of a large simulation study he conducted on synthetic data sets. Rowland and Holmes (1978), who analyzed Thiel's coefficient, U_O, also found similar results. These authors suggest that "average" to "good" models generate a value of U_O in the range of .4 to .7.

Summary

In this chapter we have focused on empirical aspects of working with systems. We have assumed that the model is formulated, many of the qualitative verification tests suggested by Forrester and Senge (1980) have been passed, and that one has gone through extensive testing of the sensitivity of both the parameters and the loop structure. In other words, from a theoretical point of view, the model seems consistent and solid. At this point, one might have either some past historical data available or might want to forecast the results of a new study or new future policy. We have argued that repeated measures designs are optimal for applying dynamic system models to psychological processes. In picking the unit of observation, we suggested following the behavior of individual cases and single small groups, organizations, communities, regions, countries, and so forth.

Once the data are collected, we described several ways to estimate the parameters of the model. First we described obtaining parameter estimates from disaggregated data, independent of the behavior one wants to eventually predict. This can be accomplished in ways analogous to using an ohm meter to calibrate a resistor one wants to use in an electrical circuit. The other estimation approach utilizes the time series data one wants to predict. These procedures include single estimation, full information, and partial-modeling methods. Although currently we are using computer implementation of Box's COMPLEX algorithm, a full information technique, there is much to recommend single equation estimation techniques. In particular, the single equation approach fits nicely into a number of activities we have covered in this book: studying the sensitivity of the model's loops (Chapter 3), estimating parameters of meaningful components, and in the use of the Thiel and Barlas statistics to assess the fit of one's model.

With respect to assessing the fit of the model to historical data, once the parameter values have been estimated, one can begin a sequence of tests of the fit of the model to the data. First we recommend, for each empirical time series, the following overall statistics: MSE, $RMSPE$, U_O, and finally U_{Mod}. These summary statistics, if interpreted correctly with respect to the purpose and requirements of the model, will give the modeler basic information concerning whether the model captured the behavior of the key variables under study. Given the amount of error, the next step is to calculate the Thiel statistics to get an idea of the relative proportions of the total MSE accounted for by bias, variance, and unsystematic random (nonpredictable) noise. Finally, at a more detailed statistical level, the last step would be to follow Barlas's sequence of comparing actual and simulated trends, periods, means, variations, and phase lags that might be inherent in one or more empirical time series.

Assessing the overall fit of the model for each variable and the proportional breakdown of the mean-squared error, followed by Barlas's sequence of tests, should provide the modeler with enough information to judge where the model has fallen down, if indeed it has. Modeling is an iterative process. These tests and statistics help guide the modeler by indicating where one should focus attention on strengthening the model in future iterations.

References

Barlas, Y. (1985). Comparing the observed and model-generated behavior patterns to validate system dynamic models. *Proceedings of the 1985 International Conference of the System Dynamics Society, 1.*

Box, M. J. (1965). A new method of constrained optimization and comparison other methods. *Computer Journal, 8,* 42–52.

Caceci, M. S., & Cacheris, W. P. (1984, May). Fitting curves to data. *Byte Magazine, 9*(5), 340–362.

Forrester, J. W., & Senge, P. M. (1980). Tests for building confidence in system dynamic models. *TIMS Studies in Management Sciences, 14,* 201–228.

Gill, P. E., Murray, W., & Wright, M. (1981). *Practical optimization.* New York: Academic Press.

Gregson, R. A. M. (1983). *Time series in psychology.* Hillsdale, NJ: Lawrence Erlbaum.

Griner, G. et al. (1978). Validation of simulation models using statistical techniques, *Proceedings of the Summer Computer Simulation Conference,* 54–59.

Hamilton, M. S. (1980). Estimating lengths and orders of delays in system dynamics models. In J. Randers (Ed.), *Elements of the system dynamics method* (pp. 168–182). Cambridge, MA: M.I.T. Press.

Homer, J. (1983). Partial-Model testing as a validation tool for system dynamics, *Proceedings of the 1983 International System Dynamics Conference,* 919–932.

Joreskog, K. G. (1979). *Advances in factor analysis and structural equation models.* Cambridge, MA: Abt Books.

Kalman, R. E. (1960). A new approach to linear filtering and prediction problems. *Journal of Basic Engineering,* Series D, *82,* 35–45.

Mass, N. J., & Senge, P. M. (1980). Alternative tests for selecting model variables. In J. Randers (Ed.), *Elements of system dynamics method* (pp. 205–225). Cambridge, MA: M.I.T. Press.

Nelder, J. A., & Mead, R. (1965). A simplex method for function minimization. *Computer Journal, 7,* 308.

Peterson, D. W. (1975). *Hypothesis, estimation, and validation of dynamic social models,* Ph.D. dissertation, M.I.T., Cambridge, MA.

Peterson, D. W. (1980). Statistical tools for system dynamics. In J. Randers (Ed.), *Elements of system dynamics method* (pp. 226–245). Cambridge, MA: M.I.T. Press.

Pindyck, R. S., & Rubinfeld, D. L. (1981). *Econometric models and economic forecasts.* New York: McGraw-Hill.

Press, W., Flannery, B., Teukolsky, S., & Vetterling, W. (1986). *Numerical recipes: The art of scientific computing.* New York: Cambridge University Press.

Rowland, J. R., & Holmes, W. M. (1978). Simulation validation with sparse random data. *Computing Electrical Engineering, 5,* 37–49.

Schweppe, F. C. (1973). *Uncertain dynamic systems.* Englewood Cliffs, NJ: Prentice-Hall.

Spendley, W., Hext, G. R., & Himsworth, F. R. (1962). Sequential application of simplex designs in optimization and evolutionary design. *Technometrics, 4,* 441–461.

Sterman, J. (1983). Appropriate summary statistics for evaluating historical fit of system dynamic models. *Plenary Session, the 1983 International System Dynamics Conference,* 203–260.

Thiel, H. (1958). *Economic forecasts and policy.* Amsterdam: North Holland.

Thiel, H. (1966). *Applied economic forecasting.* Amsterdam: North Holland.

II

Applications of Systems Approaches

6

Applications of Living Systems Theory

James Grier Miller and Jessie L. Miller

Living systems theory identifies basic principles that underlie the structure and processes of living things and relates them to the nonliving physical world, integrating and bringing order to the ever-growing mass of empirical data about them. In addition, living systems models and methodology are useful in empirical research on the great variety of systems of interest to psychology and related fields and in study of individual systems at any of the eight levels of living systems.

Research in General Systems Theory

Systems science research, whatever its particular theoretical bias, is concerned with processes of heterogeneous, complex systems. The systems of interest may be (a) living, (b) nonliving, or (c) mixed living and nonliving. The last class (c) includes both man–machine and ecological systems. These are complex not only because they have many interacting parts but also because they change over time in ways that are not necessarily predictable from their initial states. What Forrester calls

James Grier Miller • Departments of Psychiatry and Psychology, University of California at Los Angeles, Los Angeles, California 90024, and Department of Psychiatry, University of California at San Diego, San Diego, California 92093.   Jessie L. Miller • 1055 Torrey Pines Road, Suite 203, La Jolla, California 92037.

Analysis of Dynamic Psychological Systems, Volume 2: Methods and Applications, edited by Ralph L. Levine and Hiram E. Fitzgerald. Plenum Press, New York, 1992.

(1973) "the counterintuitive behavior of complex systems" results from multiple feedback loops that connect nonlinear variables in hierarchical structures.

Models in General Systems Theory

Systems scientists use models to display or discover general principles or isomorphisms (formal identities) among systems of different types or levels.

Adequate representation of complex systems, particularly living systems, requires new theoretical models in natural language and in types of mathematics capable of handling nonlinear interactions among a large number of variables.

Prigogine (1947) originated nonlinear thermodynamics, the thermodynamics of irreversible processes, to provide models for systems of this sort. Catastrophe theory, set theory, group theory, topology, fuzzy set theory, bifurcation theory, stability theory, and hierarchy theory are all used in systems models. Other mathematical approaches useful in analyzing data from this type of system or representing system processes include information theory, game theory, queuing theory, statistical decision theory, conceptual structure theory, inventory theory, factor analysis, and cluster analysis.

Computers are of great value in the study and modeling of complex systems. They can process massive amounts of data rapidly. They can also be used to simulate the interactions of thousands of variables as they develop over time from a given starting point. In this way effects of changes in initial conditions and the impact of exogenous variables on processes throughout the system can be traced. Such a simulation can be used to predict future states of systems. In addition, computerized artificial intelligence expert systems can be a part of applied work with systems of various sorts.

Models are also used to study the characteristics of a class of systems or to examine a particular system for classification, study, or diagnosis and correction of pathology.

Cross-Level Models

The following general systems models are of living and nonliving systems of many sorts. They are not based on living systems theory but are compatible with it. Living systems theory stresses the importance of cross-level research to discover general properties of living systems of all kinds. All of these general systems models are applied to systems at different levels, although they are not specifically described as "cross-level" by their authors.

Nicolis and Prigogine (1977) have used nonlinear thermodynamic models to demonstrate the phenomenon of "self-organization." Their concepts are very complex but so important that they need to be understood by psychologists and other behavioral scientists. In simple terms, they believe self-organization to be a fundamental formal identity of all systems able to maintain over time an internal order that apparently defies the second law of thermodynamics. This law of nature provides, in its broadest form, that the most probable state of systems, toward which all isolated systems tend, is an equilibrium in which molecules are randomly distributed.

Self-organizing systems are able to maintain a nonrandom and, therefore, improbable state because they are open systems that exchange inputs and outputs of matter and energy with their environment. Under proper conditions, inputs or outputs can cause fluctuations that destabilize the system so that "branching" or "bifurcation" of its states occurs. Bifurcation is a transition from the ordinary progression of thermodynamic states to new, stable, and more complex configurations; that is, order can increase in such self-organizing systems. These configurations are "dissipative" structures that can remain ordered only so long as sufficient flows of matter and energy are available.

The models of Prigogine and his colleagues illustrate the operation of this principle in many kinds of systems such as interacting chemical reagents, cells undergoing processes like glycolysis and synthesis of protein, developing embryos, neural nets, the immunity system in organisms, species in evolution, population dynamics in communities, and processes of social systems such as traffic flow in cities.

Conrad's (1983) thermodynamic models were intended to show that adaptability is a general characteristic of biological systems. Adaptability is defined as the use of information to handle environmental uncertainty.

These mathematical models are concerned with evolutionary processes in ecosystems made up of a nonliving part, the physical environment, and a living part, the biota. Conrad considers organisms the most important units in the biota. Above the organism are aggregations of organisms, including social systems and populations. Below the organism are organs, cells, organelles, molecules, and atoms.

Conrad's models describe the contribution to adaptability of several important processes which, in living systems theory, would be called *adjustment processes*. These include transformability, compensation, cycle formation, and the minimal tendency of adaptability. They operate at multiple levels of living systems.

Odum's (1983) ecological models also describe systems in thermodynamic terms of energy and entropy. He analyzes the processes of a variety of living systems, including social and economic systems as well as nonliving systems and ecological systems, to support the proposition that systems prevail that maximize the flow of useful energy (the *maximum power principle*). They ordinarily operate at intermediate rates to conserve energy but exercise maximum power when that is necessary to compete against other systems. He considers the struggle for survival to be essentially competition for free energy, that is, energy that is available for work.

The "dynamic" models of Forrester and his associates are systems of equations that are programmed in a computer language, DYNAMO, developed for the purpose. His models of a factory (1961) and an urban community (1969) apply not to any particular factory or community but to classes of systems with similar characteristics.

He and his colleagues have also developed global models that incorporate variables such as food supplies, industrial production, agriculture, pollution, and population (Forrester, 1973; Meadows, Meadows, Randers, & Behrens, 1972).

By running a model with different initial conditions and assigning different values to variables, the effects of changes in such things as prices and policies can be traced through a system. The simulation can be extrapolated into the future, and predictions of future system behavior can be made. Such models are useful to the extent that the selection of variables, processes, and data accurately reflect the real systems they represent.

Of course it is not possible to include all variables and interactions of such complex systems in even the largest simulation. Those used in a particular simulation are selected as typical of the aspects of real systems with which the modeler is concerned. Data are drawn from statistical or other sources.

Leontief and his colleagues (1953) have developed an input–output model for analyzing the flows of goods and services through an economy. He identifies various producing and consuming sectors that can be found in the economies of regions of a nation, an entire nation, regions of the world, and the world as a whole. The sectors he identifies are commonly used in economic analysis and are compatible with the *Standard Industrial Classification Manual*. The number of sectors in a model depends upon the degree and type of aggregation of data required.

Leontief's input–output models consist of sets of differential equations that represent the producing and consuming sectors of an economy. These are connected by flows of goods and services. Input and consumption coefficients for each sector describe the combination of goods and services it needs in order to produce a unit of its output, or, in the case of households, a unit of their expenditures and income. Stocks of capital goods and natural resources are also represented.

The large number of equations in his models are solved by computer. The resulting data are presented in tables that give a detailed picture of the economic system being studied.

In addition to its use as a general-purpose economic model, input–output analysis has been employed to study the possible futures of developing countries as well as world military spending. A global model (1977) has been run with alternative projections for the years 1980, 1990, and 2000. Eight sets of assumptions about, for example, growth rates of populations, growth rates of per capita incomes, and pollution abatement created eight "scenarios" for the future of the world economy with outcomes that range from bleak to optimistic.

Living Systems Theory: Basic Research and Applications

Basic Research

Any scientific theory derives its credibility and eventual validation from its correspondence to the real phenomena to which it is applied, its usefulness in solving problems and answering questions, and the degree to which it contributes to science in general. It is, therefore, critical that research be undertaken on testable hypotheses derived from it.

Basic living systems research is concerned with intersystem generalization, the search for common aspects of structures or processes among living systems of different kinds. Generalization may be among individuals of the same species or type, among systems of different species or types at the same level, or among systems at different levels.

Cross-level research is the most powerful of these approaches, although all have their place in the study of living systems. It seeks to discover isomorphisms among systems at two or more levels and to apply models based upon them to a variety of systems. This is a useful approach because it can illuminate previously undetected regularities in nature and can result in deeper understanding of basic characteristics of living systems at two or more levels. Scientists must always be alert to the fact that greater generality is a major goal of science.

Cross-Level Tests of Living Systems Hypotheses

A list of 173 testable cross-level hypotheses appeared in *Living Systems* (Miller, 1978), and others have been stated since. Some of these apply to all levels, others to two or more. Several have been tested empirically.

Formal identities across levels cannot be rigorously demonstrated unless the dimensions and units used in measurements at various levels are compatible. If this were not so, curves that plot data at different levels might appear to be identical when they were not. Conversely, such curves might appear to be of entirely different shapes and yet actually be the same. For instance, a curve with equal intervals along the abscissa and logarithmic intervals along the ordinate looks entirely different from a curve derived from the same data with logarithmic units on both coordinates.

Information Input Overload. The first multilevel experimental test of a hypothesis derived from living systems theory was research on information input overload (Miller, 1960, 1978). The hypothesis, which was tested at levels of cell, organ, organism, group, and organization was the following:

Hypothesis 5.1-1: As the information input to a single channel of a living system—measured in bits per second—increases, the information output—measured similarly—increases almost identically at first but gradually falls behind as it approaches a certain output rate, the channel capacity, which cannot be exceeded in the channel. The output then levels off at that rate, and finally, as the information input rate continues to go up, the output decreases gradually toward zero as breakdown or the confusional state occurs under overload.

Experiments at each level were conducted by specialists in relevant fields. Measurements of information input and output rates were made on single fibers (from the sciatic nerves of frogs), optic tracts of white rats (retina or optic nerve to optic cortex), human subjects working alone, human subjects in groups of three, and laboratory "organizations" made up of nine subjects.

In one form of experimental organization, two groups of three people simultaneously received information inputs. In each of these, two members served as input transducers, each member receiving a different sequence of visual inputs. The input transducer components sent electronic signals to the third member of the group who output them to a display before one member of a three-person group in another room. Each of these forwarded the signals he received to the final person in the second room, who acted as decider and output transducer. He compared the signals, made a decision about them for the total organization, and output the decision to a recording device.

Data from all the levels—cell, organ, organism, group, and organization—yielded information input-output curves alike in form. Each curve rose sharply at first, leveled off, and then, as the channel capacity of the systems was exceeded, fell toward zero. Transmission rates were lower at each higher level.

These results confirmed the hypothesis of a formal identity in this aspect of information processing at these five levels.

Nonrandom Nets. Rapoport and a group working with him did experiments on communication networks in elementary and junior high schools (Rapoport & Horvath, 1961). Rapoport (1957) had previously investigated models of neural nets through which information in the form of bioelectric pulses is propagated and conjectured that a similar model might apply to communications networks in systems at higher levels, such as organizations. They tested the following hypotheses:

The structure of the communication networks of living systems at various levels are so comparable that they can be described by similar mathematical models of nonrandom nets.

Nonrandom or "biased" nets are those in which communication over some channels is more probable than over others. They were able to predict with reasonable accuracy some aspects of the friendship patterns in schools.

Effects of Conflicting Signals on Decision Making. An experimental study by Lewis (1981) on individual human organisms and on groups tested an hypothesis that relates to the decider subsystem at those levels:

When a system is receiving conflicting command signals from several suprasystems, it intermittently being a component of all of them, the more different the signals are, the slower is its decision making.

Subjects played a computerized game in which they had to defend their own ships and destroy the enemy in war in space. A computer presented "commands" from five officers, all of the same rank and all superiors of the subjects. The subjects had to decide which command to obey. They could "fire," "warp," "scan," "dock," or "wait." Eight patterns of commands with differing amounts of conflict were used, ranging from total agreement among the officers to total disagreement. Subjects could act only on command of at least one officer. The game was scored so that subjects were more likely to succeed when more officers agreed upon a command.

The research design was the same for both groups and single subjects. In the group experiment, three subjects worked together to arrive at a joint decision. After 40 turns, a final score was calculated, based on the subject's success in defending his own ships and destroying the enemy. The results were highly significant in both the individual and the group experiments, an outcome that supported the hypothesis.

The preceding three sets of basic researches are examples of cross-level hypotheses based on living systems theory. Other cross-level hypotheses have been tested by other scientists.

Living Systems Studies of Individual Systems

Living systems theory can be used to analyze and work with individual living systems at all eight levels. It has been applied to selected systems in diagnosis of pathology, in treatment, and in efforts to improve both efficiency and effectiveness of system processes.

When a physician examines a patient, he or she uses diagnostic tests and instruments to check the structure and processes of organ systems. The living systems method for studying systems at other levels is similar. It involves observing and measuring important relationships between inputs and outputs of the total system and identifying the structures that perform each of the 19 subsystem processes discussed in Chapter 2 (Volume 1). The flows of relevant matter, energy, and information through the system and the adjustment processes of subsystems and the total system are also examined. The status and function of the system are analyzed and compared with what is average or normal for that type of system. If the system is experiencing a disturbance in some steady state, an effort is made to discover the source of the strain and correct it.

A set of symbols, shown in Figure 1, has been designed to represent the levels, subsystems, and major flows in living systems. They are intended for use in simulations and diagrams and are compatible with the standard symbols of electrical engineering and computer science. They can be used in graphics and flow charts like those used in several applications discussed in this chapter (e.g., see Figure 2, p. 177).

It is of great importance that measures or indicators be developed for structures and processes of systems at all levels and that normal values and ranges of critical variables be determined. This has been achieved for thousands of physiological variables of human beings. Normal values and ranges of many variables of interest to psychologists are also known, particularly in areas like sensation, perception, learning, and child behavior, but many are still to be established. For some other types of systems, such as organizations, such information could be obtained, but efforts to do so are rarely made.

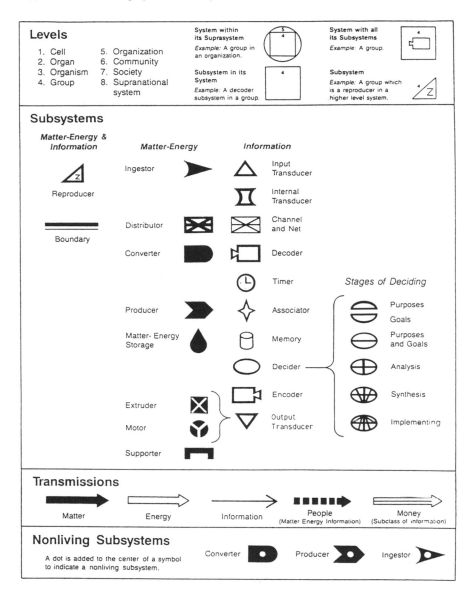

Figure 1. Living systems symbols.

The following pages contain short descriptions of studies based on living systems theory, either completed, in process, or projected. It is hoped that ultimately we or others will carry on such investigations at all levels of living systems. At present no studies have been carried out at the level of the organ.

Some of these studies consist of subsystem reviews only with little use of other living systems concepts. Others are sophisticated efforts to analyze systems and make recommendations for their improvement.

The Level of the Cell

Conceptual Structures

A new method for collecting and integrating facts has been described and computer software developed by Sowa (1983). It can be used in any scientific discipline, profession, or other area of interest in which knowledge accumulates over time. His method uses logic, mathematics, and/or simulation to display the conceptual structure of a field and show the relationships among the concepts of different authors. As data from more and more sources are added, the growing conceptual structure is amplified, refined, and made more precise. Sowa's procedures also specify the logical relationship of each new fact to preceding facts and alters the conceptual structure as necessary.

To perform an integration that reflects the current state of knowledge of a subject, it is necessary to provide the computer with a database. Parameters, variables, and specific values can be found in published research in the field of interest or related fields. Qualitative as well as quantitative statements can be included in this sort of conceptual structure and, as the field advances and knowledge increases, less precise qualitative statements can be replaced by quantitative expressions.

Cellular biology, cellular genetics, and neuropsychopharmacology have generated a vast literature that includes results of millions of experimental and clinical studies. Thousands of articles in each of these areas are published annually. As a result, no individual person can remember all the things that have been learned in his area.

We propose to attempt to achieve integration of knowledge at the level of the cell by building a conceptual structure that will include all relevant information. A small group of computer scientists, mathematicians, and cellular biologists will work together in the project. Data about cells will be drawn from the literature of several relevant fields. The levels, subsystems, and subsystem variables of living systems theory can constitute a set of parameters on which to base such a conceptual structure. It could be developed much as biologists' knowledge of cells has accumulated over the last few centuries. At the same time, the components of cells could be related to the subsystem processes of living systems theory (see Table 1).

The structure would first include facts discovered soon after microscopes were first used, such as the shape and range of sizes of cells and the presence of a nucleus in many cells. Other discoveries, like the chromosomes and other nuclear structures, the behavior of the genetic material in mitosis, and the fact that chromosomes are the sites of genes that specify the characteristics of descendants would be added successively. Thus a description of the cellular reproducer subsystem would begin to arise. It would become more precise as the structure and function of DNA and RNA and the reproducer components in the cytoplasm were included in the conceptual structure.

Descriptions of the other subsystems of the cell would be created in the same way, and ultimately the relationships among subsystems would be included in the growing representation of cellular structure and process. As it developed, the conceptual structure would become more and more precise, detailed, and congruent with current knowledge in cellular biology. Also it would be increasingly useful as a reference source for scientists and a source of suggestions as to what future studies were needed.

It seems possible that conceptual structures could be developed at all the other levels of living systems by an analysis of the relevant literature using the same method we propose to use on cells. Because the parameters used at each level would be those of living systems theory, it would be most interesting to find out whether the conceptual structures were similar at the different levels. To the extent that this was true, it would tend to confirm the evolutionary cross-level aspects of living systems theory.

The Level of the Organ

No research at this level is planned at present, but it is obvious that particular organ components, like the liver or the bladder, could be analyzed in terms of living systems theory. In addition, a living systems conceptual structure of any organ could be created using Sowa's methods.

The Level of the Organism

Expert Systems

Artificial intelligence expert systems are used as research tools or as practical devices to solve problems or replace a human interviewer. For

Table 1a. Selected Major Components of Each of the 20 Critical Subsystems at Each of the Eight Levels of Living Systems: Part 1

Level						Subsystem					
	Reproducer	Boundary	Ingestor	Distributor	Converter	Producer	Matter-energy storage	Extruder	Motor	Supporter	
Cell	DNA and RNA molecules	Matter–energy and information: Outer membrane	Transport molecules	Endoplasmic reticulum	Enzyme in mitochondrion	Chloroplast in green plant	Adenosine triphosphate	Contractile vacuoles	Cilia, flagellae, pseudodopia	Cytoskeleton	
Organ	Upwardly dispersed to organism	Matter–energy and information: Capsule or outer layer	Input artery	Intercellular fluid	Gastric mucosa cell	Islets of Langerhans of pancreas	Central lumen of glands	Output vein	Smooth muscle, cardiac muscle	Stroma	
Organism	Testes, ovaries, uterus, genitalia	Matter–energy and information: Skin or other outer covering	Mouth, nose, skin in some species	Vascular system of higher animals	Upper gastrointestinal tract	Organs that synthesize materials for metabolism and repair	Fatty tissues	Sweat glands of animal skin	Skeletal muscle of higher animals	Skeleton	
Group	Parents who create new family	Matter–energy: Insect soldiers; Information: Makers of television-use rules in family	Refreshment chairman of social club	Father who serves dinner	Work group member who cuts cloth	Family member who cooks	Family member who puts away groceries	Parent who puts out trash	Driver of family car	Birds that build nests	

Level										
Organization	Chartering group	*Matter–energy:* Guards at entrance to plant; *Information:* Librarian	Receiving department	Assembly line	Operators of oil refinery	Factory production unit	Stockroom personnel	Janitorial staff	Crew of company jet	Building repair and maintenance personnel
Community	National legislature that grants state status to territory	*Matter–energy:* Agricultural inspection officers; *Information:* Movie censors	Airport authority of city	County school bus drivers	City stockyard organization	Bakery	County jail officials	City sanitation department	City transit authority	Maintenance crew at capitol building
Society	Constitutional convention that writes national constitution	*Matter–energy:* Customs service; *Information:* Security agency	Immigration service	Operators of national railroads	Nuclear industry	All farmers and factory workers of a country	Guards at national armory	Export organizations of a country	Trucking industry	Officials who operate national public buildings and lands
Supranational system	United Nations when it creates new supranational agency	*Matter–energy:* Troops at Berlin Wall; *Information:* NATO security personnel	Legislative body that admits nations	Personnel who operate supranational power grids	EURATOM, CERN, IAEA	World Health Organization	International storage dams and reservoirs	Downwardly dispersed to societies	Operators of United Nations motor pool	People who maintain international headquarters buildings

Table 1b. Selected Major Components of Each of the 20 Critical Subsystems at Each of the Eight Levels of Living Systems: Part 2

Level	Subsystem									
	Input transducer	Internal transducer	Channel and net	Timer	Decoder	Associator	Memory	Decider	Encoder	Output transducer
Cell	Receptor sites on membrane for activation of cyclic AMP	Repressor molecules	Pathways of mRNA, second messengers	Fluctuating ATP and NADP	Molecular binding sites	Unknown	Unknown	Regulator genes	Structure that synthesizes hormones	Presynaptic membrane of neuron
Organ	Receptor cell of sense organ	Specialized cell of sinoatrial node of heart	Nerve net of organ	Heart pacemaker	Second echelon cell of sense organ	None found; upwardly dispersed to organism	None found; upwardly dispersed to organism	Sympathetic fibers of sinoatrial node of heart	Presynaptic region of output neuron	Presynaptic region of output neuron
Organism	Sense organs	Proprioceptors	Hormonal pathways, central and peripheral nerve nets	Suprachiasmatic nuclei of hypothalamus	Sensory nuclei	Unknown neural components	Unknown neural components	Components at several echelons of nervous system	Temporoparietal area of dominant hemisphere of human cortex	Larynx
Group	Lookout of gang of thieves	Group member who reports members' attitudes to group decider	Person-to-person communication channels among group members	Mother who wakens other family members on time	Member who explains rules to team	Parents who teach good behavior	Father who stores family records	Parents' family council	Writer of group communication	Jury foreman

Organization	Secretaries who take incoming calls	Factory quality control unit	All users of corporate phone network	People who operate factory whistle	Foreign language translation group	People who train new employees	Filing department	Top executives, department heads, middle managers	Annual report writers	Public relations department
Community	Representatives who report from state capital to local voters	Neighborhood watch groups	Telephone linesmen in city	Caretakers of clock on city hall tower	Attorney general of state who interprets law	City school teachers	Operators of central police computer	Governor, legislators, judges of state	Writers of city ordinances	Representatives from state to national legislature
Society	Foreign news services	Public opinion polling organizations; voters	Telephone and communications organizations	Legislators who decide on time and zone changes	Cryptographers	All teaching institutions of a country	Keepers of national archives	Voters and officials of national government	Drafters of treaties	National representatives to international meetings
Supranational system	UN Assembly hearing speaker from nonmember territory	Speaker from member country to supranational meeting	Universal Postal Union (UPU)	Personnel of Greenwich observatory	Simultaneous translation staff of supranational organization	FAO units that teach farming methods in third world nations	Librarians of UN libraries	Council of Ministers of the European Community	UN Office of Public Information	Top official who announces decisions of supranational body

instance, such systems have proved to be effective for securing admissions information in hospitals. Two diagnostic programs, Mycin (Buchanan & Shortliffe, 1984) and Internist I (Miller, Pople, & Myers, 1982) are available to suggest probable diagnoses in clinical situations.

Expert systems are of two kinds: (a) those that model the way human beings behave or solve problems and (b) those that solve problems in the optimal way regardless of whether their logical processes are like human thinking.

Living systems theory probably has little relevance to the second sort of artificial intelligence system, but it can be usefully applied to the first, the systems that are designed to operate as much as possible like human problem solvers. At present, no unifying conceptual system is in use by the developers of expert systems. Each selects her or his own space and dimensions. The living systems subsystem analysis, the designation of flows as matter–energy, information, or both, and the use of comparable units and common dimensions could integrate the field and make it possible to relate the various expert systems.

Almost all expert systems today are at the level of the organism, but the cross-level emphasis of living systems theory leads us to ask whether expert systems might some day carry out some functions of work groups on assembly lines, of certain decision-making organizations like stock markets, and even of higher level systems. The increasing availability of very large computer chips makes such developments appear possible.

Diagnosis and Treatment

Living systems theory is of potential value to physicians and mental health professionals. Although not widely accepted, it has been used as a model for diagnosing a patient's physical and psychological state (Kolouch, 1970); applied in the diagnosis and treatment of asthma (Kluger, 1969); and used in analysis of school phobia (Bolman, 1970).

A subsystem view of the sort described was used by members of a clinical seminar in an analysis of a teenage boy with clinical symptoms of anorexia nervosa. The analysis was carried out at both organism and group levels. This case illustrates how important it is to consider not only the subsystems of a system in which pathology is diagnosed but also the suprasystem, in this case, the family of the patient.

The patient was brought to the hospital by his parents, who were concerned about a major loss of weight and other symptoms brought about by his refusal to eat or drink. Physical and neurological examination showed him to have no structural abnormalities in his matter–

energy-processing subsystems, although there were functional abnormalities in the ingestor (failure to input matter–energy). Functional problems were also observed in the distributor (abnormal heart action) and extruder (urinary system and bowel malfunction). All these symptoms disappeared when he was given nourishment in the hospital.

Neurological examination disclosed no disease or structural abnormality in his nervous system (channel and net and other information-processing subsystems). Psychological examination showed normal associator and memory function for his age. Encoding and output transducing were also normal.

Certain of the information-processing subsystems were, however, functioning abnormally. He was extremely sensitive to noise (input transducer) and, unlike most patients, remained so when starvation ended. He insistently denied feeling hunger and thirst and reported feeling cold when others were uncomfortably hot (decoder). Like other anorexic patients, he perceived his body to be too fat when he was, in fact, very thin (decoder).

The major pathology in this boy appeared to be in the decider subsystem. Unlike other teenage boys, he seemed to have no personal goals but was apparently motivated entirely by reward and punishment. There was no sign of the desire for emancipation that is common among American boys of his age. He also appeared unable to control his behavior by setting limits for himself. He had become interested in collecting some time before and had filled his room with so many things that he had to move out. When he started to draw or write, he continued as long as he was rewarded for doing so. Further study of this case centered on his family.

The Level of the Group

Diagnosis and Treatment

During the course of diagnosis and treatment of the anorexic boy discussed above, a social worker had several interviews with the parents together and, infrequently, saw each parent separately. A living systems analysis of the family was based upon her report.

This family differed in many ways from others of similar cultural and educational backgrounds. The deviation was sufficiently great and the effects on family members so profound that several family processes appeared to be pathological. The primary pathology at the group level, as it had been at the organism level, was in the decider subsystem. Pathologies in several other subsystems appeared to be secondary.

Although families differ greatly in the way they make decisions, this family was clearly unusual. All decisions, including those ordinarily made by mothers, were made by the father. The mother was allowed no more independence than the children. The father insisted that the family be as isolated as possible from the community and that the parents be separated from the children only when it was unavoidable (matter–energy boundary). He censored incoming information (information boundary) by deciding what family members were allowed to read or watch on television and by prohibiting friendships. All information that entered the family was interpreted by the father in terms of his own fears and suspicions (decoder). No one could discuss personal feelings and attitudes (internal transducer). Communication among family members was discouraged (channel and net).

The results of the analyses at organism and group levels were used to make the diagnosis, decide what further treatment would be offered, and set treatment goals.

The pathological adjustment processes of this patient and his family appeared to be consequences of the father's aberrant information processing. The primary pathology was at the organism level, but in the father rather than the patient. In his interviews, the father had revealed fear that he would lose his family as he had lost his parents early in childhood.

The seminar group believed that the patient was unlikely to adjust normally unless his father was helped to gain insight into the reasons for his excessive control. It also seemed probable that the other children would experience difficulties when they reached adolescence. Treatment, therefore, would focus on him. Both individual treatment of the patient and treatment of the family appeared necessary.

Goals for treatment were (1) to help the father gain the insight that he could allow his family increased independence and still retain their love, and (2) to work with the patient toward an adjustment more normal for this age.

Family Interviews

One description of how any family could be analyzed dealt with the structure, processes, and pathologies of each subsystem as well as feedbacks and other adjustment processes (Miller & Miller, 1980).

A subsystem review of a real family (Bell, 1986) was carried out in a videotaped interview that followed a schedule designed to discover what family members were included in each of several subsystems, how the family decided who would carry out each process, how much time was spent in each, and what problems the family perceived in each process. In addition, sections of the schedule concerned how rules are made in families with children (a decider subsystem process) and how the rules are enforced and children disciplined (an adjustment process). We consider rule making to be a decider function, while enforcement and discipline are adjustment processes.

Ant Nests

For decades, analogies have been made between animal organisms and the nests of ants, termites, and other social insects. At times there was widespread skepticism among biologists as to whether such cross-level analogics are valid. In recent years, as more evidence has accumulated, the organismlike character of social insect colonies has been widely accepted. It therefore seems reasonable that the input–output flows of matter, energy, and information through the 20 critical subsystems can be observed and measured in the nests of ants and other social insects. We recognize the cross-level formal identity between organisms and insect nests, but we maintain that those systems are groups rather than societies, as they are often called.

We are planning to videotape an ant nest for several hours continuously and to make such observations and measurements from the tapes. We intend to compare our observations of these subsystem processes with the 26 kinds of behavioral acts recognized by Oster and Wilson (1978) in ant nests.

Level of the Organization

Living systems theory offers a means for analyzing the structure, function, and processes of organizations and finding disfunctions that reduce a system's effectiveness in achieving its purposes. It has been used in studies of several sorts of organizations including hospitals (Merker & Lusher, 1987). A psychiatric ward (see page 174), several public schools in a community (Banathy & Mills, 1985), and a public transportation system (Bryant & Merker, 1987) among others.

Measures of effectiveness may be related to financial variables such as total profit, profit per share of stock, or cost per unit of goods or services output. They may also relate to less tangible variables like mortality rates of hospital patients, average attendance at performances or meetings, or number of applicants for admission to universities. The factors that contribute to effectiveness in an organization can be revealed by monitoring flows of matter–energy and information through components or subsystems of the system rather than using only input–output measures of the total system.

Each type of organization is specialized for certain subsystem processes of the society to which it contributes and may be quite unlike organizations of other types in its subsystem structure and in the type of living and machine components that carry out its processes. Organizations of the same type may also differ in many aspects of structure and process. It is, therefore, necessary to observe how the individual system under consideration is structured and what living and nonliving components are involved in each process.

United States Army Project

The first large-scale application of living systems theory was a 3-year study of 41 U.S. Army battalions (Ruscoe et al., 1985). The Army's own assessments had revealed problems that affected its capacity to achieve and maintain optimal levels of training efficiency and the ability of units to accomplish their assigned missions. There was no consensus about the methods of evaluating battalion effectiveness that the Army has used for many years. These consist of many independent objective and subjective measures that do not relate to any integrated conceptual system. Although these may distinguish good and less good battalions, they do not reveal the basis for differentiating them.

The objective of this study was to use living systems process analysis to explain how battalions function and relate the quality and quantity of flows of matter–energy and information to battalion effectiveness.

In this study, three types of data were collected: (a) findings of the Army's traditional evaluation methods, which combine performance indicators, command indicators, and perceptions of personnel; (b) process perception data, which includes opinions of unit personnel on how well each matter–energy and information process was being handled, the time spent on each of the members of the unit and the unit as a whole, the importance of each process, and how well the process was being performed in terms of several process variables; and (c) process objective data that includes such things as the percentage of the unit's vehicles that are operational at a particular time.

Data sources were mass administration of standardized questionnaires, interviews with key management personnel, records, and reports available within each unit, surveys from brigade-level managers, and check sheets completed by surveyors. In all, more than 5,000 officers, noncommissioned officers, and other enlisted personnel from battalions in both the continental United States and Europe participated.

Two sets of effectiveness criteria were employed: the traditional Army criteria and a new set based on living systems theory. Although both the living systems criteria and our data collection were derived from living systems theory, we were able to avoid contamination between the two. The living systems criteria led to similar conclusions about distinctions among battalions as traditional measures of unit effectiveness, but they revealed much more about the dynamics of the units.

Among the many discoveries from the study (Ruscoe et al., 1985, pp. 45–49) was the relationship between information processing and effectiveness. The greater the unit personnel's appreciation of and skill in information processing, the more effective it was. The information variables of meaning, lag, volume, cost, and distortion were repeatedly shown to be good indicators of unit effectiveness.

IBM Study

Civilian corporations and army battalions are different in many ways, but they have many similarities in addition to carrying out the same life processes. This study is planned to use many of the same procedures as the Army study but will include several significant improvements.

1. In the IBM research, to be consistent with conventions used in studies of organizational behavior for about 25 years (Forrester, 1961), we measured five rather than three flows: materials (MATFLOW), energy (ENFLOW), communications (COMFLOW), money and money equivalents (MONFLOW), and personnel (PERSFLOW). These are basically the same matter–energy and information flows used in all other living systems research. We divided information into two classes (communications and money) and added one other flow, personnel, which are lower-level groups and individual persons, each made up of matter–energy and information.
2. We collected only subjective data in the Army study, but in this study we intend to employ objective data in addition. The objective measures will be used to check the reliability of the subjective data.
3. Subjects will give their responses to computers rather than to human interviewers in standardized interviews.
4. We shall observe the location of members of the organization and of equipment objectively, using badges that will send electronic signals to a central computer. Such data will supplement

the estimates of time spent in each activity by the people wearing the badges. Use of such badges will be voluntary.

5. Data analysis will be accomplished by computerized expert systems. It is planned to use two types: (1) ALIGN (the interactions among components within the system) and (2) IMPACT (interactions between the total corporation and the local, national, or supranational market in which it competes.

6. Three separate criteria of internal organizational effectiveness will be used: (a) Independent opinions of several judges on the best strategy in different situations. These judges will be chief executive officers. (b) Effectiveness criteria derived from the previous Army study. (c) Effectiveness and productivity data from the extensive literature on organizations.

Hospital Studies

Numerous studies have applied living systems concepts to clinical psychology and to psychiatry. Several of these specifically apply living systems theory.

1. The components, variables, and possible pathologies of the subsystems of the inpatient psychiatric unit of a university hospital were described by Chase and his colleagues (Chase, Wright, & Ragade, 1981a).

2. The same group analyzed the decision process in the case of an adult male alcoholic patient (1981b). The process included decisions by the patient's employers, his family, the patient himself, and his family physician before he reached the hospital. From his first appearance at the admission desk to his discharge to outpatient treatment 16 days later, a series of 38 decisions determined he would be admitted, the course of treatment, and the treatment plan after discharge.

The writers concluded that decision analysis of this sort can be of practical clinical significance. It can show what components are involved in the decision process, find significant feedback loops, and reveal otherwise unknown problems in a treatment situation. Superimposing decision analyses for several patients can uncover significant problems within the patient milieu.

3. Following upon two unpublished studies by Whitehead, Goldberg, and others, plans have been made for a full-scale study of hospital cost-effectiveness using methods developed in the IBM Study.

Few, if any, adequate procedures exist for objective evaluation of performance in hospitals, clinics, and health maintenance organizations. Our methods should make it possible to compare alternative approaches to cost reduction and suggest ways to improve cost-effectiveness without threatening the quality of health care.

We plan to concentrate on information processing in a hospital. There is abundant evidence that information processing in many organizations is unsatisfactory. It may be incomplete; distorted by misinterpretations, opinions, or emotions; untimely; or presented from incompatible viewpoints or in different formats that are difficult to integrate. It may also be wrongly interpreted by the receiver.

A Living Systems Theory of Accounting

Accounting is an internal transducer process that takes place at the levels of the organization and above. Living systems concepts have been used as a basis for a general theory of accounting, and a set of testable hypotheses have been generated (Swanson & Miller, 1986). This approach has been used in a quantitative analysis of the accounts of W. T. Grant, Inc., a large retail store (Swanson & Miller, 1986). Living systems theory has also been applied to marketing, an output transducer process (Reidenbach & Oliva, 1981).

The Level of the Community

A City

Communities are complex systems with many problems that could well be approached from the point of view of living systems theory, but so far few studies have been made at this level. An analysis of the city of Louisville, Kentucky (Vandervelde & Miller, J. L., 1975), was the first attempt to describe the subsystems of a community in living systems terms.

Community Health Care

An analysis of the different levels of living systems involved in community health care and the subsystems of the health care system of a community, which is a part of the producer subsystem, addressed some of the problems involved in health care delivery (Miller, 1970a).

A Community in Space

It appears likely that within a few decades a settlement will be established on a satellite in earth orbit or on the moon. In 1984, a study of such a settlement was conducted by the California Space Institute (Calspace). Representatives of NASA, including former astronauts, met with members of the Calspace staff and outside consultants to discuss the feasibility and possible program of such a project. Although it would be largely a macroengineering project, the needs of the living systems that would inhabit the space station were also considered (Miller, 1987, pp. 202–203).

Because a community of this sort would necessarily be more isolated and therefore more totipotential than a community on earth, a living systems analysis was used to determine what components would comprise each subsystem, and how flows of essential matter–energy and information to all parts of the system would be assured.

A community on the moon, set up in a location favorable for mining and extraction of the abundant useful minerals of the moon's surface, was envisioned. It would receive regular deliveries of essential inputs by shuttle from earth and send its mineral output to earth in the same way. Figure 2 is a diagram of the chief subsystem processes and flows in such a community. The symbols shown in Figure 1 are used to identify components of each subsystem.

The Level of the Society

The Administrative Side of a Government

A study for the General Accounting Office (GAO) of the United States is currently in the planning stage. It would employ the techniques used in the Army and IBM researches and the accounting concepts mentioned above. Originally the GAO was given the responsibility of making traditional accounting audits of each agency in the executive and judicial branches for Congress. A few years ago it was given the additional responsibility of performing "management audits." It is required to report to Congress how cost-effective each agency is. The GAO has been asked if management audits based on living systems theory could be conducted that might improve current methods used by that agency.

The Supranational System

In the latter half of the twentieth century a worldwide channel and net subsystem has developed that connects all but the most remote human settlements. Its components are human beings working with telephone networks, cables under bodies of water, electronic networks using ra-

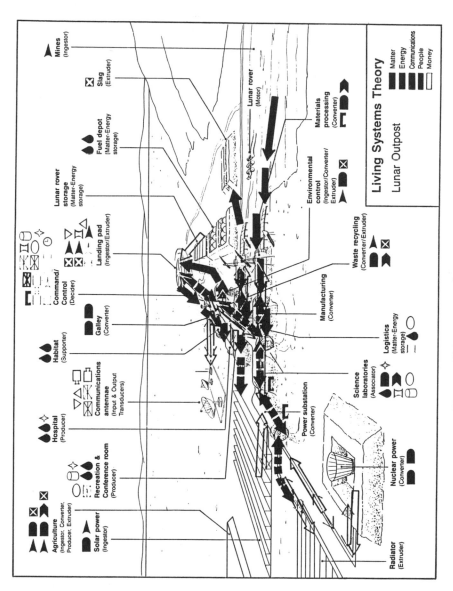

Figure 2. Living systems theory: Lunar outpost.

dio, television, and satellites in earth–stationary orbit. Ours is the first generation in which information, traveling more than seven times around the world in a second, can enable billions of people worldwide to communicate with one another.

Networks that use such virtually instantaneous transmission are in use by military organizations, banks, and other business organizations. A network for medical and health information (MEDLINE) connects the United States to many other countries. In each of the past several years, Christmas color television programs have connected religious worshipers in several countries on three continents.

A Canadian/American Transboundary Monitoring Network

The International Joint Commission of Canada and the United States has been employing living systems theory as a conceptual framework for exploring the creation of an electronic network to monitor the region surrounding the border separating those two countries (Miller, J. G., 1986). The proposed network is conceived of as a channel and net subsystem for the boundary region system. It is hoped that communication across the border on such a network can improve the adjustment processes to stresses like acid rain and provide for more effective cooperation between the two nations. A similar approach could be used to decrease tensions across many other international borders throughout the world.

University of the World

Use of international networks for educational purposes was foreseen 20 years ago (Brown, Miller, & Keenan, 1967). This vision arose from a recognition of the potential of a supranational channel and net subsystem, a concept derived from living systems theory. Conviction that such networks are possible or needed has been slow to develop, but interest in establishing such a system is growing. Plans are well advanced for such a supranational network under the auspices of a University of the World. It would use satellite transmission to participating countries and radio or television as well as other electronic technologies within countries to transmit courses taught by experts. They would travel worldwide and would particularly benefit areas that have little access to high-quality education or research. Courses would be offered from the preliterate level to a master's-degree level. Examinations would be given at central educational institutions, and degrees might be granted based on such instruction. Interest in the University of the World has been expressed by numerous countries throughout the world.

Conclusion

This has been a description of some of the uses of living systems theory at seven of the eight levels of living systems. These researches and applications should interest psychologists from a number of sub-specialties, including physiological, experimental, clinical, social, community, industrial, and educational psychologists.

Psychology today is so fractionated that it is in danger of losing its intellectual cohesiveness. Living systems theory may point the way toward reunification and integration.

References

Banathy, B., & Mills, S. R. (1985). The application of living systems process analysis in education. *ISI Monograph, 85–7.*

Bell, R. A. (1986). *Videotape script.* (personal communcation.)

Bolman, W. M. (1970). Systems theory, psychiatry, and school phobia. *American Journal of Psychiatry, 127,* 65–72.

Brown, G. W., Miller, J. G., & Keenan, T. A. (1967). *EDUNET.* New York: Wiley.

Bryant, D., & Merker, S. L. (1987). A living systems process analysis of a public transit system. *Behavioral Science, 32,* 292–303.

Buchanan, B. G., & Shortliffe, E. H. (Eds.). (1984). Rule-based expert systems: The Mycin experiments of the Stanford heuristic programming project. Reading, MA: Addison-Wesley.

Chase, S., Wright, J. H., & Ragade, R. (1981a). The inpatient psychiatric unit as a system. *Behavioral Science, 26,* 197–205.

Chase, S., Wright, J. H., & Ragade, R. (1981b). Decision making in an interdisciplinary team. *Behavioral Science, 26,* 206–215.

Conrad, M. (1983). *Adaptability.* New York: Plenum Press.

Forrester, J. W. (1961). *Industrial dynamics.* Cambridge, MA: M.I.T. Press.

Forrester, J. W. (1969). *Urban dynamics.* Cambridge, MA: M.I.T. Press.

Forrester, J. W. (1973). *World dynamics* (2nd ed.). Cambridge, MA: Wright-Allen Press.

Kluger, J. M. (1969). Childhood asthma and the social milieu. *Journal of the American Academy of Child Psychiatry, 8,* 353–366.

Kolouch, F. T. (1970). Hypnosis in living systems theory: A living systems autopsy in a polysurgical, polymedical, polypsychiatric patient addicted to Talwin. *American Journal of Hypnosis, 13*(1), 22–34.

Leontief, W., Chenery, H. B., Clark, P. G., Duesenberry, J. S., Gerguson, A. R., Grosse, A. P., Grosse, R. N., Holzman, M., Isard, W., & Kistin, H. (1953). *Studies in the structure of the American economy.* New York: Oxford University Press.

Leontief, W. (1977). *The future of the world economy.* New York: Oxford University Press.

Lewis, F. L. II. (1981). Conflicting commands versus decision time: A cross-level experiment. *Behavioral Science, 26,* 79–84.

Meadows, D. H., Meadows, D. L., Randers, J., & Behrens, W. W. III. (1972). *The limits to growth.* New York: Universe Books.

Merker, S. L., & Lusher, C. (1987). A living systems process analysis of an urban hospital. *Behavioral Science, 32,* 304–314.

Miller, J. G. (1960). Information input overload and psychopathology. *American Journal of Psychiatry, 116,* 695–704.

Miller, J. G. (1978). *Living systems.* New York: McGraw-Hill.

Miller, J. G., & Miller, J. L. (1980). The family as a system. In C. K. Hoffling & J. M. Lewis (Eds.), *The family: Evaluation and treatment* (pp. 141–184). New York: Brunner/Mazel.

Miller, J. G. (1986). A living systems analysis of a Canada/U.S. boundary region. In P. T. Haug, B. L. Bandurski, & A. L. Hamilton (Eds.), *Toward a transboundary monitoring network: A continuing binational exploration* (Vol. 1). Washington, DC: U.S. State Department, International Joint Commission, U.S.A. and Canada.

Miller, J. G. (1987). The study of living systems: A macroengineering perspective. *Technology in Society, 9,* 191–210.

Miller, R. A., Pople, H. E. Jr., & Myers, J. D. (1982). INTERNIST I, An experimental computer-based diagnostic consultant for general internal medicine. *New England Journal of Medicine, 307,* 468–476.

Nicolis, G., & Prigogine, I. (1977). *Self-organization in non-equilibrium systems: From dissipative structures to order through fluctuations.* New York: John Wiley & Sons.

Odum, H. T. (1983). *Systems ecology.* New York: John Wiley & Sons.

Oster, G. F., & Wilson, E. O. (1978). *Caste and ecology in the social insects.* Princeton, NJ: Princeton, NJ: Princeton University Press.

Prigogine, I. (1947). *Etude thermodynamique des processus irreversibles.* Liege: Desoer.

Rapoport, A. (1957). Contribution to the theory of random and biased nets. *Bulletin of Mathematical Biophysics, 19,* 257–278.

Rapoport, A., & Horvath, W. J. (1961). A study of a large sociogram. *Behavioral Science, 6,* 279–291.

Reidenbach, R., & Oliva, T. A. (1981). A framework for analysis. *Journal of Marketing,* Fall, 42–52.

Ruscoe, G. C., Fell, R. L., Hunt, K. T., Merker, S. L., Peter, L. R., Cary, Maj. J. S., Miller, J. G., Loo, Cpt. B. G., Reed, Cpt. R. W., & Sturm, Cpt. M. I. (1985). The application of living systems theory to 41 U.S. Army battalions. *Behavioral Science, 30,* 7–50.

Sowa, J. S. (1983). *Conceptual structures.* Reading, MA: Addison-Wesley.

Swanson, G. A., & Miller, J. G. (1986). Accounting information systems in the framework of living systems theory and research. *Systems Research, 4,* 253–265.

Swanson, G. A., & Miller, J. G. (1989). *Measurement and interpretation in accounting: A living systems approach.* New York: Quantum Books.

Vandervelde, K. J., & Miller, J. L. (1975). The urban grant university concept: A systems analysis. *Behavioral Science, 20,* 273–295.

7

Complex Organizations and Living Systems Theory

The Case of the U.S. Army

Gordon C. Ruscoe

Introduction

The study of complex organizations occupies a middle ground between traditional, reductionist research that stresses the simplicity of two-variable relationships and the more contemporary, probabilistic research that stresses the disorganized complexity of statistical, multivariate relationships. Research on complex organizations focuses on problems of what Weaver (1958) terms *organized complexity*—that is, on problems with many variables but with knowable relationships among these variables. Because of the stress on interrelated, interdependent variables, problems of organized complexity seem especially amenable to study within the perspective of systems science, which has frequently been called the science of complexity.

One of the problems of complexity that has persistently confronted theorists is how to account for disparities in organizational effectiveness. Such disparities occur in different types of organizations, in different organizations of the same type, and within the same organization

Gordon C. Ruscoe • School of Education, University of Louisville, Louisville, Kentucky 40292.

Analysis of Dynamic Psychological Systems, Volume 2: Methods and Applications, edited by Ralph L. Levine and Hiram E. Fitzgerald. Plenum Press, New York, 1992.

over time. In addition, because of the typical complexity of most organizations, theorists have disagreed as to the criteria by which to evaluate effectiveness (Scott, 1981), while continuing to recognize that effectiveness seems to be a key to understanding how organizations grow,
mature, and decline (Ouchi, 1980). Thus a systemic analysis of organizations and their effectiveness would seem to be called for.

There are a number of possible systems approaches to such a study.
Each of these approaches stresses, to varying degrees, the openness of
organizations as systems and the need to consider inputs and outputs
as well as the internal dynamics of these systems. Each approach also
stresses, again to varying degrees, procedures by which to categorize
and describe the internal dynamics of organizations.

But living systems theory (Miller, 1978) seems a priori especially
useful because it addresses the two key points of complexity and effectiveness. First, living systems theory argues that living systems are to be
found on seven levels, ranging from simple cells like the amoeba to
complex supranational systems like the United Nations. These levels
represent increasing degrees of complexity. They also represent not
merely increasing numbers of variables but the emergence of higher
order phenomena and properties that distinguish higher from lower
levels. Thus living systems theory intrinsically recognizes and takes
account of complexity as a distinguishing characteristic of systems.

Second, living systems theory postulates a set of 19 critical processes that describe the ways in which a system functions and that,
together, may be taken to represent an exhaustive, mutually exclusive
set of categories by which to analyze systems. These critical processes
can be used to categorize the internal dynamics of any living system,
regardless of its level of complexity. More important, these critical processes can be used to evaluate the effectiveness of the system; that is, to
be viable, a system must be able to carry out each of these processes.
Thus living systems theory seems particularly valuable in examining
effectiveness in complex organizations.

Among complex organizations, the military is an especially interesting candidate for research. The military works with both information and matter–energy, thus carrying out all the processes associated
with living systems. It is composed of a large number of echelons, thus
representing a fortiori a complex organization. It is—on paper at
least—a rationally organized system, thus inviting a rational analysis of
its functioning. And last, the military is preeminently concerned with
organizational effectiveness—especially in the sense of combat readiness, thus exemplifying a central problem of organizational theory.
Therefore, beginning in 1979, the usefulness of living systems theory
was tested by applying it to the study of effectiveness in the U.S. Army.

The Methodology of the Military Research

The application of living systems theory to the study of the U.S. Army began as a limited, exploratory study but became later a much more extensive piece of research. The preliminary study (Ruscoe, Giguet, Brown, Cary, & Burnside, 1979) was confined to examining information processing in training management activities in six armor battalions, four in the United States and two in West Germany. This study was later extended (Merker & Ruscoe, 1981; Peter & Ruscoe, 1981) to an additional 35 battalions, in which training, personnel, and logistics management activities were examined. This extended study considered both information and matter–energy processing, and the 35 battalions included combat, combat support, and combat service support units.

The single most difficult methodological problem occasioned by the research was the need to distill from living systems theory a manageable set of researchable concepts and to translate these concepts into a language that simultaneously met the canons of social science research and could be understood by the military. Because the research focused particularly on critical processes, it was necessary to develop military-oriented examples of each of the processes. Ultimately, the original names of the processes were changed in an attempt to employ terms that had a "real-world" relevance without doing too much injustice to the original meaning of the process. For example, *internal transducing* became *monitoring*, a term familiar to battalion members.

In addition, it was necessary to develop a set of measurable indicators for these processes. Living systems theory postulates that, for each critical process, there are a number of variables that can be used to assess how well the process is being carried out. Moreover, for each variable, there are a number of indicators that can be used to quantify that variable. Thus in the case of the process of input transducing (bringing information into the system), one possible variable would be the timeliness of information brought in, and one possible indicator of this variable would be the extent to which the battalion's files contain the most current training circulars issued by the Army.

The process-variable-indicator translation, because of the assistance of Army personnel who worked as researchers on the project, proceeded smoothly, but it must be recognized that only a very small sample of possible variables and indicators could be used in the research.

While the theory was being translated, it was also necessary to develop appropriate data-collection instruments. Initially, three types of instruments were used. Questionnaires were administered to all commissioned and noncommissioned officers and to a sample of enlisted personnel. These questionnaires were designed principally to assess the living systems' critical processes in terms of their importance to the battalion's activities, the time devoted to the processes in carrying out these activities, and the perceived quality in performing these processes (in terms of usefulness, volume, accuracy, timeliness, and cost).

Second, interviews were conducted with key battalion personnel to elicit their evaluations of the management of training, personnel, and logistics within the battalion. Again, the interview schedule stressed critical processes, particularly in terms of the interviewee's position within the battalion. Third, conventional Army performance and "command climate" data were collected to serve as an independent measure of effectiveness. Items such as the percentage of required equipment "ready" and the number of adverse personnel actions in the battalion were combined into Battalion Effectiveness Ranks (BERs) against which to compare the process data.

In the second phase of the research, an additional instrument was included. This instrument attempted to collect data on "objective" measures of process performance to complement the perceptual data gathered through questionnaires and interviews.

The resulting data base was, of course, very large. A total of nearly 6,000 individuals were included in the two phases of the research. For each individual who received the full battery of instruments, nearly 1,000 data items were collected. As the data analysis developed, the procedures employed came to be known as living systems process analysis (LSPA).

The Findings of the Military Research

In assessing as a research paradigm LSPA and the theory from which it was derived, two types of questions were considered. First, it was necessary to determine the *descriptive utility* of LSPA—that is, as a means of describing organizational effectiveness in terms of internal, process-related activities. Second, it was necessary to determine the *diagnostic utility* of LSPA—that is, as a means of diagnosing areas of organizational ineffectiveness in terms of these processes.

Descriptive Utility

In the case of descriptive utility, it was clear throughout the two phases of the research that the LSPA approach distinguished among the battalions in terms of effectiveness in much the same way as did the conventional measures contained within the BER evaluations. More important, however, the process analysis revealed that *differences in effectiveness* were associated with *differences in the efficiency* with which critical processes are carried out.

First, variations in effectiveness were shown to be correlated with differences in the total duty time devoted to the processing of information and matter–energy. That is, more effective battalions were characterized by a significantly greater allocation of time, on the part of upper echelon personnel, to information rather than to matter–energy processing. In the area of training, for example, key personnel in more effective battalions spent 75.8% of their time processing information and only 24.2% processing matter–energy. In contrast, in less effective battalions, the distribution of time was 61.2% and 38.8%, respectively. Even in the area of logistics, typically seen as the heart of matter–energy in the battalion, key personnel in more effective units spent 66.7% of their time processing information compared to 55.6% for their counterparts in less effective units.

The importance of information processing to organizational effectiveness is additionally confirmed in looking at other process perspectives used. More effective battalions tended to report higher evaluations of the state, importance, and performance of information processes. Whereas matter–energy—the "beans and bullets" of the battalion—is obviously important to battalion effectiveness, the processing of matter–energy hinges on the efficient processing of information.

Second, the inputting and outputting of both information and matter–energy do not fully distinguish between more and less effective battalions. To be sure, more effective battalions tended to handle incoming information and matter–energy more efficiently and to avoid being overloaded with the paperwork typically associated with outputting information and matter–energy. But such differences were relatively small, perhaps reflecting the fact that input and output processes are to a large degree inhibited by the requirements of echelons above the battalion.

In contrast, battalion effectiveness is clearly correlated with differences in the allocation of time to specific information and matter–energy throughput processes, and the patterns of these differences vary among the battalion's areas of activity. For example, in the case of training, two information throughput processes were especially important—associating and deciding. In essence, more effective battalions spent significantly *more* time in organizing training information and using information to make training decisions. In addition, these battalions spent significantly *less* time in virtually all the matter–energy processes.

In the area of logistics, organizing logistics information (associator) was again critical, as were distributing information (channel-and-net) and decoding information for use within the unit (decoder). More effective battalions spent significantly more time in each of these processes.

In the area of personnel, it was the use of information to make personnel decisions (decider) that most clearly distinguished between units. Interestingly, the *less* effective battalions spent *more* time in this process, reflecting perhaps the inability of these units to organize and routinize what should be administrative rather than policy decisions.

Third, the importance of specific information throughput processes was further illustrated by examining clusters of key duty positions (see Table 1). Battalion commanders and their staffs in more effective battalions spent significantly more time in organizing training information (associator) than did their counterparts in less effective battalions. The battalion training (S-3) section in more effective battalions concentrated on monitoring training in the battalion (internal transducer) and organizing training information (associator). Company commanders and their staffs in more effective units stressed monitoring training (internal transducer), organizing information (associator), and making training decisions (decider). And platoon leaders and their staffs emphasized monitoring training (internal transducer) and making training decisions (decider).

Similar analyses of staff in personnel and logistics again revealed distinctive patterns. In the case of logistics, for example, the logistic (S-4) sections in more effective battalions spent significantly more time on distributing information (channel-and-net), organizing information (associator), and making logistics decisions (decider) than did their counterparts in less effective battalions. Platoon leaders and their staffs

Table 1. Percentage of Workweek Devoted to Specific Information Processes by Key Personnel

Position[a]	Process[b]	More effective battalions	Less effective battalions
BnCDR and staff	AS	40.6%	22.7%
S-3 and staff	IT	34.6	9.1
	AS	31.1	17.1
CoCDR and staff	IT	28.0	16.0
	AS	23.2	12.9
	DC	28.8	11.8
PltLDR and staff	IT	36.3	18.9
	DC	23.7	13.7

Note. Derived from Ruscoe, 1981, pp. 46–53.
[a] BnCDR = battalion command; S-3 = training section leader; CoCDR = company commander; PltLDR = platoon leader.
[b] IT = monitoring (internal transducer); AS = organizing (associator); DC = decision making (decider).

in effective battalions, in contrast, devoted a significant amount of their time to monitoring logistics information (internal transducer) and to translating information for use in the unit (decoder).

Finally, this type of process time analysis permits a deeper understanding of sources of inefficiency in the ways in which personnel carry out their tasks in processing information and matter–energy. The overarching conclusion of time analysis centered on two related factors—coordination among personnel and decentralization of responsibility. That is, the flow of information (and, thereby, matter–energy) is enhanced when personnel within the battalion share in the generation and distribution of information, even if the flow thereby runs counter to the vertical chain of command traditionally associated with military decision making.

The case of training again provides an illustration of this finding. A comparison of the allocation of time among key battalion personnel in monitoring training (internal transducer) revealed a marked difference between more and less effective battalions (see Table 2). In more effective battalions, there was a fairly even distribution of monitoring responsibility, even down to the lowest levels of the unit. Especially important was the role of platoon leaders, who are in closest daily contact with individuals soldiers.

This analysis of personnel time on process should not be confused with traditional "time–motion" studies. Process time analysis is *not* an attempt to break down information and matter–energy processing into a set of simple, repetitive routines. The division of labor—or *specialization of task*—implicit in time–motion studies is in fact antithetical to process analysis. Instead, process time analysis is concerned with

Table 2. Percentage of Workweek Devoted to Monitoring Training by Key Personnel

	Percentage of workweek	
Position	More effective battalions	Less effective battalions
Battalion commander	10.4%	6.1%
Training section head	9.6	3.3
Company commander	11.5	11.8
First sergeant	10.6	14.9
Company training NCO	6.2	6.0
Platoon leader	13.9	7.8
Average percentage for company commander and below	12.2	9.0

Note. Derived from Ruscoe, 1981, p. 48.

identifying and describing an efficient *specialization of people* (Thompson, 1965). That is, process time analysis seeks to understand how individuals can best use their talents in information and matter–energy processing and how the interdependence of these individuals can produce an integrated and thereby more efficient performance of all the processes critical to the battalion. Process time allocations for training in more effective battalions is thus a clear example of optimizing the roles of key personnel.

Diagnostic Utility

In the case of diagnostic utility, it was clear throughout the research that the LSPA approach permitted an analysis of sources of ineffectiveness not readily revealed in more conventional analyses of Army data. In particular, the management of information flows lies at the heart of organizational effectiveness. The management of flows into and out from the organization determines the effectiveness of the battalion vis-à-vis the larger organization of the U.S. Army. The management of information flows within the battalion determines the efficiency of the battalion in carrying out its tasks.

Based on these general findings, LSPA can thus be used to establish measures of "health" for the various processes necessary to the battalion's functioning. These health measures, in effect, pinpoint for each process the variables that adversely affect its efficiency and, consequently, its effectiveness. For example, the process of monitoring (i.e., internal transducing) may be inefficient because the information derived from such monitoring does not reach the battalion commander in a timely fashion, even though the information is accurate and important.

An examination of these health measures yields a model of effectiveness for each area of battalion activities (Fell, 1982; Ruscoe, 1981). These models contain a set of norms critical to battalion effectiveness and a set of activities that can be employed to remediate serious deviations from these norms.

In the area of training, for example, multiple correlation analysis revealed that battalion effectiveness depended in large part on the accuracy of information—both that obtained in monitoring training and that used to make training decisions. In addition, the accuracy of information used to make decisions depended on the accuracy of the information translated for battalion use (decoder), stored and retrieved (memory), and organized (associator).

Such analysis permits the construction of a model of those processes and variables that most strongly account for variations in battalion effectiveness *and* that are susceptible to remediation. The training model (shown in Figure 1) thus represents several points at which remediation can be introduced into the battalion in order to improve training and, subsequently, overall battalion effectiveness.

Similar analyses of the areas of personnel and logistics demonstrated that more effective battalions carried out more efficiently those processes necessary to each of these activities. In personnel, effective battalions exhibited significantly greater accuracy of information translated (decoder) and organized (associator) for the battalion and significantly more timely acquisition (input transducer) and transmission (channel-and-net) of information. In logistics, effective battalions were characterized essentially by the timeliness of the information they brought into the battalion (input transducer), transmitted (channel-and-net), stored and retrieved (memory) within the battalion, and translated into codes appropriate to send out from the battalion (encoder).

The establishment of norms for these critical process variables permits battalion leadership to monitor, on a regular basis, a battalion's critical processes. One means by which to carry out this monitoring is to examine the values of the unit's process variables relative to those values obtained from the entire data base. Such a comparison can be accomplished by using a "performance indicator" (see Figure 2) that serves to describe the battalion's process variables in relation to norms based on all battalions in the study. By locating the unit's values on this indicator, battalion leaders can determine the extent and direction of

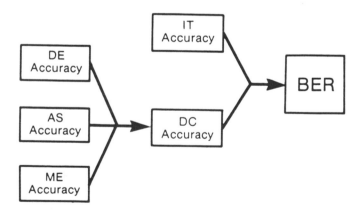

Figure 1. A model of training performance and battalion effectiveness. Derived from Ruscoe, 1981, p. 21. Key: DE = decoder; AS = associator; ME = memory; IT = internal transducer; DC = decider; BER = battalion effectiveness.

Figure 2. An example of a performance indicator. Derived from Ruscoe, 1981, p. 22.

unit deviations from the norms. These deviations then can be subject to remediation.

The suggested forms of remediation are based on and refer to specific Army documents and procedures. These procedures assure that remedial efforts fall within acceptable Army guidelines for battalion management. That is, each process is discussed, first, in terms of its importance to battalion effectiveness, and, second, in terms of its relevance to Army procedures. To assist Army personnel in understanding the relationship between system processes and Army procedures, each discussion is accompanied by a "publication index" that contains the information necessary to remediate process inefficiency in accordance with standard Army procedures.

Living Systems as a Research Paradigm

In assessing living systems theory as a research paradigm, it is necessary to take into account three types of considerations—the particular methods employed in the Army research, the underlying theory itself, and the peculiar problems of doing research on complex organizations. Each of these considerations will be examined in turn.

Methodological Considerations

The Army research proved to be an important case study of the application of living systems theory at the level of organizations. The research provided a body of data that differed significantly in emphasis from that conventionally gathered on Army operations and their relationship to military effectiveness. It is important to recognize, however, that the research methodology contained certain inevitable limitations. Because of the synchronic nature of the research, it was difficult to obtain a "dynamic" picture of the organization or of changes in effectiveness as a result of changes in process efficiency. This problem is especially acute in an organization, such as the Army, which has an extremely high rate of personnel turnover. Changes in personnel occur rapidly, but changes made by personnel while in a particular position are likely to be more enduring. Thus a "snapshot" approach to organizational research may mistakenly attribute to current individuals' actions organizational characteristics inherited from predecessors.

Second, the use of largely perceptual measures of organizational efficiency and effectiveness may raise concern about the essentially ephemeral nature of perceptions. To be sure, given the large amount of data collected from a large number of respondents and the persistent uniformity of perceptions, this problem is somewhat alleviated in the Army research. Furthermore, the addition in the second phase of the research of more objective process data, even if only partially successful, lends support to the findings based on perceptual data. Further efforts to develop and test objective measures of process efficiency are, however, critical.

Third, and perhaps most important, the Army research tapped only a very small portion of the possible variables and indicators that might be used to characterize living systems processes. We have discussed elsewhere (Ruscoe, 1982) the problem of the "inferential leaps" that have to be made in moving from living systems processes, as described in the theory, to empirically based variables and indicators meaningful to specific organizations. Indeed, attempts to use the basic format developed for the Army research to study transportation and hospital systems (Merker, 1985) demonstrate the difficulty both of selecting appropriate measures and of obtaining results that can be compared to those from the Army study.

Theoretical Considerations

The Army research is also limited in terms of the degree to which it adequately tests the utility of living systems theory. Most obvious is the fact that the research examined only one level of living systems—the organization—and did not attempt to generate and test from this research any cross-level hypotheses. Data obtained in the research could conceivably be reanalyzed in order to examine process efficiency at the platoon level, thereby including at least some information on the level of the group. But this level of analysis was not a part of the original research design.

In addition, the research did not consider either the theoretical adequacy of the postulated 19 critical processes or the possibility of another schema for categorizing and analyzing these or similar processes. Already an additional critical process has been accepted by

Miller (Bosserman, 1982). Thus the original assumption of the research that the 19 critical processes represent an exhaustive and mutually exclusive set has been successfully challenged. Moreover, it is possible to identify competing schemata for examining processes, at least at the level of organizations (e.g., Katz & Kahn, 1966; Mintzberg, 1973). These competing approaches, however, do not claim to be—and certainly do not appear to be—generalizable to other levels of living systems. And clearly, the appeal of living systems theory is precisely its claim for applicability at all levels of living systems, regardless of their complexity.

Organizational Considerations

Quite apart from methodological and theoretical considerations, which hinge largely on the utility of living systems theory per se, there are considerations that arise from the suspicion that complex organizations are basically "recalcitrant." That is, complex organizations as an object of study seem to display a perverse resistance to rational analysis, from a systems or any other point of view. Describing organizations as recalcitrant is deliberately anthropomorphic and arises from a consideration of the difficulties in achieving a rational picture of the how and why of complex organizations, particularly as characterized by conventional organizational theory.

Organizational theory has been essentially a theory of order (Pondy & Mitroff, 1979). Organizational disorders have typically been seen as mere pathologies or dysfunctions that need to be remediated or eliminated. Yet it is quite possible that such disorders reflect the often overlooked fact that complex organizations are created by and populated with human beings, whose recalcitrant behavior has been frequently noted. Cohen and March (1974), for example, have suggested that at least some complex organizations can best be described as "organized anarchies." Attempts to describe such organizations rationally are likely to be frustrated because these descriptions fail to account adequately for the recalcitrance that the human actors themselves display.

In order to understand organizational recalcitrance, it is useful to appeal to the concept of "loose coupling." A loosely coupled system (Glassman, 1973) is one in which the various subsystems or elements function relatively independently of one another: Although they are responsive to each other and to central authority, they maintain their own identities to a large degree. Attempts to change the system often fail because the subsystems are themselves able to resist change. Simple cause–effect relationships do not always hold. In effect, loose coupling provides less of the "organizational glue" usually associated with tightly controlled systems.

In a loosely coupled system, moreover, control is problematic precisely because it is difficult to exercise. Indeed, some have warned of the difficulty in predicting the effects of planning and change in a loosely coupled organization and have urged that tighter coupling is necessary (e.g., Brown, 1980). Others, however, have argued that, in order to achieve some degree of management in complex organizations, it is in fact necessary to promote decoupling. Such decoupling, it is assumed, encourages greater manageability of the organization by making it more easily understood (Connolly & Pondy, 1980) and by promoting adaptability and, ultimately, survival of the organization (Weick, 1976).

In either case, living systems theory—and other systems approaches to the study of organizations—are perhaps more likely to detect and appreciate loose coupling than are more traditional, reductionistic approaches to research, especially those that emphasize organizational structure. The stress in living systems theory is upon critical processes and their role in determining both the efficiency and ultimate viability of living systems. Such an emphasis encourages a view of organizations as dynamic flows of information and matter–energy rather than as rigid structures. And, as the U.S. Army research demonstrated, it is processes, not structures, that determine how well the system is functioning.

The U.S. Army research, of course, explored only one small part of the theory and only within the context of one type of complex organization. The research did, however, reveal some of the potential contributions that living systems theory can make to our understanding of complex organizations and other living systems.

References

Bosserman, R. W. (1982). Internal security processes and subsystems in living systems. In R. Trappi (Ed.), Cybernetics and systems research (pp. 113–119). New York: North-Holland Publishing Co.

Brown, L. D. (1980). Planned change in underorganized systems. In T. G. Cummings (Ed.), Systems theory for organization development (pp. 181–203). London: John Wiley & Sons, Ltd.

Cohen, M. D., & March, J. G. (1974). Leadership and ambiguity. New York: McGraw-Hill.

Connolly, T., & Pondy, L. R. (1980). General systems theory and organization development: A dialectical inquiry. In T. G. Cummings (Ed.), Systems theory for organization development (pp. 15–34). London: John Wiley & Sons, Ltd.

Fell, R. L. (1982). Remediation of organizational information process performance: A *case of the U.S. Army battalion.* Unpublished master's thesis, University of Louisville, Louisville, Kentucky.

Glassman, R. B. (1973). Persistence and loose coupling in living systems. *Behavioral science, 18,* 83–89.

Katz, D., & Kahn, R. L. (1966). *The social psychology of organizations* (2nd ed.) New York: John Wiley & Sons.

Merker, S. L. (1985). Living systems process analysis: A comparison of three studies. *Systems Inquiring,* Proceedings of the Annual Meeting of the Society for General Systems Research, Vol. I, pp. 488–496.

Merker, S. L., & Ruscoe, G. C. (1981). *Application of living systems process analysis to critical activities of the U.S. Army battalions: Personnel and logistics.* Technical Report, U.S. Army Research Institute, Alexandria, Virginia.

Miller, J. G. (1978). *Living systems.* New York: McGraw-Hill.

Mintzberg, H. (1973). *The nature of managerial work.* New York: Harper & Row.

Ouchi, W. G. (1980). Framework for understanding organizational failure. In J. R. Kemberly & R. H. Miles (Eds.), *The organizational life cycle* (pp. 395–429). San Francisco: Jossey-Bass Publishers.

Peter, L. R., & Ruscoe, G. C. (1981). *A living systems theory analysis of Army battalions impacted by the Battalion Training Management System.* Final Report, U.S. Army Training Board, Fort Eustis, Virginia.

Pondy, L. R., & Mitroff, I. I. (1979). Beyond open system models or organization. In B. M. Straw (Ed.), *Research in organizational behavior* (pp. 3–39). Greenwich, CT: JAI Press.

Ruscoe, G. C. (1981). *Application of living systems analysis to the establishment of process norms in the United States Army.* Final Report, U.S. Army Training Support Center, Fort Eustis, Virginia.

Ruscoe, G. C. (1982). Application of living systems process analysis to Army organizations: A review and critique. *Behavioral science, 27,* 203–213.

Ruscoe, G. C., Fell, R. L., Hunt, K. T., Merker, S. L., Peter, L. R., Cary, J. S., Miller, J. G., Loo, B. G., Reed, R. W., & Sturm, M. I. (1985). The application of living systems theory to 41 U.S. Army battalions. *Behavioral science, 30,* 7–52.

Ruscoe, G. C., Giguet, L. R., Brown, B. R., Cary, J. S., & Burnside, B. (1979). *Application of living systems theory to the evaluation of critical processes in the armor battalion: An exploratory analysis.* Technical Report, U.S. Army Research Institute, Alexandria, Virginia.

Scott, W. R. (1981). *Organizations: Rational, natural, and open systems.* Englewood Cliffs, NJ: Prentice-Hall, Inc.

Thompson, V. A. (1965). *Modern organization.* New York: Alfred A. Knopf.

Weaver, W. (1958). A quarter century in the natural sciences. *The Rockefeller Foundation Annual Report* (pp. 7–122).

Weick, K. (1976). Educational organizations as loosely coupled systems. *Administrative Science Quarterly, 23,* 541–552.

8

General Systems in Psychiatric Disorder

Irwin M. Greenberg

Introduction

During the third decade of this century, the Shroedinger equation of quantum mechanics was developed and its theoretical and practical application elaborated (Semat, 1944). Modern physics was already well developed. Moreover, the higher mathematics that were later to become the bases of more recent advances were already published and understood by workers in appropriate fields (Greene, 1986). Biological sciences and medicine, despite major advances in infectious diseases and tissue pathology at the end of the nineteenth century, still had not advanced so far in terms of theory and application. The possible role of insulin in diabetes mellitus was the most notable advance in biological science in 1922 (Banting & Best, 1922). Major health advances at the time often resulted from public health measures such as insuring sanitation, a clean water supply, and adequate housing, all still luxuries in the Third World today (*The Economist*, 1986). Sulfa drugs were not to be available for another decade, and penicillin was not yet even a laboratory scientist's dream.

There were, however, significant theoretical advances in biology despite the lag of biological technology behind that of the physical

Irwin M. Greenberg • Department of Psychiatry, Harvard Medical School–Cambridge Hospital, 1493 Cambridge Street, Cambridge, Massachusetts 02139.

Analysis of Dynamic Psychological Systems, Volume 2: Methods and Applications, edited by Ralph L. Levine, and Hiram E. Fitzgerald. Plenum Press, New York, 1992.

sciences. Walter Cannon (see Cannon, 1932; Brobeck, 1973) espoused the theory of homeostasis—the steady state—in *The Wisdom of the Body*, following the lead of Claude Bernard (Bernard, 1878; Brobeck, 1973) and advanced the concepts of fight, flight, and fright as biological phenomena. Freud (Brenner, 1955), in his efforts to develop a biological psychology, endeavored to use a biological model based on the theory of the biogenetic law of ontogeny (i.e., the development or embryology of the individual) recapitulating phylogeny (i.e., the development of the species). He strove to utilize a systems theory in his efforts to demonstrate that the *interaction* of civilization and biological need produced emotional disorder. In addition, investigators in microbiology overtly or covertly believed that microorganisms in themselves did not necessarily "cause" disease but interacted with hosts of varying degrees of disease resistance to produce varying degrees of pathology (Bennett, 1966).

Within this intellectual climate, general systems theory took root, originally as a *biological theory* as advanced by Ludwig Von Bertanalffy (1968, 1972), himself a biologist. Medicine, as well as biology, had utilized some of his ideas in terms of describing *organ systems* within the organism, a superordinate biological system. Medical science, however, persisted in looking for specific individual "causes" or "etiologies" of diseases, or more sophisticatedly, aberrant target enzyme systems.

Although psychiatry was well represented in the systems approach by the psychobiological models of Adolph Myer (Cameron, 1963, Muncie, 1948), the conceptual framework did not enjoy widespread application by his colleagues. Remarkably, application of general systems theory flourished among social scientists, engineers, and later, among computer scientists. Fortunately, psychology, psychiatry, medicine, and biology have now caught up, with gratifying results.

Current methods of clinical diagnosis in psychiatry require four factors to be considered other than the basic diagnostic picture. These include character structure or pathology, medical or surgical illness (considered to be any physical disorder contributing to the clinical picture), the degree of external sociofamilial stress, and the level of the patient's global functioning during the previous year (American Psychiatric Association, 1980). Although not specifically described in systems terms, such considerations imply a theoretical framework that goes beyond a cause-and-effect concept.

In the following development, the theoretical framework of the systems approach will be formulated in terms of a more specific set of subsystems and index systems (Figure 1; Greenberg, 1976; Gross, 1981, 1982).

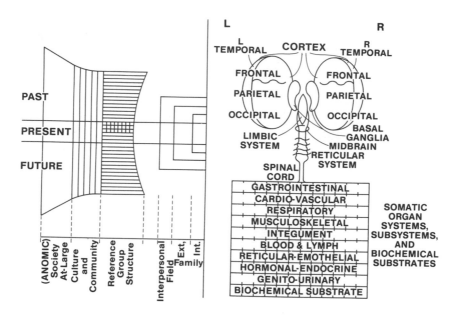

Figure 1. Schematic diagram of systems structures. (Note that the entire configuration should be considered as being embedded in a general physical environment, itself subject to systems analysis.)

The organism, in this case the human organism, shall be taken as an index system, in turn to be considered as being the anatomical and functional sum of its component subsystems (i.e., the organ systems usually described in the study of mammalian and human biology). Although the organism's biological systems would require extensive elaboration for a fuller understanding of this approach, this discussion will deal only with the central nervous system (CNS) and some of its subsystems. The human organism, in turn, shall be considered to be embedded in a set of nested social systems as superordinate systems.

Subsystems and Index Systems

The specific CNS subsystems and superordinate social systems that shall be considered are the following:
> I. CNS subsystems (the prefix "sub" shall be omitted in the following description).
>> A. *The reticular system:* This system has its cell bodies, localized into nuclei, located in the archaic portion of the brain stem, and deals peripherally with activation, inhibi-

tion, sleep and wakefulness, and integration of all nervous system functions. Its most typical pathological manifestations may be found in attention deficit disorder, delirium, and some comatose states.

B. *The limbic system:* This system articulates with the reticular system by means of a very great number of synapses in the midbrain, is the anatomical underpinning of emotion, motivation, olfaction, hormonal control, and short-term memory, as well as the central control station for autonomic function. Typical pathological states are found in classical bipolar illness and in partial complex seizures (temporal lobe epilepsy).

C. *The basal ganglia:* These are also archaic brain structures, found in premammalian vertebrates, and constitute most of the cerebral cortex in such animal forms. They deal principally with motor *power* and synchronous movement. The most common pathological manifestations occur in Parkinson's syndrome (from degenerative or vascular disease, as well as from neuroleptic drug toxicity), although there are numerous other disease states of the basal ganglia.

D. *The cerebellum:* This structure deals with balance, and although of great importance, will not be discussed here.

E. *The cerebral cortex:* The cortex can be viewed in terms of lobar functions (which would itself require several textbooks), but for the present purposes, subsystems, defined by (1) the dominant (i.e., verbal and usually left-sided) hemispheric functions together with the nondominant (i.e., spatial and usually right-sided) hemispheric functions and (2) the pre-Rolandic (i.e., motor, planning, and social comprehension) functions as well as the post-Rolandic (i.e., sensory and cognitive–intellectual) functions. The four functional subsystems are then represented approximately by the (1) pre-Rolandic dominant cortical quadrant (usually the left frontal lobe); (2) the pre-Rolandic nondominant cortical quadrant (usually the right frontal lobe); (3) the post-Rolandic dominant cortical quadrant (usually the left temporal, partial, and occipital lobes); and (4) the post-Rolandic nondominant cortical quadrant (usually the right temporal, parietal, and occipital lobes). Because the discussion that follows will not require specific details of these functions, the reader is referred to standard neurological and neuropsychological texts for further elaboration. An important point concerning cortical function, as viewed in this communication, is that *all ideational processes*, including thinking about intrapsychic, interpersonal, and social memories, as well as judgments and other thought processes, are basically cortical functions. The specific anatomic localization and integration of each thought process depends on its particular components.

F. Other CNS subsystems such as the spinal cord, the covering of the brain and cord (i.e., the meninges), the blood supply, and the like, will not be discussed here and are beyond the limited scope of this discussion.

Having described, however briefly, the principal CNS subsystems to bear in mind, three additional sets of major factors require definition.

A. For each system or subsystem, four parameters will be considered. These parameters are (1) anatomic or structural intactness of each subsystem; (2) functional capacity of each subsystem; (3) *quality* of function; and (4) *time* variations of functions.

B. In addition, the actual *interaction* between and among systems and subsystems requires the same four parameters to be considered.

C. Furthermore, it can be stated at this juncture that the same structural, functional, time interactive and time reaction factors pertain to psychosocial systems and subsystems as well as to biological systems.

II. Before proceeding with case illustrations, the *sociofamilial systems* considered should also be stated. These include (in addition to the intrapsychic system mentioned previously) the following:

A. *Nuclear family:* This system has been redefined several times in the past decade, principally to conform to changing social standards. In this communication, any interpersonal system of two or more people related by enduring biological, physical (sexual or otherwise), or emotional closeness could be accepted as a nuclear family, although the cases described fit into the more classical definition of nuclear family.

B. *Extended family:* This shall be taken to include more than two generations or branched lineage, with defined role functions, of a group of related people, with or without a nuclear core family as defined previously.

C. *Interpersonal field:* This field consists of the set of other people with whom the index person has meaningful, or relatively meaningful, interpersonal relations.

D. *Reference group systems:* This superordinate system consists of all relevant groups to which the index person relates,
or to which he or she *must* relate by necessity or obligation.
The reference group super system shall be considered to
include the community at large as well as the cultural background, in this communication.

E. *The anomic society at large:* This social structure is individually defined and consists of those social (as well as
political and economic) structures which affect the index
person, but on which he or she has no perceived influence.

Clinical Cases

The following three clinical case presentations are intended to demonstrate some aspects of a systems approach to diagnosis and as a consequence, treatment. There is deliberate overlap of some of the presenting
problems in order to illustrate the interaction of method and concept in
arriving at diagnosis. Two of the cases are 10-year-old children, both
girls and both presenting with the chief complaint of anorexia. One was
very disturbed by her symptoms, whereas the other was almost indifferent (item in Table 1, "anxiety concerning symptoms"). The child who
experienced distress and was very adept at making it evident to others
was suffering from a physiological disorder that was basically a variant
form of seizure activity. She presented affectively and interpersonally
in a manner very similar to a 52-year-old widow whose chief complaint
was nocturnal panic attacks and who also was discovered to have
atypical seizure activity, among several other physical disorders. To
add a further factor, the relatively indifferent child behaved as if she
were four decades older.

The individual case summaries are as follows.

Case 1

The patient was a 52-year-old widow of Mediterranean origin who said
she suffered from "nerves." By this she meant that she went to bed
feeling fine and woke up about 1 hour later feeling very frightened and
experiencing palpitations. These events had occurred for about 5 years
but had become worse during the previous 2 years with frequency of
attacks increasing from monthly to two or three times per week. In
addition, she said that "my legs freeze on me; I can't stand on them."
She also was experiencing some difficulty breathing and had been eval-

Table 1. Comparison of Symptoms

Clinical variable	Case 1	Case 2	Case 3
Sex	F	F	F
Age	52	10	10
Ethnicity/religion	Mediterranean/Catholic	Mediterranean/Protestant	Anglo/Dutch/Protestant
Medical–surgical	History of hysterectomy	Abdominal pain	Severe abdominal pain
Symptoms or history	History of bleeding ulcer, calcium deficiency, cervical arthritis, thrombophlebitis, trouble breathing	Anorexia, chest pain, frequent feedings in first year	Anorexia, foot blisters, symptoms began after sore throat, persistent abnormal laboratory tests
Central nervous System signs and symptoms	Dizziness Episodic fearfulness, sleep disturbance, mildly abnormal E.E.G.	Dizziness History of fetal distress requiring delivery by Caesarean section	Sleep disturbance History of fainting, abnormal E.E.G.
Intrapsychic conflict	Unresolved grief concerning husband and mother	Retreat from maturation, identification with depressed grandmother	Difficulty meeting expectancies of self and family
Familial relations	Reliance on daughter	Parents often absent working, more time with grandmother	Gets along very well with mother
Loss	Husband, age 55—1 year prior to onset of patient's symptoms; mother, age 77—concurrent with symptoms worsening	Death of grandfather 5 years earlier	None noted
Peer relations	Good	Fair	Good
Work or school performance	Good	Good	Very good
Dislike of effect of illness on social function	Moderately high	Moderately high	Moderately high
Discomfort with respect to symptoms	High	Moderately high	Very high

(continued)

Table 1. (*Continued*)

Clinical variable	Case 1	Case 2	Case 3
Anxiety concerning illness	Minimal	None	Some
Concern about the illness	Moderately high	Little	Moderately high
Anxiety			
Patient's rating	Moderate	None	Moderate
Examiner's rating	Moderate	None	High
Depression			
Patient's rating	Little	None	High
Examiner's rating	Moderate	Moderate	High
Agitation			
Patient's rating	Moderate	None	High
Examiner's rating	Moderate	None	High
Paranoia			
Patient's rating	None	None	None
Examiner's rating	None	None	None

uated for "low calcium, arthritis, and left-sided headaches" on a medical inpatient service earlier that year.

Past medical history revealed that she had a hysterectomy at age 25 for a "bleeding tumor" and had a polyp on her vocal cord removed surgically at age 37. In addition, she had a bleeding duodenal ulcer at age 41, but an operation had not been required; the patient adhered to her special diet. Her personal physician was treating her for spinal arthritis and calcium deficiency as well as for thrombo-phlebitis in her leg veins. The vein inflammation had also been worse for the past 2 years, and her leg arteries were beginning to be blocked.

Despite these difficulties, the patient continued working at a local factory, as she had for 25 years. The family history was contributory, in that her husband had cancer for 3 years before he died. The onset of his cancer, however, did not coincide with any exacerbations of the patient's medical illnesses, however. He died at the age of 55, six years prior to the patient's request for psychiatric assistance. The patient's panic attacks appeared to begin within a year after his death. In addition, her mother had also died of cancer earlier in the current year, shortly after the patient herself had been in the hospital.

The patient had five sisters and two daughters. No one in the family had required psychiatric care. She said she did not like living alone and was afraid to fall asleep because she might wake up with an attack. Although she did not like her symptoms and clearly stated that they bothered her, she was not otherwise fearful and functioned well.

A brain-blood flow study had been done earlier in the year and had shown no abnormalities. In keeping with a systems approach that would consider a physiological basis for the patient's disturbance, an analysis of all the medical conditions was made.

Although it was clear that the patient suffered from several real and somewhat painful conditions, there did not seem to be any physiological basis for her clinical symptoms of palpitations and fearfulness after one hour of sleep. The patient said she was not depressed, but she appeared to be experiencing some depression on clinical interview. She also appeared to be excessively concerned with her arthritis and leg symptoms, despite very careful monitoring by her personal physician.

Because no medical or physiological basis for the psychological symptom could be ascertained and because there appeared to be intact social functioning both in the long- and short-term, the sleep disturbance and occasional anxiety were diagnosed initially as having an intrapsychic basis, with a possible interpersonal component. The latter appeared to consist of an effort to establish a more dependent relationship with her daughters.

A course of psychotherapy was begun, based on a working diagnosis of an unresolved grief reaction concerning her mother's death, which, in turn, had aroused unresolved grief, and possibly conflict, over her husband's death.

The hypothesis was made that her nocturnal palpitation episodes were manifestations of anxiety, itself generated by conflict about the unresolved grief. The patient was articulate, cooperative, and spoke about her husband, her mother, her relationships with them, and how she thought she would meet them in heaven, but her symptoms persisted.

At this point, there appeared to be a reasonably well-developed therapeutic alliance, no ascertainable transference resistance, and good developing insight. Inasmuch as the cerebral blood-flow structure as well as the patient's cognitive presentation clinically gave no reason to suspect structural organic pathology, the failure of psychotherapy to relieve the patient's symptoms was considered to reflect an error in original diagnosis. Therefore, on the further hypothesis that there might be an endogenous biological depression operative, a trial of tricyclic antidepressant medication was begun. This succeeded in increasing the frequency of her attacks.

When the patient responded to tricyclic antidepressant treatment in this manner, another reconsideration of diagnosis was made. Clearly, the biochemical effect of the medication had been to increase activation (i.e., had adversely affected the sleep function of the reticular system) and had worsened the episodic fear concomittently (i.e., had exacerbated episodic limbic dysfunction). An unusual side effect of tricyclic agents in this patient's age group is that of exacerbation of seizure activity. Thus, although there were no other signs or symptoms of such activity, the time-related episodic nature of the symptom, itself exacerbated by tricyclic medication, as well as the lack of symptom remission with psychotherapy, prompted an investigation into the possibility of seizure activity. It should also be noted that a very small focus of cerebral (cortical or subcortical) irritability would not appear on a blood-flow study. Therefore, at this point, an electroencephalogram was performed and interpreted as follows:

> This tracing is an essentially normal one. However, rare, high voltage, spikelike waves are seen at the right temporal lead, which is somewhat suggestive of seizure potentiality, but not diagnostic of it.

The antidepressant medication was stopped, and the patient begun on a regimen of phenytoin, an anticonvulsant medication. Her attacks of palpitations improved. The regimen succeeded in eliminating her symptoms of nocturnal palpitations and fearfulness entirely, although she persisted in complaining about symptoms related to her arthritis and to her legs. She was followed psychotherapeutically for 6 months with excellent results.

Case 2

The patient was a 10-year-old girl, native born of Mediterranean origin, who was seen for psychiatric consultation in a pediatric inpatient setting "cause my stomach hurts me." She complained of burning pains in her abdomen and sharp pains in her chest. The pains had been moderate and present occasionally for 1 year but had become severe in the previous month. Sometimes the pain appeared suddenly, but at other times there was gradual onset. When her stomach hurt her, the patient felt sad, but otherwise she was generally happy unless she had suffered an injury. She rarely became sad for any other reason and would not be happy if she had to stay in the hospital for a long time.

She slept well but had frequent dreams. Some dreams were frightening, of two-headed people, and of monsters. She did not get frightened easily, had no enemies, and got angry only if someone was bad to her (e.g., bothering her by tugging at her when she wanted to play).

At times she did not like tests at school. Although she did not like coming home when no one was there, she was not afraid. She had one brother, aged 9, and got along well with him. She also had friends and frequently played with them.

The patient's father drove a truck and was often away from home but had been at home more frequently during the past year. Her mother worked in a grocery store in the evening, and at those times the child was frequently in the back of the store. If her mother worked during the day, she would be able to play with friends outside the store. Her maternal grandfather had died when she was 5, but she saw her other three grandparents regularly.

Past history indicated that the child was delivered as a result of a complicated pregnancy. She was a breech presentation, and Caesarean section had to be performed after 12 hours of labor. She required feeding every half-hour to every hour during her first year of life, but after that, she became an obedient, quiet, and well-mannered child who gave no trouble to her parents.

On interview, the patient had constricted affect, with rare episodes of appropriate smiling. She was remarkably unconcerned about her situation and offered no complaints about her family. She was thin and emaciated looking. There was no cognitive disturbance, no difficulty thinking, and no conscious depression.

The initial diagnostic impression was that of an organically based anorexia nervosa, with a familial component characterized by a difficulty in allowing the patient to mature. Diagnostically, psychological testing was suggested.

Test results revealed a Full Scale IQ of 116 on the Wechsler Intelligence Scale for Children, a need for caution, and "strong compulsive mechanisms." The projective tests revealed some somatic and sexual phobic material. There was no indication of thought disorder or any other psychotic process.

Clinically, the relatively sudden onset of symptoms, the episodic intense nature of the abdominal pain, the propensity to be somewhat clinging interpersonally, and the frightening dreams all were consistent with a component of cerebral dysfunction that might respond to an anticonvulsant drug, even in the absence of frank seizures. Therapeutically, a trial of phenytoin was suggested, based on the reasoning that there was no significant familial, intrapsychic, or cognitive disorder, coupled with the significant history of a very difficult delivery. In addition to phenytoin treatment, continuing evaluation in a psychotherapeutic interview situation with the family was indicated.

When the mother was interviewed in an out-patient setting, she was dressed fashionably and seductively. The patient, on the contrary, looked as if she were a 50-year-old immigrant who had just arrived from abroad. She was dressed in a little artificial fur coat and hat, a long dress, and looked very serious. Mother said that the patient's manner and speech had changed and she had become a different kind of person during the past year. This is also characteristic of developing cerebral dysfunction, with a seizure activity component.

During joint interviews, mother encouraged the patient to express her anger, and within 2 weeks, the patient began feeling hungry and was experiencing less pain.

During the third week after discharge from the hospital, the patient related a happy dream she had had of her grandfather "a long time ago," and when she had told her mother about it, 3 days before the interview, her stomach hurt. Both mother and child discussed the grandfather and his death. Toward the end of the interview, the mother stated that the patient had been spending a lot of time with her grandmother, who was still mourning the grandfather's death, showing the patient pictures of the grandfather and telling her that it would be nice to see him in heaven. The grandmother ate very little and was very thin. At the next interview 1 week later, the patient said she felt much better than last week *and* since the onset of her illness in terms of no longer having abdominal pain. She was doing well in school, dreamt good dreams, and even ate lunch at home. She also said that although she had been concerned about dead relatives in the past, she no longer was. She continued to take phenytoin for another month, but it was discontinued because of a side effect of hair appearing on her back.

The patient continued to do well in terms of eating but experienced an increase in abdominal pain. She visited her *paternal* grandfather, who was still living, and experienced worse pain, from which she recovered. Continued discussion with mother, child, and occasionally father appeared to resolve the issues of depression, dealt with some death anxiety, and reinforced social adjustment, despite the persistance of some abdominal pain.

The final diagnostic impression was that of an hysterical identification with a depressed grandmother, complete with symptoms of *folie a deux* and possible minimal cerebral dysfunction secondary to birth trauma. Only a family therapy intervention would appear to have served the principal therapeutic purpose in this case.

In this case, the organic component originally hypothesized was not considered to be a sufficient explanation for the presenting symptoms. Continuing investigation into nuclear and extended family dysfunction was done from the outset, with the thought that such dysfunction was contributing a significant component to the patient's distress. In actuality, this was the case, and it is very likely that patient's dependency and need for intense identification was a product of both minimal cerebral dysfunction and disruption of the familial structure and function.

Case 3

The patient was a highly intelligent 10-year-old girl who presented with the chief complaint of abdominal pains that she described as "gas" and that were relieved by "burping." She had more severe pain after eating and therefore ate less. She had a history of frequent colds and had developed severe infections several months prior to the onset of her gastrointestinal symptoms. In addition, she had inflammatory blisters on her feet. There had been a persistent low-grade fever, and the significant pains had been present for 1 month at the time of hospitalization and psychiatric consultation.

Sleep disturbance was also present, such that the patient could fall asleep at 9 or 9:30 P.M. when she went to bed only once or twice a week; otherwise, she could not fall asleep until 11 P.M. or even 1 or 2 A.M. because of pain. She did not like being away from the mother because of anxiety about the abdominal distress. She experienced some crying spells at night, had begun to have disturbing dreams, and felt anxious about not feeling well.

In addition, she began experiencing periods of being angry at herself. She could not understand why she cried or why she was angry but related her anxiety to being ill.

Family relations were good, according to the patient. Her brother, aged 12, and she, got along reasonably well. Her relationship to her father was described as good, and the one to her mother as very close and somewhat dependent. Recently, according to the patient's mother, the patient's brother had been wondering about life and reading the Bible intensively but had recovered from this episode uneventfully. Both parents were highly educated professionals with high expectancies of their children as perceived by their daughter.

The patient did very well in school but disliked one teacher in second grade and experienced some loss when two other teachers whom she liked left school in the middle of the previous schoolyear. The patient had no behavior or emotional problems previously. She also had many friends in school, well-defined goals, and specific reading interests.

During the first interview, the patient appeared to be in moderately severe physical distress, somewhat agitated and depressed, and episodically tearful. She was thin, looked physically ill, and felt distressed that her symptoms were viewed by staff as emotionally or psychologically induced. Although she was not pleased by the interview, she was cooperative, elaborated answers to questions, was thoughtful, and was able to communicate well. She had a good range of affect, was highly intelligent, and after some initial difficulty, related well to the interviewer. There was no evidence of any delusional thinking or of body image distortion. She was very clear about not liking to be alone because of her pain, not liking her physical discomfort, and was well aware of her anxiety.

Laboratory examinations were only minimally contributory. There were some minor indications of residual inflammation found in the examination of blood cells and thyroid function studies but no specific findings indicative of any single disease entity. She had a mildly abnormal electroencephalogram that showed some spike activity in the left temporal area. This was considered to be consistent with, but not diagnostic of, seizure activity.

The Wechsler Intelligence Scale for Children demonstrated a Full Scale IQ of 138 with minimal scatter. The projective tests concurred with the clinical picture and showed some anxiety, perfectionism, and compulsive defenses, with some surfacing of anxiety and anger. There was no indication of any serious psychopathology consistent with disturbance of a body image or a psychotic process.

Although this young lady presented with anorexia, there was neither physiological evidence of a basic gastrointestinal or metabolic disorder, nor of serious psychological disorder. There was, however, some evidence of seizure activity on the electroencephalogram, episodic abdominal pain, and fairly sudden onset after a febrile episode. A trial of anticonvulsant medication, phenytoin, was begun, and the patient then followed in an out-patient setting with her mother, and occasionally her father, for individual and family psychotherapy. The symptomatic goals of medication treatment were to relieve the pain and its concommitant anxiety, as well as to assist the patient in falling asleep by means of pain reduction. The goals of psychotherapy were to deal with the loss of self-esteem consequent to performance failure induced by her illness, both in terms of intrapsychic attitude change and familial expectancy change.

The symptoms of abdominal pain and anorexia remitted to some extent, although the pain was still present at times. Six months after treatment began, the patient wrote, stating at the end of her letter that "the stomach ache isn't like it used to be though, it's not the same and it's not as bad. Thanks a lot for your help." She still experienced short episodes of unexpected depression and had developed sensitivity to "noises and other little things." At that time, the patient had not been seen for 4 months, the initial psychotherapeutic intervention having been once weekly for 2 months. She was free of chronic depression and anxiety and was followed every 3 months for the next year, and then every 6 months.

The patient continued to have persistent fatigue, with laboratory data suggestive of several different hormonal conditions involving the thyroid, and possibly sex hormones. No laboratory test was specifically significant or indicative of a disease process. However, the blisters on her feet, which had recurred, were successfully treated with steroid hormones (cortisone). Four years after the initial febrile illness, the patient appeared stronger physically and better able to cope. The discrepancy between her physical stamina and her expectations continued, however.

By the age of 17, the patient felt well enough to attend college away from home. At that point, she no longer required medication, was handling one-half to three-quarters of a full program, and had a boyfriend. Tragically she and her boyfriend both died in an automobile accident that year. Her mother wrote to say how the patient had gained weight at college, had finally surmounted her health problems, and was becoming independent. She expressed her thanks and grief and enclosed a short self-description of the patient written shortly before her death.

Discussion

It should be taken as axiomatic, for the purposes of the discussion, that appropriate and successful treatment depends on accuracy of diagnosis. The purpose of the systems approach is to provide a framework and method of analysis. The framework endeavors to enlarge the field of diagnostic possibilities, whereas the method provides an approach to focusing critical and ancillary subsystem dysfunctions that lead to the clinical picture. The subsystems that are brought into focus as dysfunctional elements then provide the appropriate diagnostic indicators.

Three Principal Points

Specifically, the three cases presented can be used to illustrate three principal points:

1. Presenting symptoms do not lead *a priori* to diagnosis. Even a second order analysis of dynamics, whether of psychological or physiological processes, does not necessarily lead to correct diagnostic formulation.
2. The same underlying pathological process does not necessarily lead to the same clinical picture in all cases. Although this is taken as axiomatic for psychopathological processes, it is not

always evident when the underlying pathology is of physiological origin.

3. Value system assumptions, overt or covert, conscious or unconscious, on the part of staff, can obscure correct diagnosis, and lead to what can be termed *sociogenic* or *ideologenic* countertransference difficulties. The latter can be described as staff attitudes aroused by staff's incorrect assumptions that the patient is motivated by a particular set of values either held by staff members or believed by staff members to be held by the patient.

Symptoms not Leading A Priori to Diagnosis

An analysis of Case 1 and a comparative analysis of Cases 2 and 3 can be used to illustrate this point.

In Case 1, that of the 52-year-old widowed woman who had recently lost her mother and who had multiple physiological problems, there was a clear precipitant for her distress, as well as a clinically evident hysterical character structure.

There was, however, a history of competent functioning as a mother and employee for 25 years and an absence of a history of emotional disorder. Nevertheless, because the patient's interpersonal field had been radically altered by her mother's death, it appeared reasonable to make a working diagnosis of an anxiety reaction, the symptom of which appeared to induce the immediate consequence of restoring an interpersonal field. In addition, a high likelihood of unresolved grief was reasonable to assume in view of the clinical picture. The clinical evidence against this diagnostic formulation was her long-term life adjustment, as well as her presentation on interview. As noted, she appeared reasonably concerned, was not hypochondriacal, although she had many medically corroborated complaints, and had a very good range of affect. There was no convincing evidence of internalized unresolved conflict with her mother or husband.

Consequently, her lack of improvement with a psychotherapeutic approach led to a change in working diagnosis to that of an *endogenous* biologically induced depression precipitated by her mother's death. The worsening of her symptoms with tricyclic antidepressant medication, which frequently relieves endogenous depression, coupled with the episodic and sleep-induced nature of her panic attacks, led to the investigation of seizure activity, which she had and which responded to anticonvulsant therapy.

Although it is unlikely that the diagnosis could have been made sooner, it is of importance that it was made, even after several months. The systems analysis required examining intrapsychic, interpersonal, and social functioning, as well as medical, neoclassical neuropharmacological, and classical neurological approaches. The patient's seizure focus was probably part of the same disease process that had already given rise to peripheral blood vessel disease and calcium metabolism difficulties. Nevertheless, she continued to do well psychologically for the next 10 years. The comparative analysis of Cases 2 and 3, two 10-year-old girls, is almost self-evident. They both presented with the same set of symptoms, namely with abdominal pain, anorexia, and separation anxiety, with one major difference: One child appeared clinically to be almost indifferent to her difficulties, whereas the other was in great distress. Not too surprisingly, the patient who was not in overt psychological distress, actually had made an hysterical identification with her anorectic and depressed grandmother. Each child had one brother, and sibling rivalry was considered by staff in both cases. In actuality, this was of little import. It is also important to note that both children had good psychosocial adjustments and presented with acute ego-alien illness. Continued interviewing and exploration of family process was required to ascertain the importance of familial function in one case and its relative lack of importance in the other. As in Case 1, pursuit of data concerning *both familial process and neurological disorder* was imperative in establishing correct diagnosis and treatment. Interestingly, the child with seizure activity required more prolonged psychotherapeutic intervention, particularly in terms of family therapy, because of her physiological disorder, even though family dynamics were not of etiological significance. Staff countertransference difficulties in establishing diagnosis will be discussed in another section.

Pathology Productive of Different Symptomatologies

This idea may almost be self-evident at this point of the development of the systems approach in diagnosis. The presentation of Cases 1 and 3, when compared, are excellent demonstrations of this point. *Both* patients had underlying seizure activity; both were concerned about their illness, had good premorbid adjustments, and saw their symptoms as ego-alien. Neither had significant structured psychopathology. Their clinical presentations, however, that were different could feasibly have been very similar but were not. Specifically there is no obvious reason why the child did *not* present with night terrors, other than she did not fall asleep *until* she felt better. The finding of seizure activity on the electroencephalogram is not unusual in children and adolescents who present with anorexia, but the concept of seizure activity giving rise to anorexia is *not* commonly held. Moreover, there was no family history or medical history that would have indicated the possibility of seizure activity in the child, although there were some mild indications in the older woman.

Lastly, a comparison of the two children of Cases 2 and 3 in terms of *medical history* would have made the child presented in Case 2 (i.e., the patient who had identified with her anorectic, depressed grandmother), a more likely candidate for seizures, inasmuch as she had a significantly difficult birth. However, she did *not* have seizures. The separation anxiety present in each of the two children was derived from two different origins. The child with seizure activity wanted her mother present because of intense physical discomfort and consequent fearfulness. The other child was somewhat fearful and perfectionistic, possibly as a consequence of mild birth injury and was therefore more dependent on her adult family members. This fostered a propensity for hysterical identification. The anxiety was reinforced by the frequent physical absence of her parents.

"Sociogenic" Countertransference Considerations

Possibly the most important concept in terms of permitting appropriate diagnostic procedures and effective treatment is that of attitudinally induced (i.e., "sociogenic") countertransference. Such countertransference difficulties are well illustrated in these cases. Patients with multiple, and seemingly minor, physical difficulties, especially older patients, are often viewed as seeking attention, replacing lost objects, and binding anxiety. Their apparent dependency needs run counter to cultural expectations, and hostility is thereby often aroused in staff. The active pursuit of diagnostic accuracy serves both as a conceptual and practical protection against the effects of such countertransference.

The children present similar, as well as somewhat different, countertransference problems as the adult patient. Similar issues are raised by invoking "attention seeking" and by the hostility aroused in staff by patients whose illnesses are very difficult to diagnose. Different issues are involved by the almost automatic assumption that sibling rivalry must be involved. It is likely that "sociogenic" countertransference varies in its content both ethnically and geographically, but the principal task of the systems diagnostician may be to dissolve countertransference with clinical accuracy and precision.

Conclusion

The systems approach, by providing a framework for concept and a blueprint for method, allows diagnostic accuracy and may provide an instrument for the prevention of countertransference induced by socially expected norms.

References

American Psychiatric Association. (1980). *Diagnostic and statistical manual of mental disorders* (3rd ed.). Washington, DC: Author.

Banting, F. F., & Best, C. H. (1922). The internal secretion of the pancreas. *Journal of Laboratory and Clinical Medicine, 7*, 265.

Bennett, I. L., Jr. (1966). An approach to infectious disease. In T. R. Harrison, R. D. Adams, I. L. Bennett, W. H. Resnik, G. W. Thorn, & M. M. Wintrobe (Eds.), *Principles of internal medicine* (5th ed.; pp. 1470–1476). New York: McGraw-Hill.

Bernard, C. (1887). *Lecons sur les Phenomenes de la Vie Commune aux Animaux et aus Vegetaux.* Paris: Bailliere.

Brenner, C. (1955). *An elementary textbook of psychoanalysis.* New York: International Universities Press.

Brobeck, J. R. (Ed.). (1973). *Best and Taylor: Physiological basis of medical practice* (9th ed.). Baltimore: Williams & Williams.

Cameron, N. (1963). *Personality development and psychopathology, a dynamic approach.* Boston: Houghton-Mifflin.

Cannon, W. B. (1932). *The wisdom of the body.* New York: W. W. Norton.

The Economist. (1986). Primary health care is not curing Africa's ills. Vol. 229, No. 7448, May 31–June 6, 91–94.

Greene, M. B. (1986). Superstring. *Scientific American, 225*, 48–60.

Greenberg, I. M. (1976). *General systems: Social and biological interactions.* Unpublished manuscript.

Gross, D. A. (1981). Medical origins of psychiatric emergencies. *International Journal of Psychiatry in Medicine, 11*, 1–24.

Muncie, W. (1948). *Psychobiology and psychiatry* (2nd ed.). St. Louis: Mosby.

Semat, H. (1944). *Introduction to modern physics* (2nd ed.). New York: Rhinehart.

Von Bertanalffy, L. (1968). *General systems theory.* New York: George Brazzilier.

Von Bertanalffy, L. (1972). General system theory. In S. Arieti (Ed.), *American handbook of psychiatry* (Vol. 4, pp. 1095–1112). New York: Basic Books.

9

A Systems Approach to Understanding and Enhancing Grassroots Organizations
The Block Booster Project

Paul Florin, David Chavis,
Abraham Wandersman,
and Richard Rich

Introduction

This chapter reports on the use of systems concepts in an action-research project studying and intervening with voluntary community organizations. Voluntary community organizations include block or neighborhood associations, tenant associations, church volunteer groups, youth groups, and merchant associations. They may stand independently or be connected through federations or coalitions.

Whatever their type, voluntary community organizations (VCOs) share several characteristics that define their distinctive place in our

Paul Florin • Department of Psychology, University of Rhode Island, Kingston, Rhode Island 02881.   **David Chavis** • School of Social Work, Rutgers University, New Brunswick, New Jersey 08903.   **Abraham Wandersman** • Department of Psychology, University of South Carolina, Columbia, South Carolina 29208.   **Richard Rich** • Department of Political Science, Virginia Polytechnic Institute and State University, Blacksburg, Virginia 24061.

Analysis of Dynamic Psychological Systems, Volume 2: Methods and Applications, edited by Ralph L. Levine and Hiram E. Fitzgerald. Plenum Press, New York, 1992.

society. They are (1) *geographically based*, representing residents who live in a particular area; (2) *volunteer driven* where the primary resource is the time, skills, and energy of their members; (3) *locally initiated,* formed by local residents responding to local conditions; (4) *human scale* where decisions are made face-to-face and operations are informal; (5) *problem solving* where getting specific tasks accomplished is important; and (6) *multipurpose and flexible,* able to address a variety of issues simultaneously and adjust strategies when necessary.

A variety of case studies and a small but growing body of empirical research has documented a variety of positive impacts of voluntary community organizations (see Florin, 1989, for a summary). These organizations have many impacts on the physical and social conditions in the neighborhoods where they exist. VCOs may engage in cleanup and beautification projects, mount home maintenance and repair programs that improve the existing housing stock, and renovate abandoned buildings that add to the housing stock. They may also change the social relations within an area, reducing isolation, and increasing neighboring and a sense of community among residents.

VCOs often strengthen, facilitate, or substitute for social services needed by local residents. They organize and deliver their own services, often provided entirely by volunteers, including day-care and baby-sitting cooperatives, employment services for teens, food-buying clubs for working-class families, food pantries for the hungry, and temporary shelter for the homeless. Some organizations have entered formal and informal "co-production" relationships with municipal governments in which the city and a VCO assume joint responsibility for a service (Rich, 1979, 1981). Co-production arrangements with city governments have been established in areas as diverse as health care, housing rehabilitation, and job counseling, and can sometimes provide services cheaper than the public sector (Cardwell, 1987).

The activities and projects of VCOs can have a stabilizing effect on urban neighborhoods by increasing satisfaction, reducing fear of crime, and promoting confidence and investment in the community. Participants can develop competencies and confidence that is "empowering" to them, leading to decreases in feelings of helplessness and increases in the sense of citizen duty (Ahlbrandt, 1984; Rappaport, 1981; Zimmerman & Rappaport, 1988). These organizations can also collectively be "empowered," obtaining increased mastery over the affairs of their neighborhood by altering the distribution of power and decision-making authority within the community (Crensen, 1983; Henig, 1982; Rappaport, 1987; Zimmerman, in press).

The number of VCOs has been growing during the past two decades. Examples of successful grassroots groups can be found in all regions of our country, among all income, racial, and ethnic groups. In 1979, the National Commission on Neighborhoods identified more than 8,000 neighborhood organizations in the United States. The Federal Office of Neighborhoods, Voluntary Associations and Consumer Affairs found nearly 15,000 citizens groups concerned with rural and urban problems. The Community Information Exchange, an information and referral service for community groups, maintains a list of 4,500 neighborhood associations, and the National Association of Neighborhoods has a mailing list of 15,200. Although detailed distributions are unavailable, concentrations of such groups occur in a number of older, low, and moderate income neighborhoods in our larger cities. The numbers reflect only organizations large enough to contact national organizations. The small and informal nature of most VCOs makes obtaining accurate counts difficult. The Citizens Committee for New York City has identified over 7,000 block, tenant, youth, and other VCOs in New York City (Clark, Chavis, & Glunt, 1988). These numbers are almost certainly not typical of other cities, but it is safe to assume that every large city has many such organizations.

The growth in VCOs has been given impetus by several large-scale trends within our society. One of them is what has been called a "rooted distrust of bigness" (Glazer, 1983). The alienating qualities of big business and big government have led to mounting cries to empower people and strengthen "mediating structures" such as the family, church, volunteer associations, and neighborhood groups (Berger & Neuhaus, 1977). Simultaneously, the search for community involves many Americans and has been growing (Bellah, Madsen, Sullivan, Swidler, & Tipton, 1985). Pollster Daniel Yankelovich reports that in 1973 approximately 32% of Americans felt an intense need to compensate for the impersonal and threatening aspects of modern life by seeking a community. By the late 1970s, that number had increased to 47% (Yankelovich, 1981). In areas where grassroots community organizations exist, they receive a ready response from residents and participation rates range from 15% to 22% of the residents. This number also represents only a portion of the potential participants. A 1980 Gallup Poll showed a striking 69% of the urban population willing to devote an average of 9 hours per month to neighborhood activities, including "the performance of some neighborhood social services." Participation also seems to increase as the size of the group's turf decreases. The Block Booster Project discussed in this chapter found that membership in block associations averaged 62% of the residents.

Clearly, people appear ready to seize opportunities for the kind of connection and sense of control provided by VCOs. Their numbers and potential for positive effects make them a potent vehicle for the maintenance and revitalization of American urban communities. Yet our understanding of these important social entities remains limited, and our documentation of their effects underdeveloped. Perhaps most important, our ability to foster, nurture, and sustain VCOs is hampered by lack of systemic knowledge of the important dimensions of their functioning. The questions most frequently raised are: Why do some organizations thrive while others decline and die out? What can be done to help VCOs survive and become more effective? How can large numbers of VCOs be supported in a cost-effective and efficient manner given the tremendous demand? The Block Booster Project, using a systems framework to promote understanding and systems concepts in the design of an intervention, has provided some initial answers to these questions.

The Block Booster Project

The Block Booster Project, funded by the Ford Foundation and conducted in conjunction with the Citizens Committee for New York City, was the most comprehensive empirical investigation of small voluntary community groups on record. Over 1,000 residents on 48 blocks from three culturally and economically different neighborhoods in New York City participated in the study. Information was gathered through telephone surveys, self-administered questionnaires, in-depth interviews, environmental observations, and archival records.

The 2½-year study was undertaken to clarify the relationship between volunteer block associations and overall community development and crime. Block Booster documented the impacts of block associations, identified the ingredients of successful block associations, and developed and field-tested a technical assistance method to help such groups remain active and vital. The Block Booster Project assessed the impacts of block associations by comparing blocks with block associations to demographically similar blocks without associations and comparing members in block associations with nonmembers from the same blocks. Block associations were analyzed for characteristics that distinguished viable groups from those that eventually declined into inactivity. Block associations were chosen at random to receive the technical assistance package. These groups were later compared to groups that received no assistance in order to test the method's effectiveness.

Findings from the Block Booster Project concerning the impacts of block associations and participation in them on a variety of individual, social, and physical outcomes can be found elsewhere (Chavis, Florin,

Rich, & Wandersman, 1987; Florin, 1989; Perkins, Chavis, Florin, Rich, & Wandersman, 1985; Perkins, Florin, Wandersman, Chavis, & Rich, 1985; Rich, Chavis, Florin, Perkins, & Wandersman, 1986). Here we outline the systems framework used to conceptualize the functioning of voluntary community organizations, summarize the findings related to organizational viability, describe and report the outcome of our intervention efforts, and explore implications for nurturing large numbers of VCOs.

A Systems Framework and Organizational Functioning

The Block Booster Project used a systems framework and concepts to gain an understanding of block associations and to develop an intervention aimed at improving their functioning. In this perspective, block associations and other types of voluntary community groups, although small scale and informal, are nevertheless organizations that should function according to systems principles. The departure points for our framework were basic systems concepts from general and social systems theory (Berrien, 1968), Katz and Kahn's (1978) open systems model of organizations, a prototype of such a model used previously (Prestby & Wandersman, 1985; Wandersman, Florin, Chavis, Rich, & Prestby, 1985), and earlier research that identified correlates of active individual participation (Florin & Wandersman, 1984) and social climates associated with viability (Giamartino & Wandersman, 1983).

Following general systems theory (Berrien, 1968; Von Bertalanffy, 1950), we conceive of the block association as exhibiting principles and characteristics isomorphic with systems at levels as varied as molecules and the solar system. A basic concept is that the block association, as a system, receives various "inputs" from its environment. This input (e.g., matter–energy or information) is transferred across an organizational boundary that separates the block association from its environment. The input is then transformed by organizational components (e.g., structures) and processes (e.g., organizational functions) into products (or "output"). Output is discharged back across the boundary, from the system into the suprasystem or environment. These products or output are either useful to the suprasystem, or they are waste products and useless. "Output" can, in turn, impact various aspects of the environment that "feedback" into a new cycle of inputs into the organization.

The block association, as all systems, must combat a tendency toward disorder or entropy by accessing sufficient resources and developing structures and functions that maintain operations and produce products. Organizational boundaries that are too rigid prevent the identification or acquisition of resources from the environment, whereas overly fluid boundaries weaken structures and dissipate internal resources. The "state of the system" is the particular pattern of relationships among components and the particular filtering condition of the boundary and tends toward homeostasis (Berrien, 1968). Our intervention to improve block association functioning (described below) was an attempt to decrease entropy in block associations through changing the "state of the system."

Katz and Kahn called their systems view of organizations an "open-systems" model. It is true that voluntary community groups are influenced by large-scale social forces such as demographic trends, industrial out-migration, and general economic downturns over which they have no control (e.g., the forces are exogeneous to the system) and therefore function as "open." However, these organizations work to have as much control and impact as possible over their immediate environment. Feedback and reciprocal causation between the organization and environment is thus involved as the organization seeks to try to "close" the system, making environmental factors endogeneous to the system. Resources are accessed and environmental factors controlled as much as possible. For example, when a neighborhood organization opposes a municipal decision to close a firehouse in their neighborhood (an exogeneous variable), an attempt is being made to "close" the loop between organization and environment and bring the decision under the control of the organization (e.g., transform it to an endogeneous variable).

Figure 1 outlines the general components of our systems framework and their relationships. Within the organization, on-hand resources in the form of characteristics of members (e.g., knowledge, skills, time, political connections, etc.) and externally accessed resources (e.g., funds, power, support of other organizations) must be "organized" into structures and functions, because proximity alone is insufficient to create a system (Berrien, 1968). Structural components and the functions they perform must also be linked or bonded in a way that permits the components to function together. Insufficient linkages or bonding of the components leads to structural disintegration. The particular pattern of organization of structures and functions within the block association and the linkage or bonding between them creates an organizational climate or group atmosphere for the participating members. The group climate within the organization influences the degree to which internal resources are mobilized, that is, the degree to which members are satisfied and committed to the organization and devote their time and energy to participate in the organization. The extent of

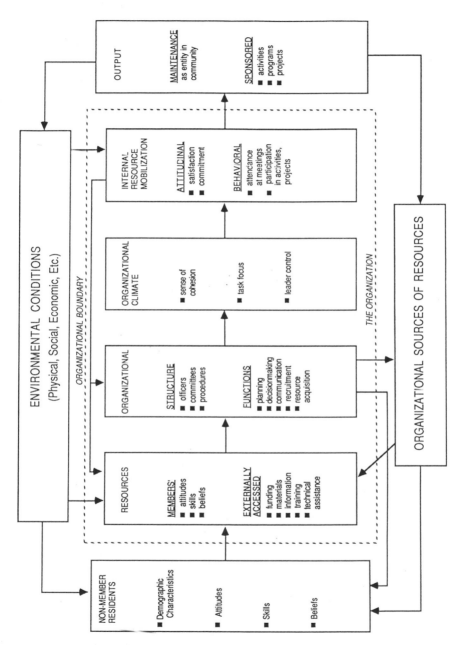

Figure 1. A systems framework.

internal resource mobilization determines the organization's ability to maintain itself as an organizational entity within the community and successfully mount various projects and programs it hopes will impact the environment. For example, several skilled members in an association organize themselves into structured committees that systematically engage in both production and maintenance functions. This creates an organizational climate high in task focus and sense of cohesion. Satisfaction and commitment thrive in this atmosphere, and high levels of participation are generated. The high level of participation is used to mount home improvement and youth recreation projects presented to the community as services of the organization.

The presence of the feedback loop *internal to the organization* indicates how the *quantity* and *quality* of the participation that is mobilized can, in turn, affect other organizational elements. The *quantity* of participation mobilized (e.g., number of members) can impact organizational structure and functions (e.g., committees added or dropped, the degree of formalization necessary, the ability to adequately perform several functions simultaneously). The *quality* of members' participation (e.g., extent of involvement, degree of responsibility) can impact upon members' skills and attitudes, a resource of the organization. For example, a committee chair develops leadership skills, increasing the pool of potential officers for the organization; another member, who feels she is being asked to do "busy work" beneath her skills and abilities, develops a negative attitude toward the organization and withdraws, thus reducing the resources of the organization.

The external feedback loops in the framework indicate how the functioning of a voluntary neighborhood group like a block association impacts on environmental factors that will, in turn, affect its own functioning. For example, the "output" (e.g., the presence of the association as an entity and the projects it undertakes) may impact positively upon the physical and social aspects of the block (e.g., a new sense of optimism among residents, a cleaner block with fewer rundown houses; youth basketball leagues instead of groups of teens hanging out on corners, more neighboring, etc.) This in turn influences the attitudes or beliefs of nonmember residents and may lead some to join and contribute their skills to the association. These environmental impacts also feedback to members, expanding their sense of the possible, and amplifying their commitment and contributions to the association. For example, Chavis and Wandersman (1991) demonstrated how changes in the physical and social environment can affect factors influencing the level of one's participation in a block association. Outputs can also influence the direct support supplied by other organizations in the environment when the programs or activities of the association are perceived as related to their own mission (e.g., police department support to buy recreation equipment for youth, use of church van for transporting elderly residents to a clinic, low-interest home improvement loans from a local savings and loan). These are examples of the output of the block association being seen as useful to the "suprasystem" outside of its boundary and therefore supported by organizations in that suprasystem. Organizations may indirectly contribute resources to the block association by influencing their own membership to become involved with the association (e.g., a business sponsors a baseball league, and its employees serve as coaches or referees; a minister encourages participation in the organization). Resource-acquisition functions of the association may also directly access resources from external organizations (e.g., donations of tools for home improvement projects from a local hardware store, funding sources identified by a municipal official, training in meeting management from a support organization), or increase membership directly through a recruitment campaign.

Most volunteer community organizations want to create positive spirals in their feedback loops: more resources mobilized and accessed from the environment that are transformed into a broader array of positive output that builds member competence and attracts new members to further increase organizational capacity and so on. Some voluntary neighborhood organizations mobilize around a specific problem and disband when the association successfully resolves the issue that led to its formation, a negative feedback loop from the environmental change leading to a decrease in members' belief that an organization is necessary. Still others are content to exist in a relatively quiescent state, producing little output other than holding periodic social activities that maintain a sense of community on the block and the potential for participation should an environmental threat necessitate other activities.

All too often, however, voluntary community organizations get caught in a downward spiral while they still want to continue to exist and mount projects. An organization risks a downward spiral where it becomes increasingly difficult to maintain the organization and produce programs and products if it fails to (1) obtain adequate resources, (2) develop an organizational structure with functions for effectively producing outputs and accessing more resources, (3) create a supportive organizational climate, and (4) mobilize acquired resources. This leads to decreased participation and even less likelihood of recruiting new members or accessing additional resources. Such a downward spiral may be halted by an exogeneous environmental force or event (e.g., a new municipal program targets the neighborhood with an infusion of grant money; the discovery of a toxic waste dump nearby rouses an angry citizenry), but often the outcome is a decline into inactivity and the loss of potential benefits to residents, leaving leaders feeling frustrated and defeated.

The possibility of one other potential feedback loop should be briefly mentioned, one in which a positive loop eventually changes the character of the organization itself. Occasionally, resource acquisition through grant writing and/or contracted service delivery is very successful, leading to increased competence in this realm and even more funding for the organization. Through this positive feedback loop, a VCO sometimes transforms itself into a neighborhood development organization or community development corporation. However, the considerable resources and presence of paid professional staff demand bureaucratization with more complex administrative structures and centralized decision making (Milofsky, 1987). Although these organizations maintain their commitment to locale-based development and can have enormous positive impacts on the entire community (Mayer, 1984; Peirce & Steinbach, 1987), they are by necessity organizations that are qualitatively different than more informal and participatory community organizations where the primary resource is the energy and involvement of volunteer members (Rich, 1983). Greater material effectiveness may come at the price of less flexibility and reduced capacity for promoting the social integration of the community.

Organizational Characteristics and Organizational Viability

There are several possible ways to think about organizational viability or effectiveness, and researchers have disagreed about the best way to measure viability. We define *viability* as the capability for living and growth. The primary indicator of viability in any organization is its ability to continue to exist or maintain itself. Whatever other fine distinctions can be made, organizations that cease to function are obviously less effective and viable than those that continue to exist. As suggested, the demise of a block association deprives residents of important benefits and may also have the negative effect of dampening people's willingness to take collective action in the future (Rich, 1988). Unfortunately, the inability of the association to maintain itself is a common problem for many block associations.

Organizers and community leaders have long recognized that maintaining a VCO is often a great challenge. Rapid decline is possible because voluntary organizations like block associations rely on the energy and expertise of people who are volunteering their time and are free to withdraw at any time. Formation of such organizations may, in fact, be relatively easy in comparison to maintaining the association after the initial enthusiasm and excitement has faded (Miller, Malia, & Tsembersis, 1979). Researchers have confirmed that it is difficult to keep members involved and active and a block association viable. Yates (1973) found more than 50% of the block associations he studied became inactive after they had performed initial simple tasks. In another study of 17 block associations, only 8 were found to be functioning 1 year after the initial contact (Prestby & Wandersman, 1985; Wandersman et al., 1985).

The Block Booster Project was also witness to the fragility of block associations. As part of the project, data from a variety of sources was gathered on 28 active block associations during February–May 1985. By May 1986, eight of these block associations had lapsed into inactivity and ceased operations. "Inactivity" was operationalized as having had no meetings or activities during the preceding 6 months. Block association leaders did not believe that the inactivity of their association was due to lack of any need for an association. Instead, they attributed inactivity primarily to lack of involvement and participation by members and the failure of the association to accomplish much. The data gathered from leader interviews and member surveys 12 to 15 months earlier was used to distinguish characteristics of those block associations that maintained operations from those which ceased functioning.

The search for organizational variables associated with maintenance presents a wide array of choices. Maintenance could be a function of leadership, the composition of membership, assistance from outside organizations, the way the organization is structured, or a host of other possibilities. There is no tight theory of organizational maintenance in voluntary organizations that postulates a small and manageable number of variables. We used our systems framework to guide our efforts and clustered the variables we examined into categories. Figure 2 is a simplified, flow-type framework, derived from our systems framework, that shows the operationalized variables related to maintenance and viability that we examined. Our goal here was not to examine the systemic relationships among these groups of variables or between any of these variables and the environment but only to examine the relationship of each of these separate groupings of variables to maintenance—that is, to treat each as an independent variable. The simplified framework thus shows no feedback loops, only the presumed flow process. All of these independent variables were gathered when all 28 blocks were active, a minimum of 12 months before 8 of the blocks declined into inactivity. Our objective was to determine which of these

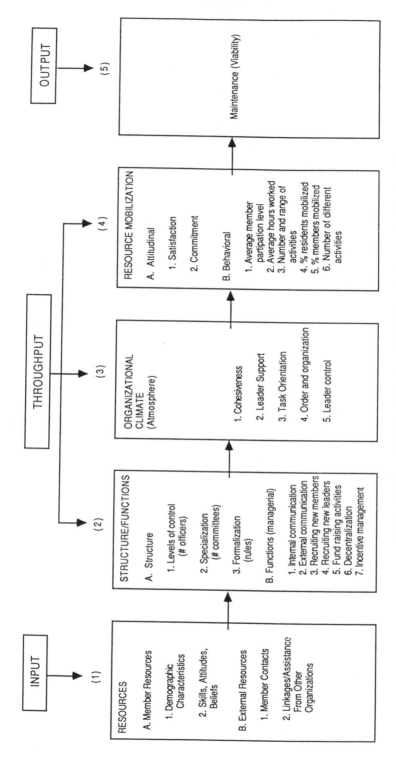

Figure 2. Framework of variables used in Block Booster study to investigate organizational maintenance.

variables would distinguish the group of inactive from the group of active block associations.

Here we report preliminary results on variables in each category that successfully distinguished between the group of associations that remained active and the group that became inactive. Each was statistically significant at the .05 level or below. It should be noted, however, that with such a small N in the two groups (20 active and 8 inactive associations) this significance level is very conservative and vulnerable to Type II errors of not identifying differences where they exist. That is, additional variables might be identified as significant with a larger data set or a less conservative significance criteria. (A complete description of all variables and statistics can be found in Florin & Chavis, 1991.)

Resources

Members themselves are one of the primary resources of any organization, bringing with them abilities, values, and beliefs. We examined 18 demographic and attitudinal member variables that included (a) demographics, (b) variables related to block attachment, (c) length of membership, (d) extent of participation in other community organizations, (e) perceptions of the block environment such as problems and sense of community, and (f) expectations for success and perceptions of participation-related skills. We found no significant differences between the group of active block associations and the group of inactive block associations on any of these 18 member-resource variables.

In addition to the internal resources of member attributes, block associations establish ties or linkages with sources of external resources. The external resources can provide knowledge, facilities for meetings, mailing lists, additional personnel for special projects or funds in the form of grants, loans, or donations. Active block associations had made significantly more linkages and received more external assistance than inactive block associations. Sixty-seven percent (67%) of the active associations received help from six or more external organizations, whereas only 17% of the inactive associations had received assistance from six or more organizations. Of particular interest is the fact that active block associations were more likely than inactive associations to receive support and assistance from other block associations. Thirty-one percent (31%) of active block associations reached out to sister organizations versus 14% of inactive associations. An assessment of the environmental supports available to all associations showed that these differences were not due to differences in the availability of support in the association's environment but rather the degree to which such support was actively identified and accessed.

Organizational Structure and Functions

Structure refers to the way an association organizes its human resources for goal-directed activities. It is the way the members of an organization are arranged into relatively fixed relationships that largely defines patterns of interaction, coordination, and task-oriented behavior (Steers, 1977). We examined three simple structural variables: (a) *dispersion of authority* or the number of officers (e.g., president, vice-president, treasurer, etc.) within the association, (b) *specialization* or the degree to which activities were divided into specialized committees within the organization, and (c) *formalization* or the degree to which rules and procedures are written and precisely defined (e.g., presence of by-laws, agendas at meetings, written minutes, and written responsibilities for officers). Active block associations were significantly different than inactive block associations on all three of these variables. Active block associations, on the average, had a greater number of officers (dispersion of authority), twice as many committees (specialization) and were more likely to operate with rules and procedures written and precisely defined (formalization).

Organizations also perform functions that focus their activities both quantitatively and qualitatively. One organization may devote more energy to one activity or function than another organization or do so in a different way. For example, one organization may not pay much attention to the recruitment of new members and leaders, whereas another may undertake more activities in this area. Organizations also differ with respect to how the decision-making process is conducted. One organization may vest considerable authority in a small group of individuals, whereas another may seek to involve more of its members in decision making.

We examined seven functional dimensions for differences between active and inactive block associations. Block associations that remained active differed significantly from those block associations that became inactive in that they (a) used a greater number of methods (e.g., newsletters, personal contact, telephone calls, etc.) to communicate with members (internal communication), (b) used more personalized outreach strategies to recruit new members, (c) took a more proactive stance in the recruitment and preparation of new leaders, (d) used more participatory decision making and delegated more responsibilities (decentralization), and (e) attempted to provide benefits and reduce costs for participation (incentive management).

Social Climate

Assessing organizational climate assumes that members within a group are participant observers in the group milieu and are uniquely qualified to appraise it. The members' perceptions of organizational characteristics such as relationships among members, leadership, and structural characteristics can be used to describe and contrast different group settings.

We compared the organizational climate of active and inactive associations along five dimensions: (a) cohesiveness ("team spirit"), (b) leader support (the amount of help and concern displayed by the leader to members), (c) task orientation (the degree of emphasis on practical, concrete "down-to-earth" tasks), (d) order and organization (the degree to which the activities of the group are formalized and structured), and (e) leader control (the degree to which leaders direct the group and enforce rules). Although active block associations, on the average, did have more task orientation, order and organization, and leader control than inactive block associations, only on the cohesiveness dimension did active block associations have a statistically higher rating than inactive block associations.

Resource Mobilization

Organizations that are able to mobilize their internal resources are more likely to survive (Prestby & Wandersman, 1985). We examined attitudinal variables and participation-related variables for differences between active and inactive block associations.

The average level of member satisfaction and member commitment in the group of 20 active blocks did not significantly differ from the level of satisfaction and commitment for the group of 8 inactive blocks. Active blocks also did not significantly differ from inactive blocks in the average level of a member's participation nor in the number of hours the average member contributed to the association each month outside of meetings. In other words, active block associations did not remain active because their average member was participating to a greater degree or "doing more" for the association.

Block associations that maintained operations did, however, provide significantly more participation choices for members because they worked simultaneously on a significantly wider variety of activities. All of the associations that remained active sponsored five or more different activities compared to only half of those associations that eventually ceased operations.

Block associations that maintained operations recruited a significantly greater proportion of the residents on their blocks to become members and mobilized a greater percentage of nominal members into being active members. This ability to mobilize resources and not actual block size accounted for larger numbers of members in active associations.

How do systemic principles inform these results? First, the Block Booster study of maintenance tells us that concerns about block association maintenance are realistic. Data on the ages of the block associations point to the fact that the block association is an adaptive system subject to entropy, a natural process that must be continuously averted by importing and storing energy (resources). Thus maintenance is not something done once and then forgotten. It is not the case that maintenance difficulties occur only within the first year or two. In our study, the average age of block associations that lapsed into inactivity was 6.9 years (versus 6.6 years for the associations that remained active). Maintenance takes *ongoing* efforts, no matter what the age of the block association.

Second, maintenance is directly related to the block organization performing particular functions that relate to identifying, accessing, and mobilizing resources both from within and without its organizational boundary. Active block associations recognized the variety of supports available to them in their environment and made more use of these external resources. Particular functions such as internal communication, recruitment of new members, and preparing for leader succession were also related to maintenance. These findings lead to a characterization of active block associations as being more proactive and intentional in their continual "fueling" of their organization. Block associations with rigid boundaries that did not identify and access resources across their boundaries and block associations that did not perform maintenance functions were more likely to decline.

Third, the maintenance of block associations was *not* associated with the characteristics or qualities of the component parts (members) but was strongly related to the manner in which these component parts were systemically organized. The average member of the two groups of block associations did not differ significantly on demographic or attitudinal variables. They also did not differ in terms of their levels of satisfaction, commitment, level of participation, or amount of time devoted to the association. Rather, the active block associations differed from the inactive block associations in how these component parts were related to one another through their structural arrangements, a systemic principle.

Fourth, reciprocal influences, a systemic principle, are implied by the structural and functional variables identified. More officer positions and committees provide a greater variety of opportunities to attract participants and mobilize members. The same can be said of the number of different activities undertaken. More participants generated through mobilization functions can also lead to more offices, committees, and a formalization of roles. We have found some insight into this "chicken-and-egg" issue—successfully self-maintaining associations implement more structure *while* they are engaged in more mobilization-related functions.

Finally, differentiation of structure and integration of function, another systemic principle, is implied in the findings. Structural variables such as dispersion of authority and specialization indicate differentiation of structures, yet more decentralization of responsibilities and participatory decision making in active associations point to important integrative functions. Structural differentiation must be integrated through coordinated functions and integrative participatory mechanisms. This enables the organization to function efficiently as a total system, prevents structural fragmentation, and builds social cohesion within the group. A proper balance of both structural differentiation and functional integration would thus seem to be optimal for organizational maintenance.

In sum, block associations that successfully resisted entropy revealed a "system state" different from those that declined into inactivity. Viable associations showed a pattern of relationships among components that was more structured and organized, and they performed a wider variety of specific maintenance-related functions. These structures and functions were linked through mechanisms of delegation of responsibility and participatory decision making that integrated operations and created a strong sense of cohesion within the group. Viable associations also attended to the environment outside their boundary and identified and accessed more external resources. In essence, viable associations were more adaptive systems and, through continual self-adjustment and self-renewal, survived in the face of the forces of entropy. We now turn to whether such adaptive system functioning can be enhanced through purposeful intervention.

The Block Booster Process

A major goal of the Block Booster Project was to develop and evaluate a technical assistance intervention that would promote viable block associations. We wanted to develop an approach in which (1) the information would be useful and usable by small voluntary organizations, (2) the technique would be affordable to voluntary organizations, and

(3) would be cheap and efficient for technical assistance organizations to provide assistance to a large number of voluntary associations with minimal staff time.

In general, the intervention, called the Block Booster Process, was targeted at organizational *capacity building*, which involves increasing the competence and resources of the organization. Several systems principles guided the development and provision of this technical assistance intervention. First, we wanted to influence the overall "state of the system," that is, the pattern of relationships among components and the particular filtering conditions of the organizational boundary. Second, we wanted to increase viability and decrease the possibility of entropy. Here we used the principle that entropy is proportional to the negative of information available to a system. The larger the total input devoted to information in a system the more likely its survival (Berrien, 1968). Third, we wanted to take advantage of the concept of a block association as an adaptive system that can use information for self-corrective feedback. When the information is accompanied by suggested interventions, the system can engage in self-guided actions for system change. Finally, the particular content of the information to be supplied to the system would be related to two kinds of inputs: (1) maintenance inputs or those that energize the system and make it ready to function and (2) signal inputs that provide inputs to be processed and directed to particular outputs. Both maintenance inputs and signal inputs are important for the system to successfully produce output (Berrien, 1968).

Our technical assistance invention thus assessed the "state of the system" for block associations. We assessed selected variables from the systems framework described in Figure 1 and operationalized in Figure 2 that our earlier (pre-Block Booster) research had associated with active individual participation and organizational viability. Some variables would provide information related to maintenance inputs, such as perceived skill levels among members, members' perceptions of organizational climate dimensions such as cohesiveness and task focus, and members' perceptions of obstacles and benefits to participation. Other variables would address the filtering conditions of the organizational boundary such as nonmembers' perceptions of the openness of the block association and the degree and utilization of members' contacts with organizations outside the block association. Still others, such as members' expectations of what the block association would accomplish, would serve as signal input that directed the block association toward producing certain outputs. Taken together, they assessed the "system state" of the block association. Feeding this assessment back to the block association was seen as a way of increasing the amount of information available to the organization (thus decreasing the possibility of entropy) and allowing the adaptive self-correcting process within the organization to work with increased maintenance and signal input. A variety of suggestions for interventions targeted to each of these variables were also supplied. These suggestions pointed to potential changes in structures (e.g., creating subcommittees), the introduction of particular functions (e.g., recruitment, resource acquisition), changes in boundary conditions (e.g., conducting an inventory to identify and access external resources), or the formulation of output goals (e.g., targeting those highest on the list of members' expectations). To the extent that these variables were systemically interconnected, we felt changing even one would have reverberating effects on the state of others and changing several simultaneously might be expected to have a synergistic effect that could boost the system to a new "system state" (e.g., changed relationships among components or changed boundary conditions) that would be more conducive to successful maintenance.

Our technical assistance intervention was, from an organizational perspective, essentially an organizational development technique. *Organizational development* consists of a system self-study and derived actions taken to improve organizational functioning and production, which is totally compatible with general systems principles. Organizational-development interventions have been used in private corporations, public sector bureaucracies, and large-scale voluntary organizations. Our effort was an example of extending organizational development to small-scale community development groups (Keys & Frank, 1987).

The Block Booster Process falls between two extremes of technical assistance and organizational feedback. At one extreme is the *generic workshop,* where members of different organizations come to a time-limited (e.g., 1 day) workshop on a specific topic (e.g., leadership skills). One of the limitations of this approach is that typically the workshop is attended by members of organizations that are at different stages of development and have different strengths and weaknesses. Therefore, the workshop provides generic information that must be sifted, filtered, and translated by each attendee to make it relevant to his or her level of skills and needs and to his or her organization's structure and needs. At the other extreme is the corporate *organizational consultation* model in which a particular unit, branch, or company hires a consultant or organizational development expert to come into the organization for an extended period of time to perform an organizational assessment, give feedback, and facilitate recommendations for change. This model is intensive, the information is directly relevant to the particular organization, and it is expensive in time and money. Our approach to intervention and feedback was somewhere in the middle. It provided each organization with information about its own strengths and weaknesses and with handbooks containing possible suggestions for change. But this information and the intervention plans developed from it were presented in group workshops, not individually to each organization.

The initial design of the Block Booster Process was refined through consultation with 30 successful block association leaders and community organizers from New York City who lived outside of the neighborhoods in the study. The leaders participated in one of three focus group meetings. These meetings allowed us to present the rationale, process, and content of our interventions to the types of leaders who would be participating in our project. The results suggested that they viewed the intervention as a commonsense approach that would generate interest. There were several design suggestions around presentation and content that strengthened our design and made it more "user friendly." The final design of the intervention was thus a product of scientist–citizen collaboration (see Chavis & Florin, 1991, for a complete description and evaluation).

The Block Booster Process consisted of the following components (see Figure 3):

1. *Assessment–information gathering*—Our systems model of organizational functioning indicated that problems with viability ultimately result from weaknesses in mobilizing internal resources and accessing external resources. Through our organizational-assessment questionnaires, we assessed both member characteristics (e.g., perceived participation-related skills, expectations for success, benefits from and obstacles to participation, and contacts outside the organization) and social climate dimensions (e.g., cohesiveness, leader support, etc.). Using telephone surveys, we assessed the perceived skills of nonmember block residents and incentives for and barriers to their participation in the block association.

2. *Development of "Block Booster Profiles"*—The information gathered in Step 1 was developed into 17 graphs of the strengths and weaknesses of each association.

3. *Workshop for data feedback and suggested interventions*—A 2½-hour workshop was held for leaders of participating block associations. "Block Booster Profiles" were presented to these leaders so they themselves could assess the viability of their block association. This process used models of behavioral feedback on the individual level and survey guided feedback on the organizational level. Theoretically, the process integrates two paradigms in community development that have until now remained relatively separate: the "social facts" paradigm and the "social behavior" paradigm (Blakely, 1980).

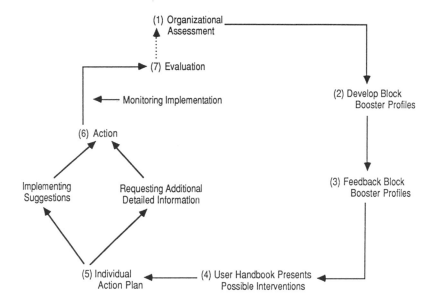

The process, if successful, can continue to be used by the block
association as they will have feedback about what areas they have made
improvements in and what areas continue to need work.

Figure 3. The Block Booster Process.

The "social facts" paradigm usually involves citizens themselves
gathering information or "facts" about their community (e.g., problems,
power distribution, actors, etc.); the "social behavior" paradigm usually involves scientists studying participants in community development (e.g., identifying characteristics, motivations, organizing dynamics, etc.). In our hybrid approach, we, like other scientists using the
social behavior paradigm, gathered information relevant to participatory behavior and processes. Unlike the usual social behavior approach,
however, we returned this information so that it became "social facts"
for the participants. These "social facts" are about the participants
themselves and their organizations. They can act upon these "social
behavior facts" to influence the participation of members and their
organization's functioning. Facts about the participants and their organizations thus become more than the property of the researcher. The
research information can be used by the participants themselves, along
with further technical assistance information, to improve their organization and better serve their community.

   4. *Matching profiles with interventions in Users' Handbook*—At

the same workshop where the profiles were presented, each block leader received a copy of the *Block Booster Users' Handbook*. This handbook is cross-referenced to the Block Booster Profiles and supplied the leader with initial practical, concrete suggestions to address areas needing improvement within their block association. For example, if the organizational assessment indicated that members believed that the association was low on task focus and needed more directed problem-solving efforts, the *Users' Handbook* supplied several practical tips on structuring the association to increase task focus. The suggested interventions addressed each category presented in the profiles (e.g., providing training or modeling to develop member skills) and also often pointed to how the leader could use structural changes and the introduction of specific functions as means to change the targeted category. As mentioned, interventions were directed to several systemic levels within the organization: developing individual competencies, changing organizational structures and processes, modifying the social climate of the organization and altering organizational boundaries. The overview of the Block Booster Profiles and *Users' Handbook* were presented in groups, yet the individualized profiles allowed each leader to concentrate his or her energies on their own organization. In addition, the identification of areas of strength was both reinforcing to the leader and members and useful as a resource for addressing weaknesses.

5. *Individual action plans*—During the workshop, leaders were aided in completing an individualized action plan that helped them identify the next steps they would take to implement their own organizational intervention, the resources they needed, and whose help they needed. These plans were completed at the end of the workshop. They were based on the strategy of learning contracts that have been found to increase the staying power of "low-dose" training efforts such as ours.

6. *Action*—Over the next several months, the leaders put their action plans into practice, drawing upon the suggested interventions in the *Users' Handbook*. In addition, detailed information on any of the topics covered in the *Handbook* was available from the Citizens Committee for New York City.

7. *Evaluation*—Block association leaders were encouraged to assess the impacts of their efforts by "checking back" with their members after the implementation of their organizational interventions. They were also instructed in how a simplified version of the process could be used to create a periodic "checkup" of organizational functioning. In this way the technical assistance intervention could be adapted for purposes of periodic "self-renewal" in the organization (Bartunek & Keys, 1979; Friedlander, 1980).

We evaluated the Block Booster Process in two ways. First, at the completion of the workshop, participants were asked to complete a brief questionnaire. Overall, the block association leaders who attended the Block Booster workshop rated them highly, found the information useful, and said they were quite likely to use the materials and ideas developed in the workshop. Second, the real test came 10 months later. We had randomly assigned 27 block associations to either a comparison ($N = 9$) or test ($N = 18$) group. All blocks, both comparison and test, received a "Block Conditions Report" that provided feedback from our telephone survey on selected variables such as residents' perception of block problems, satisfaction with the block, crime victimization, perception of crime on block, feelings of safety, and perception of the block association. The comparison block associations, which were essentially identical to test block associations on organizational and member characteristics at the outset of the project, received only the usual telephone-based technical assistance offered all block associations by the Citizens Committee for New York. The test block associations received the Block Booster Process as technical assistance: the Block Booster Profile for their association, *Block Booster Users' Handbook*, and the workshop orienting them to their use in planning organizational interventions. All 27 block associations were contacted 10 months after the workshops to determine if their associations were still active, using the operationalized definition of "inactivity" described earlier (e.g., no meetings or sponsored activities for the previous 6 months). We found that only 22% of the test blocks became inactive in contrast to 44% of the comparison group. Although it appears that the Block Booster Process had helped reduce the attrition rate by 50%, we are reluctant to attribute the higher "survival" rate of test blocks exclusively to the specific effects of our intervention. Nonspecific effects such as a "Hawthorne" effect or increased support through contact with other block association leaders at the workshop might have contributed to the lower attrition rate. The Hawthorne effect attributes better performance to respondents knowing they are being observed and measured. This effect should be equally true for members of both test and comparison block associations, because they received the same number of measurements. Still, we cannot rule out the impact the "special treatment" of having a workshop for test blocks might have had, at least in communicating the message that the association was important. Increased support from other block associations, however, should not be considered only an artifact of a research project as is the Hawthorne effect. Rather, increasing contact with other, similar organizations as a means to provide additional sources of external support should be built into all efforts to maintain VCOs. Whatever the exact combination of specific and nonspecific mechanisms, the Block Booster Process demonstrated that systematic support and technical assistance for voluntary community organizations such as block associations can make the difference between ongoing activity and going out of business.

Implications of the Block Booster Project: "Enabling Systems" for Community Development

The Block Booster Project demonstrated that concerns about the viability of voluntary community groups over time are realistic. But it is unrealistic to expect grassroots volunteer groups to address all of their maintenance issues on their own over many years. As they nurture and sustain their communities, they themselves must be nurtured and sustained. Support such as training, written information, providing hands-on assistance and bringing such groups into contact with one another increases the likelihood that organizations like block associations will survive. Results from the Block Booster Project on organizational viability show that, without assistance, many VCOs decline into inactivity.

In fact, if our overall goal is to make the benefits of block association membership available to as many residents as possible, maintenance interventions may be a most efficient means of doing so because when a block association becomes inactive, potential benefits are lost for all residents of that block. Therefore, cutting block-association decline in half, something the Block Booster Process demonstrated is possible, would mean that many more residents would continue receiving benefits than is the case if the usual rate of attrition occurs.

The Block Booster Project identified categories of variables that predicted maintenance longitudinally. This research, combined with the success of the Block Booster Process, has straightforward implications for technical assistance for voluntary community organizations. Training and technical assistance should strive to bring voluntary community organization leaders and emerging leaders into contact with each other, whether through informal means or the more structured vehicles such as federations or coalitions. The sharing of common problems and experiences should be a central component of these encounters. Technical assistance needs to be offered to all voluntary community groups so that they can do a maintenance "checkup" that can serve as an early warning system for organizational problems. Continuous attention to maintenance activities must be emphasized as a strategy for the prevention of later inactivity. It is also important to convey the message that the members of successfully maintained organizations are not themselves different or better but only organized differently.

Our research results on organizational maintenance identified specific structures and functions that distinguished which organizations remained active from those that became inactive. Technical assistance and training materials can be organized simply and concisely into four domains based on the findings: (1) structure of the organization, (2) functional activities related to mobilization of member resources, (3) promotion of decentralization and leadership development, and (4) identification and access of external resources. Interventions for maintenance need not be complex or involve the more extensive and expensive survey-feedback method employed in the Block Booster Process. Simple self-administered "checklists" in each of these domains may be more practical and can take advantage of the systemic self-correcting ability of voluntary community organizations by indicating the presence of "risk factors" associated with the organization's "health" and "vitality." The checklist can be cross-referenced to sections of a workbook/reference manual where recommendations for organizational change are found, as in the *Block Booster Users' Handbook.*

Perhaps the broadest implication of the Block Booster Project is that "enabling systems" should be established for the nurturance of voluntary community organizations of all types. Wherever substantial numbers of voluntary community organizations exist or the potential for them exists, an intermediary organization should be established whose primary mission is the organization, support, and maintenance of these groups (Florin, 1989). Such an intermediary organization must establish an "enabling system" for community development that is able to respond to large numbers of organizations with minimal staff, is easily accessible, provides assistance in a variety of "dose strengths," is multifaceted and multiculturally adaptable and is affordable and manageable over a long period of time. The major components of this system include:

1. *Training programs for skills development.* The training programs may focus on topics such as leadership and organizational development, resource acquisition, coalition building or specific content issues such as housing or crime.
2. *Telephone and on-site consultation* around organizational development issues such as team building and conflict resolution.
3. *Information and referral services* to appropriate sources of assistance such as city departments, agencies, or other sources of technical assistance.
4. *Mechanisms for creating linkages* among key community institutions and "actors," including conferences, coalitions, coor-

dinating structures, and networks. This would include the establishment of public and private sector (government, voluntary organizations, and business) partnerships.

5. *Incentive grants and methods of recognizing group achievement* in order to stimulate leadership and support and highlight model programs. Incentive grants can "seed" activities within VCOs that serve as "incubators" for innovative programs. Promising innovations can then be "harvested" and disseminated while they receive appropriate recognition which is reinforcing to the group.

6. *Publications that promote voluntary neighborhood initiatives* such as how-to guides, a newsletter that networks groups and keeps them up to date, and public information materials to educate citizens about the potentials of voluntary community organizations.

The mission of the intermediary support organization is itself systemic. No single component or type of support is sufficient by itself. Rather the previously mentioned components must be integrated into a coordinated "enabling system" that produces, through its synergistic effect, a climate of support and encouragement for voluntary community organizations (Florin, 1989). The intermediary organization creates, through these components and the organizations and institutions linked by them, a dynamic system from heretofore unconnected elements. As a true system, different components may actually be "housed" in different organizations, with the coordinating structure among them the responsibility of the intermediary organization. The important point is to provide a coordinated system with as many components as possible.

The intermediary organization and the "enabling system" it establishes can also be used for broader purposes than supporting the indigenous activities of voluntary community organizations. Voluntary community organizations have often emerged as the result of indigenous action on the part of their membership. Increasingly, however, as the potentialities of voluntary community organizations are being realized, their formation and nurturance is being encouraged by legislative action, foundation initiatives, and policies and programs from governmental bureaucracies. Social policy has begun to consider the use of "mediating structures" such as VCOs as vehicles to be engaged in social problem solving efforts (Berger & Neuhaus, 1977; Glazer, 1988; Meyer, 1982). Here the intermediary organization and "enabling system" would create linkages between the more formal sectors of communities such as government, businesses, and bureaucratic social service agencies and the informal sector represented by voluntary community groups. In this way, the output of voluntary community groups would be useful to the "suprasystem" of these more formalized organizations and thus nurtured and supported by these organizations. Enabling systems that are oriented to larger social issues can be used to promote community development approaches to health promotion (Chavis, 1988), substance abuse prevention (Florin & Chavis, 1989), crime prevention (Florin, 1989), and a host of other pressing social problems.

Conclusion

The Block Booster Project found several uses for systems concepts. They provided the foundation for a framework to view voluntary community organizations and their environments in dynamic interaction. They guided the categorization of variables used to identify crucial ingredients in block association maintenance and informed our understanding of the results. An intervention to build organizational capacity and promote viability in block associations was designed around systemic principles. These principles were also employed to outline the character of "enabling systems" that can nurture voluntary community organizations and promote their use in social policy initiatives. Obviously, this project found systemic principles to be a rich heuristic to guide investigation, intervention and broader initiatives.

References

Ahlbrandt, R. S., Jr. (1984). Neighborhoods, people and community. New York: Plenum Press.

Bartunek, J., & Keys, C. (1979). Participation in school decision-making. Urban Education, 14, 52–75.

Bellah, R. N., Madsen, R., Sullivan, W. M., Swidler, A., & Tipton, S. M. (1985). Habits of the heart: Individualism and commitment in American life. Berkeley: University of California Press.

Berger, P. L., & Neuhaus, R. J. (1977). To empower people: The role of mediating structures in public policy. Washington, DC: American Enterprise Institute.

Berrien, F. K. (1968). General and social systems. New Brunswick, NJ: Rutgers University Press.

Blakely, E. J. (1980). Building theory for CD practice. In J. A. Christenson & J. W. Robinson, Jr. (Eds.), Community development in America. Ames: Iowa State University Press.

Cardwell, D. (1987). Community groups help lower costs of public services. The Privatization Report, 6, 1.

Chavis, D. M. (1988). Concept paper on a support center for community enabling systems. Proposal to H. J. Kaiser Family Foundation. New Brunswick, NJ, Center for Community Education, Rutgers University.

Chavis, D. M., & Wandersman, A. (1991). Sense of community in the urban environment: A catalyst for participation and community development. *American Journal of Community Psychology, 18,* 55–81.

Chavis, D. M., Florin, P., Rich, R., & Wandersman, A. (1987). *The role of block associations in crime control and community development: The Block Booster Project.* Final Report to the Ford Foundation. New York: Citizens Committee for New York City.

Chavis, D. M., & Florin, P. (1991). *The Block Booster Process: An intervention to promote maintenance in volunteer neighborhood organizations.* Unpublished manuscript, Rutgers.

Clark, H., Chavis, D. M., & Glunt, E. K. (1988). *1988 state of the neighborhoods report.* New York: Citizens Committee for New York City.

Crensen, M. A. (1983). *Neighborhood politics.* Cambridge: Harvard University Press.

Florin, P. (1989). *Nurturing the grassroots: The role of volunteer neighborhood organizations in urban problem solving.* New York: Citizens Committee for New York City.

Florin, P., & Chavis, D. (1989). *Community Development and Substance Abuse Prevention* Santa Clara, CA: Bureau of Drug Abuse Services, Health Department, County of Santa Clara.

Florin, P., & Wandersman, A. (1984). Cognitive social learning and participation in community development. *American Journal of Community Psychology, 12,* 689–708.

Florin, P., & Chavis, D. M. (1991). *Organizational viability in voluntary community organizations.* Unpublished manuscript, University of Rhode Island.

Friedlander, F. (1980). The facilitation of change in organizations. *Professional Psychology, 11,* 520–530.

Giamartino, G. A., & Wandersman, A. (1983). Organizational climate correlates of viable urban block associations. *American Journal of Community Psychology, 11,* 529–541.

Glazer, N. (1983). Towards a self-service society? *Public Interest, 70,* 66–90.

Glazer, N. (1988). *The limits of social policy.* Cambridge: Harvard University Press.

Henig, J. R. (1982). *Neighborhood mobilization.* New Brunswick, NJ: Rutgers University Press.

Katz, D., & Kahn, R. L. (1978). *The social psychology of organizations* (2nd ed.). New York: Wiley.

Keys, C. B., & Frank, S. (1987). Community psychology and the study of organizations: A reciprocal relationship. *American Journal of Community Psychology, 15,* 239–253.

Mayer, N. S. (1984). *Neighborhood organizations and community development.* Washington, DC: The Urban Institute Press.

Meyer, J. A. (Ed.). (1982). *Meeting human needs: Toward a new public philosophy.* Washington, DC: American Enterprise Institute.

Miller, F. D., Malia, G., & Tsembersis, S. (1979, September). *Community activism and the maintenance of urban neighborhoods.* Paper presented to the 87th Annual Meeting of the American Psychological Association, New York City.

Milofsky, C. (1987). Neighborhood-based organizations: A market analogy. In W. W. Powell (Ed.), *The nonprofit sector: A research handbook* (pp. 277–296). New Haven: Yale University Press.

Peirce, N. R., & Steinbach, C. F. (1987). *Corrective capitalism: The rise of America's community development corporations.* New York: Ford Foundation.

Perkins, D., Chavis, D., Florin, P., Rich, R., & Wandersman, A. (1985). Block crime with block organizations. *Citizen Participation, 5,* 7–9.

Perkins, D., Florin, P., Wandersman, A., Chavis, D., & Rich, R. (1985, August). *Block organizations in community development and crime control.* Paper presented at the annual meeting of the American Psychological Association. Los Angeles.

Prestby, J., & Wandersman, A. (1985). An empirical exploration of a framework of organizational viability: Maintaining block organizations. *Journal of Applied Behavioral Science, 21,* 287–305.

Rappaport, J. (1981). In praise of paradox: A social policy of empowerment over prevention. *American Journal of Community Psychology, 9,* 1–25.

Rappaport, J. (1987). Terms of empowerment/exemplars of prevention: Toward a theory for community psychology. *American Journal of Community Psychology, 15,* 121–148.

Rich, R. C. (1979). The roles of neighborhood organizations in urban service delivery. *Urban Affairs Papers, 1,* 81–93.

Rich, R. C. (1981). Municipal services and the interaction of the voluntary and governmental sectors. *Administration and Society, 13,* 59–76.

Rich, R. C. (1983). Balancing autonomy and capacity in community organizations. *Policy Studies Review, 3,* 96–102.

Rich, R. C. (1988). A cooperative approach to the logic of collective action: Voluntary organizations and the prisoner's dilemma. *Journal of Voluntary Action Research, 17,* 36–50.

Rich, R., Chavis, D., Florin, P., Perkins, D., & Wandersman, A. (1986, November). *Do community organizations reduce crime and fear?* Paper presented at the annual meeting of the American Criminological Society, Atlanta.

Steers, R. M. (1977). *Organizational effectiveness: A behavioral view.* Santa Monica: Goodyear Publishing Company.

Von Bertalanffy, L. (1950). An outline of general systems theory. *British Journal of Philosophical Science, 1,* 134–165.

Wandersman, A., Florin, P., Chavis, D., Rich, R., & Prestby, J. (1985). Getting together and getting things done. *Psychology Today, 19,* 64–71.

Yates, D. (1973). *Neighborhood democracy.* Lexington, MA: Heath.

Yankelovich, D. (1981). *New rules: Searching for self-fulfillment in a world turned upside down.* New York: Random House.

Zimmerman, M. A. (in press). Empowerment: Forging new perspectives in mental health. In J. Rappaport & E. Seidman (Eds.), *Handbook of community psychology.* New York: Plenum Press.

Zimmerman, M. A., & Rappaport, J. (1988). Citizen participation, perceived control and psychological empowerment. *American Journal of Community Psychology, 16,* 725–750.

10

A Multivariate Causal Analysis of Teenage Child Rearing

Hiram E. Fitzgerald and Mark W. Roosa

Studies of teenage parenting have failed to demonstrate strong relationships between multivariate predictors of development and most measures of developmental outcome (Furstenburg, 1976; Quay, 1981; Roosa, Fitzgerald, & Carlson, 1982c, 1985). One reason for this failure is that most of the relationships investigated are confounded by other relationships with the independent and/or dependent variables used. For example, confounding often occurs with socioeconomic status (SES), race, marital status, or age of mother (cf. Blank, 1976; Sameroff, 1975; Social Research Group, 1977).

It is reasonable to suppose that the developmental events (reciprocal transactions between the organism and its environments) that predict parenting behavior in the teenage mother differ from those that predict parenting behavior in the older mother. Developmental pathways that predict parenting behavior may vary as a function of such variables as geographic region of residence, urban versus rural residence, ethnic/racial composition of the community and family, socioeconomic status of the community and family, religious values of the family/individual, personal standards held by the parents as individuals, and peer group standards. It is because so many variables may

Hiram E. Fitzgerald • Department of Psychology, Michigan State University, East Lansing, Michigan 48824.   **Mark W. Roosa** • Department of Family Resources and Human Development, Arizona State University, Tempe, Arizona 85287.

Analysis of Dynamic Psychological Systems, Volume 2: Methods and Applications, edited by Ralph L. Levine and Hiram E. Fitzgerald. Plenum Press, New York, 1992.

influence parenting behavior that we decided to use a different approach to examine the complex linkages among the variables contributing to parenting by teenage and older mothers (Roosa et al., 1982a,b).

In this chapter we present an overview of our attempt to test a theoretical model consisting of the major causal variables thought to influence adolescent parenting. The first step in such modeling is to generate a theory about the causal relationships among the variables in the model. Structural-equation models require "clear, precise, and explicit model or conception of the network of causal relationships among variables." (Baltes, Reese, & Nesselroade, 1977, p. 240). The theoretical model should be based upon current theory and previous empirical research.

Theoretical Model

By applying systems concepts to the extant literature on teenage parenting, we generated a theoretical model that included SES, pregnancy performance, neonatal status, maternal attitudes, home environment, and mother–infant interaction (see Figure 1). The pathways indicate the mechanisms through which each variable's influence is transmitted, rather than linking each variable directly to the final outcome variable. By doing so, we emphasized the processes of the influences, thereby reducing the importance of any single variable. This approach is consistent with suggestions that there is a need to shift the emphasis in studies of teen pregnancy from assessing changes in outcome measures to identifying the processes underlying the changes (Social Research Group, 1977).

Consider, for example, the position of prenatal preparation in the model. The literature suggests that both maternal age and SES are related to the month in which prenatal care begins and to the number of prenatal visits. In turn, the relationship of prenatal care to neonatal status should be evident both in the medical indicators of birth status and in the results of the Brazelton examination used to assess neurological and behavioral status. Both prenatal care and neonatal status are related to the mother's attitudes toward her infant and to the infant's temperament. The mother's attitudes will affect her behavior with her child. Thus it is only through this complex network that prenatal care is seen to have an effect on child development. It is difficult to imagine prenatal care having a direct effect upon child development as the results of correlational studies often imply. Unlike prenatal care, however, we expected both maternal age ξ_1 and SES ξ_2 to have a broad range of direct and indirect effects upon child development.

Notice that the model does not include any measure of develop-

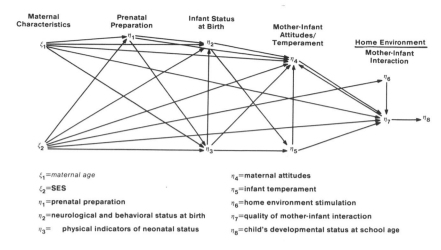

Figure 1. Theoretical model for examining the effect of maternal age upon child development. Reprinted from M. W. Roosa, H. E. Fitzgerald, & N. A. Carlson (1982a). A comparison of teenage and older mothers: A systems analysis. *Journal of Marriage and the Family, 44,* 367–377. Copyrighted (1982) by the National Council on Family Relations, 3989 Central Ave. N.E., Suite #550, Minneapolis, MN 55421. Reprinted by permission.

mental status at the 1- to 3-month level. We excluded a measure of infant developmental status from the model because of the notoriously poor predictive validity of infant developmental examinations for later outcome. Socialeconomic status and caretaking environment are much more potent influences on development than the infant's developmental status at 3 months (Honzik, 1976; Kagan, 1979; Korner & Grobstein, 1976; Lewis, 1976; McCall, 1979, McCall et al., 1973; Sameroff, 1975). Notice that not every potential causal pathway is drawn; for example, there is no arrow from maternal age (ξ_1) to infant temperament (η_1). There is no theoretical or empirical basis for hypothesizing such a causal linkage. However, the model does show that maternal age, insofar as it influences infant status at birth $(\eta_2$ and $\eta_3)$, will have indirect influences on infant temperament.

Analysis

The primary analysis for this study consisted of an estimation and evaluation of the theoretical model using a computer program called LISREL, the analysis of linear structural relationships by the method of maximum likelihood (Joreskog & Sorbom, 1984). LISREL is specifically designed to handle causal models similar to that in Figure 2, especially those with cases of mutual causality.

LISREL has a number of advantages for the analysis of complex causal models. First, LISREL allows one to use models with two parts: a measurement model and a theoretical model. The measurement model allows the researcher to specify how the theoretical or unobserved variables will be measured using the observed variables. For each theoretical variable that uses more than one observed variable as an indicator, a factor analysis is performed to define the theoretical variable by means of the common underlying variance of the indicators. The reliabilities for the measurements of the observed variables (e.g., interrater reliabilities) can be supplied by the researcher, or they can be calculated by the program. The theoretical model or the structural equation model indicates the causal effects (Joreskog & Sorbom, 1984).

Second, LISREL allows the researcher to evaluate not just the specific causal linkages but also the entire causal model. LISREL performs a chi-square goodness of fit test to determine how well the estimated

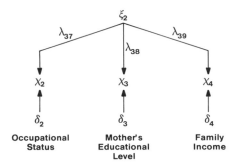

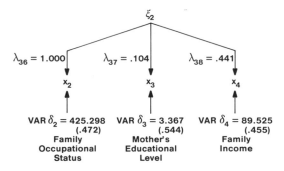

Figure 2. (A) Measurement model for socioeconomic status. (B) LISREL estimates for the measurement models for the socioeconomic status variables.

Table 1. Structural Equations for the Theoretical Model in Figure 1

$$\eta_1 = \gamma_1\xi_1 + \gamma_2\xi_2 + \zeta_1$$
$$\eta_2 = \gamma_3\xi_1 + \gamma_4\xi_2 + \beta_1\eta_1 + \beta_2\eta_3 + \zeta_4$$
$$\eta_3 = \gamma_5\xi_1 + \gamma_6\xi_2 + \beta_3\eta_1 + \zeta_3$$
$$\eta_4 = \gamma_7\xi_1 + \gamma_8\xi_2 + \beta_4\eta_1 + \beta_5\eta_2 + \beta_6\eta_3 + \beta_7\eta_5 + \beta_8\eta_7 + \zeta_4$$
$$\eta_5 = \beta_9\eta_2 + \beta_{10}\eta_3 + \zeta_5$$
$$\eta_6 = \gamma_9\xi_2 + \zeta_6$$
$$\eta_7 = \gamma_{10}\xi_1 + \gamma_{11}\xi_2 + \beta_{11}\eta_4 + \beta_{12}\eta_5 + \beta_{13}\eta_6 + \zeta_7$$

model explains the variance of the data. For exploratory research, this goodness-of-fit test provides the researcher with important information about which causal linkages are most or least useful.

Third, LISREL is a full-information maximum likelihood statistical routine. This means that LISREL uses the information that is available for all variables in the model to make each estimate. It also means that LISREL's estimates of population parameters are those values of the population parameters that are most likely to have generated the observed sample data. For these two reasons, the LISREL estimates are both consistent and efficient (cf. Fink, 1980).

The model presented in Figure 1 is a theoretical model. The structural equations for the theoretical model are presented in Table 1. These equations are read much like regression equations with γs (gammas) as coefficients for all exogenous variables (ξs), and βs (betas) as coefficients for all endogenous variables (ηs).

With the exception of maternal age, each theoretical variable in Figure 1 has more than one observed variable as an indicator. Therefore, each theoretical variable except maternal age has a measurement model associated with it. The measurement model for socioeconomic status is shown in Figure 2a as an illustration of a measurement model. The equations for the measurement model for socioeconomic status are shown in Table 2. Again these equations are read like regression equations except, in the case of measurement models, the λs are equivalent to factor loadings and the ζs are unique variance or error terms. Thus ξ_2 will be created from the common underlying variance of occupational status, mother's education level, and family income.

Table 2. Equations for the Measurement Model for Socioeconomic Status Shown in Figure 2a

$$X_2 = \lambda_{37}\xi_2 + \sigma_2$$
$$X_3 = \lambda_{38}\xi_2 + \sigma_3$$
$$X_4 = \lambda_{39}\xi_2 + \sigma_4$$

Identification of the Model

One of the critical issues in the estimation of such a complex model is that of identification (see Chapter 2, this volume). Identification refers to whether or not there is enough information available to make the estimates requested. Models that are underidentified cannot be estimated (when the number of equations is not adequate to provide solutions to some of the unknown parameters of the model). Models that are just identified can be estimated but not tested because solutions are not testable (parameters that are uniquely resolved due to the right number of equations). Overidentified models are constrained to have unique solutions, and the degree of overidentification (the amount of excess information available upon which to base the estimates) is related to the reliability of the results. The combined measurement and structural model in this example meets the necessary conditions for overidentification with 679 degrees of freedom (Joreskog & Sorbom, 1984).

Subjects and Descriptive Results

The sample consisted of 62 primiparous (first-time) mothers between the ages of 15 and 32 and their well, non-twin infants. The women were recruited from a variety of sources including physician's offices, family planning clinics, birth preparation classes, and schools having programs for pregnant teenagers in both rural and urban areas of central Michigan. Fourteen of the subjects were teenagers; all were volunteers (see Roosa et al., 1982b). The mothers were interviewed in their homes, the medical records of their deliveries were reviewed, their infants were tested shortly after birth, the mothers and their infants were observed monthly for 3 months, the home environments were evaluated, and the mothers completed scales assessing their infants' temperaments and their own attitudes toward their infants.

A comparison of those mothers below 20 years of age with those over 20 revealed a number of differences (Roosa et al., 1982b). The teenage mothers were living in relatively low SES conditions, and almost half of them were on public assistance. Only half of the teenagers maintained a relationship with the infant's putative father, and half of the teenage mothers continued to live with their parents or another relative. Though the teenage mothers began prenatal training later and were less likely to participate in birth preparation classes than were the older mothers, the younger mothers had easier births.

As has been reported in numerous other studies, the infants of teenage mothers in this study tended to have slightly shorter gestations and to be slightly smaller at birth than the infants of older mothers. There were no differences between the Apgar scores of the two groups of infants and only one of seven Brazelton Scale summary scores (motor performance) differentiated between the two groups, with the infants of older mothers doing better. No between-group differences were found for maternal attitudes or ratings of infant temperament.

The homes of teenage mothers were more crowded than those of older mothers and were less likely to offer a quiet place where the infant could be put during naps to escape from stimulation. Teenagers' homes also provided fewer nonhuman sources of audiovisual stimulation than those of older mothers. The differences in the home environments were strongly related to the SES differences between the mothers.

Although there were numerous similarities in the mother–infant interaction patterns of the two groups, the older mothers spent more time talking to their infants and talking to their infants during mutual gaze than the younger mothers did. The older mothers were also more responsive to their infants' distress signals than the younger mothers, which is consistent with the contention that teenage mothers are less sensitive than older mothers to an infant's signals. There were no significant differences between the behaviors of the two groups of infants.

Results of the Structural Analysis

The parameter estimates of the measurement model for the exogenous variable ξ_2 socioeconomic status, are shown in Figure 2a. As in all the measurement models, one of the coefficients relating the theoretical variable ξ_2 to its indicators, the observed variables X_2 (family occupational status), X_3 (mother's education level), and X_4 (family income), was set equal to 1. Without this (or other) restriction, each measurement model would be underidentified, and an infinite set of consistent estimates would be possible, with no way to choose among them (Long, 1983). Thus, each measurement model has a unique solution, and the estimated coefficients are calculated relative to the restricted parameter (Fink, 1980).

For the analysis presented here, the variance–covariance matrix for the 39 observed variables was analyzed. Therefore, the SES measurement model in Figure 2b has the following interpretations:

Var. X_2 = 1.000 Var. ξ_2 + 425.298
Var. X_3 = .104 Var. ξ_3 + 3.367
Var. X_4 = .441 Var. ξ_4 + 89.525

where Var. = variance.

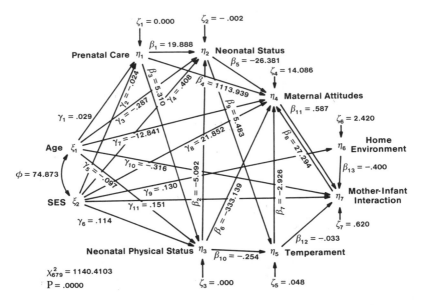

Figure 3. LISREL estimates for the theoretical model for comparing the childbearing and child rearing experiences of teenage and older mothers.

That is, ξ_2 is created by factor analyzing X_2, X_3, and X_4. The coefficients indicate how the variance of ξ_2 would be used to recreate the observed variables. The error of measurement terms (δs), indicate how much of the variance of the observed variables is not accounted for or recreated by the decomposition of ξ_2. The size of the error terms depends on both the measurement scales that were used and upon the amount of common underlying variance (degree of intercorrelation) of the indicator variables. A large measurement error does not indicate that the measurement of a particular indicator variable was poorly done, only that the measurement of that indicator as it relates to the theoretical variable is poor. Thus if the indicator variables are not highly intercorrelated, the measurement errors will be large.

One way to evaluate the significance of an estimate is to compare it to its standard error. If an estimate is at least twice as large as its standard error, it is significant at the $p < .05$ level (Hanushek & Jackson, 1977; Wheaton, Muthen, Alwin, & Summers, 1977). For example, in the measurement model for socioeconomic status, $\lambda_{37} = .104$ and the standard error for $\lambda_{37} = .021$; for λ_{38} the estimates were .441 and .084. The ratio of estimate to standard error is approximately 5 in each case indicating that these estimates are significant. However, in the whole measurement model only one-third of the estimates were significant by this criterion (Roosa et al., 1982a).

Another important aspect of the measurement model is the interpretation of the directionality of the created or unobserved variables. By assigning a coefficient of 1.0 to the first indicator, one forces the unobserved variable to vary in the same direction as the reference indicator. In Figure 2b, increasing occupational status is indicative of increasing SES; thus the positive coefficient means that large values of ξ_2 indicate high SES. One must always pay close attention to this relationship between the reference indicator and the unobserved variable when interpreting the analysis of the model.

For instance, prenatal preparation η_1 was measured by two variables each of which was scored such that they are negatively associated with quality prenatal care. By using 1.0 as the coefficient for the reference indicator, the prenatal preparation variable is actually a measure of poor prenatal preparation (large values indicate poor prenatal care, and small values indicate higher-quality prenatal care). Thus interpretation of the causal relationship of prenatal care to infant status at birth will have to take this fact into consideration. Similarly, η_5 infant temperament is a measure of negative infant characteristics or difficult temperament because of the choice of reference indicator. Careful choice of reference indicators can make interpretation of unobserved variables and causal models more straightforward.

The estimates of the parameters of the theoretical model are presented in Figures 3 and 4. The gammas γ and betas β are regression coefficients that indicate the relationships among the exogenous ξ and endogenous η theoretical variables. The size of a regression coefficient is sensitive to the measurement scales of each of the two variables it relates, to the strength of the relationship between them, to the effects of other variables in that equation, and to the structure of the whole model. For these reasons, one cannot directly compare the coefficients in Figure 3 to one another. In order to compare regression coefficients, the standardized solution to the theoretical model, which is an accurate description of the relationships in the model when the model is not scale free (Wheaton et al., 1977), is presented in Figure 4. In the standardized solution, the variance of each theoretical variable is 1.0.

In Figure 3, one of the first points to notice is the opposing direction of the effects of the two exogenous variables ξ_1 (maternal age) and ξ_2 (SES). ξ_1 and ξ_2 were highly positively correlated as indicated in Figure 3 where ϕ represents the covariance of ξ_1 and ξ_2 and in Figure 4 where ϕ represents the correlation of ξ_1 and ξ_2. The gammas γ relating the exogenous variables to the endogenous variables indicate the independent effects of maternal age and of SES; that is, the effect of each exogenous variable after the effect of the other has been partialled out.

Thus, with the effect of SES accounted for, maternal age had a positive impact on prenatal care η_1 and a negative impact on both neonatal status variables η_2 and η_3, on maternal attitudes η_4 and on

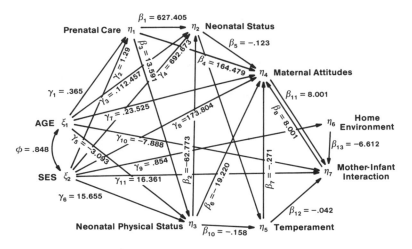

Figure 4. Standardized estimates for the theoretical model for comparing the childbearing and child rearing experiences of teenage and older mothers. Adapted from M. W. Roosa, H. E. Fitzgerald, & N. A. Carlson (1982a). A comparison of teenage and older mothers: A systems analysis. *Journal of Marriage and the Family, 44,* 367–377. Copyrighted (1982) by the National Council on Family Relations, 3989 Central Ave. N.E., Suite #550, Minneapolis, MN 55421. Reprinted by permission.

mother–infant interaction η_7. Because high scores on the prenatal care variable η_1 are indicative of poor prenatal care, the positive relationship between ξ_1 and η_1 shows that given equal SES, younger mothers had better prenatal care than older mothers. In fact, all the coefficients for ξ_1 indicate that young mothers or their offspring did better than older mothers and their offspring on each of the endogenous variables, when SES effects are accounted for or controlled.

With the effects of maternal age partialled out, SES had a negative impact upon prenatal care and a positive impact upon all other endogenous variables it affected. In other words, high SES mothers or their infants did better than low SES mothers or infants on each of the predictors of child development incorporated in the model. The impact of SES was considerably larger than that of maternal age, or indeed, of any other variables (Figure 4).

According to the estimates in Figure 3, prenatal care η_1 had a positive impact upon both neonatal status variables η_2 and η_3 and upon maternal attitudes η_4. Because of the inverted interpretation of η_1 these relationships mean that, when combined with the other predictors in the model, poor prenatal care (high values of η_1) contributes to good neonatal status and to positive maternal attitudes.

Brazelton neonatal status η_2 had a negative impact on maternal attitudes η_4 and a positive impact on infant temperament η_5. That is, in combination with the other predictors, the high Brazelton status scores contributed to lower maternal attitude scores and to higher (more difficult) temperament scores. Both of these influences are very small (see Figure 4).

Neonatal physical status η_3 had negative impacts upon the Brazelton neonatal status variable η_2, maternal attitudes η_4 and infant temperament η_5. High levels of physical status thus contributed to lower levels of Brazelton neonatal status η_2, lower or negative levels of maternal attitudes η_4, and to lower (more positive) infant temperament scores η_5, when used as a predictor with other variables as indicated in the model.

In combination with other predictors, high levels of infant temperament η_5, which are indicative of a difficult child, had a negative impact upon both maternal attitudes and mother–infant interaction. Similarly, high levels of the home environment variable (η_6—associated with more space and toys) made a negative contribution to the level of mother–infant interaction η_7. Finally, the level of mother–infant interaction made a positive contribution to maternal attitudes, whereas maternal attitudes made a positive contribution to mother–infant interaction. (Note that the relationship between maternal attitudes η_5 and mother–infant interaction is reciprocal as indicated by the double arrows between them.)

Several of the individual causal linkages described seem contradictory to both theory and logic. However, one should keep in mind that each coefficient represents the unique contribution of each theoretical variable after all other influences have been considered. Rarely in studies of teen parenting do we look at regression coefficients after the influences of maternal age and social class have been removed. Thus it is hard to compare these coefficients to those we are more familiar with from multiple regression. Still, some of these results are sufficiently paradoxical, especially the strong relationship between poor prenatal care and positive maternal attitudes, that one must begin to question certain aspects of the model (Joreskog & Sorborn, 1984).

The ζ are the errors of prediction or the amount of the variance of the endogenous variables that are not linearly predicted by the variables associated with it. Comparing the values of the ζ with the variances of the endogenous variables, we see that two of the errors of prediction ζ_4 and ζ_7 have values larger than the variance of the variable with which they are associated. This may indicate that we were wrong in assuming that the error terms for the indicators of η_4 and η_7 are uncorrelated, thus indicating that the model may be misspecified.

Test for Goodness of Fit

To evaluate the significance of the complete measurement model, a χ^2 goodness-of-fit test was performed (Figure 3). This test compares the original variance–covariance matrix of the 39 observed variables with the estimated variance–covariance matrix, which is based upon the paths or specifications of the model. The goodness-of-fit test determines how well the original matrix is approximated by the estimated matrix.

The goodness-of-fit test evaluates the null hypothesis that the model as specified is the true model of the relationships among the variables. The alternative hypothesis is that the estimated matrix is any positive definite matrix (Joreskog & Sorbom, 1984). Contrary to most significance tests, the goal here is not to reject the null hypothesis. The probability level that is given in Figure 3 is defined as the "probability of obtaining a χ^2 value larger than the value actually obtained given that the model is correct" (Joreskog & Sorbom, 1984, p. I-38). Thus small values of χ^2 with probability values approaching 1.0, under the assumption of multinormality and in large samples, are indicators of models that fit the data well.

In Figure 3, the results of the goodness-of-fit tests were $\chi^2_{679} = 1140$, $p = .0000$. If χ^2 was the only option available to test the model, we would have to conclude that the model as specified in Figure 1 (and in the measurement models) does not fit the data well. However, the goodness-of-fit test is sensitive to sample size and must be interpreted cautiously (Joreskog, 1974, pp. 4, 14; Wheaton et al., 1977, p. 99). Instead of using the χ^2 value to accept or reject the model as a whole, the results of the goodness-of-fit test can be used to compare the fit of various models to the data (Joreskog, 1974, p. 4; Wheaton et al., 1977).

An inspection of the residuals matrix (the S-sigma matrix) suggested that the model did a fairly good job of reproducing the original variance–covariance matrix. Most of the residuals are quite small, especially in comparison to the size of the corresponding variance or covariance terms. Furthermore, the χ^2/df ratio of 1:679 suggests that the model is a reasonable representation of the data, though it might be improved by the addition of theoretical variables and/or paths between variables (Wheaton et al., 1977, p. 99). This suggests the need for a specification research, an exploratory process designed to improve fit of the model consistent with theory (Long, 1983). However, such an exercise is exploratory, and any model that results from a specification research must be confirmed with data from other samples.

The present model is highly overidentified, that is, the number of degrees of freedom is large. This degree of overidentification is the result of numerous assumptions that were made or restrictions that were placed on the model (e.g., the error terms of the measurement model are uncorrelated or the error of prediction terms are uncorrelated). Because of the status of theory in the social sciences, there were no theoretical guides to the placement of these restrictions (cf. Lands & Felson, 1978, p. 288). Therefore, after inspecting the residual matrix, there are numerous submodels that can be created by relaxing some of the overidentifying assumptions. These submodels can then be compared to the original model to determine which one best fits the data.

Nevertheless, as the inspection of the standard errors above indicated, most of the parameter estimates of the model were not significant.

Although the estimates are consistent (a large sample property), with the sample size studied, many standard errors are so large that often the hypothesis that the estimate is zero cannot be rejected. Thus, one cannot have faith that the estimates obtained are the true values of the parameters, reducing the usefulness of the model for making predictions with any degree of accuracy. In fact, it is argued that one of the requirements for useful maximum likelihood estimation is very large samples. Thus the sample size available for this study may have produced biased estimates (Long, 1983).

Suggestions for Alternative Models

There are several possible explanations for the large standard errors and large χ^2 value reported above and several ways to improve the model. First, as reported above, ξ_1 (maternal age) and ξ_2 (SES) were highly correlated. Highly correlated exogenous variables result in multicollinearity, a situation in which there is only limited information to separate out the independent effects of each variable. Multicollinearity reduces the precision (increases the standard errors) of the estimates. Furthermore, because the estimates are based upon so little information (only the variance that is left after the joint effects are removed), they become very sensitive to slight modifications of the model specifications. Thus, in cases of multicollinearity, the estimates may prove to be artifacts of the sample used and of the model specifications (Hanushek & Jackson, 1977). One possible solution to multicollinearity is to create a single theoretical variable out of ξ_1 and ξ_2.

Second, the values of the variance–covariance matrix for the theoretical endogenous variables show that the variances for η_1, η_2, η_5, and η_7 are quite small, with the variance of η_2 being ".000." This indicates that the sets of indicators for these variables were not good choices; that is, the amount of common underlying variance is too small. The intercorrelations for each set of indicators supports this conclusion. This situation may result from the lack of precision of the scales used (not precise enough to show variability) or from the use of orthogonal (perfectly distinct) factors as indicators. It also may result from the use of small scaling metrics, for example, scales with maximum values between 0 and 1.

For instance, each of the Brazelton summary scores is an indicator of neonatal well-being. However, the correlation matrix shows that the intercorrelations among the indicators are sometimes quite small, suggesting that these variables are not indicators of a single construct. The extremely small measurement reliabilities for these indicators adds strength to this contention. Similar conclusions can be reached for the indicators of η_3 η_4 and η_5. Two possible solutions to the problem of weak indicators are to (1) create additional unobserved variables or (2) reduce the number of indicators used for the unobserved variables (Costner & Schoenberg, 1973; Duncan, 1975; Fink & Mabee, 1978). The resultant sets of indicators should be characterized by relatively strong intercorrelations and by theoretical relevance. To avoid problems within measurement models during structural-equation analysis, it is now recommended that one first analyze one's measurement models using LISREL and resolve any such problems before evaluating the entire model (Long, 1983).

The choice of indicators for maternal attitudes η_4 poses another type of problem. Because these indicators are the results of repeated measurements, one would expect to find the measurement errors to not be truly random but to reflect factors left out of the model that may be correlated (Long, 1983), a possibility that was not reflected in the model. This possible misspecification is probably the cause of the inflated error of prediction for maternal attitudes ζ_4. The model should be modified to allow the autocorrelation of the error terms.

The standard error is also influenced by the quality of the measurement instruments. If the instruments do not have a high level of validity, we cannot be certain that we are measuring the construct of interest, that is, the relationship between that which is measured, and the theoretical concept may be weak. If the instruments do not have a high level of reliability, a certain amount of the variation in the data will be due simply to the unreliability of the instruments that were used. Low reliability leads to larger measurement errors and to larger estimation errors.

If a model is "empirically underidentified," large standard errors will result (Fink, 1980). Though the number of degrees of freedom of the model (calculated by subtracting the number of estimates to be made from the number of elements in the data matrix) is large and nonnegative, this is only a necessary condition for being overidentified. However, "empirical" identification depends, for example, upon various covariances not being equal to zero (Fink, 1980). Some covariances were small, possibly contributing to the size of certain standard error terms.

Finally, if some of the relationships in the model are not linear (one type of heteroskedasticity), the assumption for the estimates (multivariate normality) is violated, and the model is misspecified. By choosing appropriate transformations, the specification problem can be corrected, for example, the relationships made linear, and such transformations often produce normal distributions (Fink, 1980). Where heteroskedasticity exists in the model, appropriate transformations could be found by using residual analysis of ordinary least squares regressions.

In summary, the research model does not completely account for all of the variance that appears in the data set. It is possible that adding theoretical variables or causal linkages between variables would strengthen the model. The large standard errors for the estimated parameters reduce one's confidence in the research model. These large standard errors may be reduced by using the observed exogenous variables as indicators for a single unobserved exogenous variable. Improved measurements would help to reduce the standard errors also as would analyzing data from a much larger sample.

Relationship of Estimated Model to Theory

Given the results of the goodness-of-fit test and the large standard errors, it is difficult to provide an interpretation of the results with any degree of confidence. However, a few observations from the preceding analysis need to be made. For instance, one of the more interesting aspects of the results presented is the finding that maternal age and SES have opposing effects upon the endogenous variables in the model. The fact that maternal age and SES have causal effects in opposite directions challenges much of what has been reported about teenage parenting. Even an examination of the correlation tables (not shown) suggests that the causal effects of these two variables would be similar, that is, the size and direction of the correlations of each with the other variables are quite similar. It is the high positive correlation between maternal age and SES ($r = .85$) that creates this illusion and confounds the results of studies dealing with teenage childbearing.

It appears that the negative effects of adolescent childbearing upon child development are not due to maternal age per se. Instead, the negative effects are due to the correlates of early childbearing: truncated education, limited job opportunities, and reduced earning power for the mother and her male companion. Teenage parents have fewer material and nonmaterial resources to invest in their child's development. However, within the age range studied and given equal SES levels, the more optimal status of the young mother's reproductive system (at least for the first birth), and her enthusiasm and energy may provide more optimal conditions for child development than that provided by older mothers.

It is also important to consider the relative size of the contribution of age and SES to the various aspects of childbearing and child rearing (Figure 4). In every case, the influence of SES is several times the size of the influence of age. This finding suggests that the positive aspects of early motherhood are probably more than offset by the negative impact of the low SES that is associated with early motherhood. Once again, note that this remains a tentative interpretation of the analysis because of problems that were found in the model. Further, the sample studied was small and nonrepresentative, thus precluding making generalizations from this study to other groups. A systematic model specification research to find a stronger model followed by reanalysis with data from another, larger, sample should lead to a model with more useful theoretical and practical implications. Accordingly, the purpose of this chapter has been to illustrate the usefulness and power of causal modeling and analysis for those in child development. Simultaneously, we have tried to point out some of the costs and restrictions that meaningful causal analysis requires.

Finally, one of the more intriguing implications of the present study is the dire need for more multivariate, systemic analysis in child development. A number of relationships reported were difficult to explain and contrary to previous findings in studies that were primarily bivariate. Were these unusual findings simply artifacts of the model and sample used, or do they represent our ignorance about the transactions that occur among variables in multivariate causal models? Only by applying both bivariate and systemic analyses to our research data can we begin to answer this question. When we begin to understand the complexities of the transactions among the variables used in child development research, our ability to predict developmental outcome will increase accordingly.

Acknowledgment

The research described in this chapter was supported in part by the Grant Foundation.

References

Asher, H. B. (1976). *Causal modeling*. Beverly Hills: Sage Publications.

Baltes, P. B., Reese, H. W., & Nesselroade, J. R. (1977). *Life-span developmental psychology: Introduction to research methods*. Monterey, CA: Brooks/Cole.

Blank, M. (1976). The mother's role in infant development: A review. In E. N. Rexford, L. W. Sander, & T. Shappiro (Eds.), *Infant psychiatry: A new synthesis* (pp. 91–103). New Haven: Yale University Press.

Costner, H. L., & Schoenberg, R. (1973). Diagnosing indicator ills in multiple indicator models. In A. S. Goldberger & O. D. Duncan (Eds.), *Structural equation models in the social sciences* (pp. 167–199). New York: Seminar Press.

Duncan, O. D. (1975). *Introduction to structural equation models*. New York: Academic Press.

Fink, E. L. (1980). Unobserved variables within structural equation models. In P. R. Monge & J. Cappella (Eds.), *Multivariate techniques in human communication research* (pp. 111–141). New York: Academic Press.

Fink, E. L., & Mabee, T. I. (1978). Linear equations and nonlinear estimation: A lesson from a nonrecursive example. *Sociological Methods & Research, 7*, 107–120.

Furstenburg, F. F., Jr. (1976). *Unplanned parenthood: The social consequences of teenage childbearing*. New York: The Free Press.

Hanushek, E. A., & Jackson, J. E. (1977). *Statistical methods for social scientists*. New York: Academic Press.

Honzik, M. P. (1976). Value and limitations of infant tests: An overview. In M. Lewis (Ed.), *Origins of intelligence: Infancy and early childhood* (pp. 59–75). New York: Plenum Press.

Joreskog, K. G. (1974). Analyzing psychological data by structural analysis of covariance matrices. In D. H. Krantz, R. C. Atkinson, R. D. Luce, & P. Suppes (Eds.), *Contemporary developments in mathematical psychology* (pp. 1–56). San Francisco: Freeman.

Joreskog, K. G., & Sorbom, D. (1984). *LISREL VI: Analysis of linear structural relationships by the method of maximum likelihood*. Mooresville, IN: Scientific Software.

Kagan, J. (1979). Overview: Perspectives on human infancy. In J. D. Osofsky (Ed.), *The handbook of infant development* (pp. 1–25). New York: Wiley.

Korner, A. F., & Grobstein, R. (1976). Individual differences at birth: Implications for mother–infant relationship and later development. In E. N. Rexford, L. W. Sander, & T. Shipiro (Eds.), *Infant psychiatry* (pp. 68–78). New Haven: Yale University Press.

Lands, K. C., & Felson, M. (1978). Sensitivity analysis of arbitrarily identified simultaneous equation models. *Sociological Methods & Research, 6*, 283–307.

Lewis, M. (1976). What do we mean when we say "infant intelligence scores?" A sociopolitical question. In M. Lewis (Ed.), *Origins of intelligence: Infancy and early childhood* (pp. 1–17). New York: Plenum Press.

Lewis, M., & Starr, M. D. (1979). Developmental continuity. In J. D. Osofsky (Ed.), *The handbook of infant development* (pp. 653–670). New York: Wiley.

Long, J. S. (1983). *Confirmatory factor analysis* (Sage University Paper series on Quantitative Applications in the Social Sciences, 07-033). Beverly Hills: Sage.

McCall, R. B. (1979). The development of intellectual functioning in infancy and the prediction of later I.Q. In J. D. Osofsky (Ed.), *The handbook of infant development* (pp. 707–741). New York: Wiley.

McCall, R. B., Appelbaum, M. I., & Hogarty, P. S. (1973). Developmental changes in mental performance. *Monographs of the Society for Research in Child Development, 38* (3 Serial No. 150).

Quay, H. C. (1981). Psychological factors in teenage pregnancy. In K. G. Scott, T. Field, & E. Robertson (Eds.), *Teenage parents and their offspring* (pp. 73–89). New York: Brune & Stratton.

Roosa, M. W., Fitzgerald, H. E., & Carlson, N. A. (1982a). A comparison of teenage and older mothers: A systems analysis. *Journal of Marriage and the Family, 44*, 367–377.

Roosa, M. W., Fitzgerald, H. E., & Carlson, N. A. (1982b). Teenage and older mothers and their infants: Descriptive results. *Adolescence, 17*, 1–17.

Roosa, M. W., Fitzgerald, H. E., & Carlson, N. A. (1982c). Teenage parenting and child development: A literature review. *Infant Mental Health Journal, 3*, 4–18.

Roosa, M. W., Fitzgerald, H. E., & Crawford, M. (1985). Teenage parenting, delayed parenting, and childlessness. In L. L'Abate (Ed.), *Handbook of family psychology and therapy* (pp. 623–659). Homewood, IL: Dorsey Press.

Sameroff, A. J. (1975). Early influences on development: Fact or fancy? *Merrill-Palmer Quarterly, 21*, 267–294.

Social Research Group. (1977). *Toward interagency coordination: FY '76 federal research and development on early childhood, sixth annual report.* Washington, DC: George Washington University.

Wheaton, B., Muthen, B., Alwin, D. F., & Summers, G. F. (1977). Assessing reliability and stability in panel models. In D. R. Reise (Ed.), *Sociological methodology* (pp. 84–136). Washington, DC: Jossey-Bass.

11

A Theory of Service Delivery and Its Application to the Dropout Phenomenon

Gilbert Levin and Edward B. Roberts

System dynamics permits one to construct explicit theories or models of sufficient size and complexity to encompass the rich and unwieldy phenomena encountered in psychology and other social and behavioral science disciplines. Moreover, a well-constructed model has the potential to guide practical application of such knowledge. For about 10 years, from the mid-1960s to the mid-1970s, we and our colleagues had the opportunity to develop several such models aimed at making sense out of some of the more vexing phenomena encountered in the delivery of health and human services.

The work joined the access to services and expertise in health and mental health management resident in an urban academic medical center with the systems analytic capability of the industrial dynamics group at MIT. Our hope was that this collaboration would produce new

Portions of this chapter appeared previously in Levin, G., Roberts, E. B., with Hirsch, G. B., Kliger, D. S., Roberts, N., and Wilder, J. F., *The Dynamics of Human Service Delivery*, Cambridge, Ballinger, 1976. © 1976 Gilbert Levin and Edward B. Roberts. Figures, tables, and runs are reprinted here with the permission of the authors.

Gilbert Levin • Department of Epidemiology and Social Medicine, Albert Einstein College of Medicine, 1300 Morris Park Avenue, Bronx, New York 10461.   **Edward B. Roberts** • Alfred Sloan School of Management, Massachusetts Institute of Technology, Cambridge, Massachusetts 02139.

Analysis of Dynamic Psychological Systems, Volume 2: Methods and Applications, edited by Ralph L. Levine and Hiram E. Fitzgerald. Plenum Press, New York, 1992.

insights into the nature of health and mental health systems and that prescriptions for improving the performance of those systems might result. We were able to show how the psychological, social, and economic variables interact to cause the urban heroin problem and to evaluate a range of policies aimed at controlling it (Levin, Hirsch, & Roberts, 1972; Levin, Roberts, & Hirsch, 1975). Later this work was expanded to the larger context of substance abuse as a whole (Levin, Roberts, & Hirsch, 1981). Hirsch and Miller (1974) applied the method to develop a model for planning health maintenance organizations. A methodology for evaluating the performance of mental health programs in meeting patient needs was developed based upon the human service work described in this chapter (Levin, Wilder, & Gilbert, 1978; Levin & Lewis, 1977a,b). Also guided by the human service model were empirical studies of the patient flows between components of mental health services (Hertzman & Levin, 1974a,b), several analytic pieces on the same topic (Levin, 1974; Roberts & Hirsch, 1976), and applications to the elementary-school classroom (N. Roberts, 1974) and to dental manpower needs (Hirsch & Killingsworth, 1975).

A Theory of Service Delivery

The purpose of this chapter is to show how a simple and relatively noncontroversial, or at least face-valid theory about the system of forces that engage service programs and their clients with one another can be elaborated into a full-scale model that illuminates some of the substantial problems that are familiar to anyone involved in delivering human services.

We begin with the assumption that the interaction between a service program and its clients or patients is at the heart of the service delivery process. The two sectors of the system, one representing the service program, the other representing the client or client population, are linked together through services that are rendered by the program in response to demands generated by the client as shown in Figure 1. The circles that represent the two sectors are, in reality, highly complex and differentiated; each contains many variables that operate through time and may cause the client to demand services and the program to respond by rendering these services in varying quantity.

A variety of circumstances may call the system into active operation. Forces either within the client or impinging upon him or her from outside the system, such as economic emergency or serious illness, have the potential to produce a significant need for services. This need becomes expressed as some degree of demand for service—if the de-

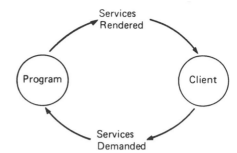

Figure 1. A first approximation of the theory.

mand is large enough to stimulate the program to respond to the client demand by allocating resources. As a result, service is delivered that can satisfy the client's needs, and when satisfied the demand for service shuts off.

A service program is thus viewed as a negative feedback system governed by its operation to satisfy client need. The greater the client need, the greater the demand; the greater the demand, the greater the program activity; and the greater the services rendered, the less the client need. The system answers needs with services that reduce these needs and consequently lessen the demand for services. The addition of another causal link further elaborates the theory. The client not only signals that a service is needed but also cooperates actively in the service transaction, as shown in Figure 2. She telephones the dentist for an appointment and later appears at the dentist's office willing to cooperate. He not only applies for income assistance but cashes and uses the check on mutually agreed-upon occasions.

Thus the rendering service is an interactive process, requiring mutual actions on the part of program and client. Without this mutuality or contact, service does not occur. Allocation of resources for the client only results in service when the client cooperates. And client need for service is only met when the program perceives the need and responds to it.

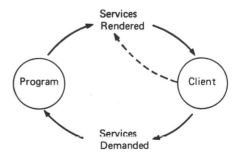

Figure 2. A causal link is added.

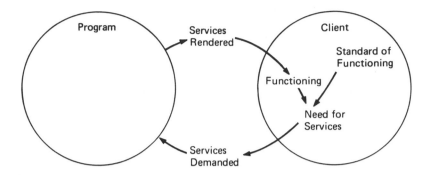

Figure 3. Each of the two major sectors of the system is composed of a number of interacting variables.

Figure 3 makes explicit three factors that have a prominent influence on client behavior throughout a large range of human service systems. As Figure 3 illustrates, this need for service is itself dependent upon two other factors: the client's *level* of functioning and the client's *standard* of functioning. When the level of functioning is low, the client's need for service tends to increase. However, this increase in need does not necessarily trigger a visible demand for service. Demand for service increases only when the client's level of functioning is significantly at variance with his or her standard of functioning. Suppose, for example, that as a result of an illness, the client's level of functioning has declined rapidly from some arbitrary value, say 100, to a new level of 80. His or her demand for service will increase only if at that point in time his or her standard of performance is sufficiently greater than 80 to materially affect the demand mechanism. The client's demand for service at any moment is based on the difference between his or her actual level of functioning and his standard of functioning.

In reality, the standard reflects not only the expectations of the individual client but also those of his family, his friends, and others in his immediate social environment. Although a man in the act of assaulting his wife and disturbing his neighbors may not be violating his own present standard of functioning, he may be described as needing service on the basis of the discrepancy between his behavior and the expectations others have of him.

A demand for services can generate several possible results, depending on the manner in which the program meets the demands upon it. Consider the three scenarios shown in Figure 4. Starting at Time zero and extending for an initial period of time, the actual functioning and the standard are identical, generating no need for service. Then a crisis (the person loses his job, develops cancer, is injured in a car accident,

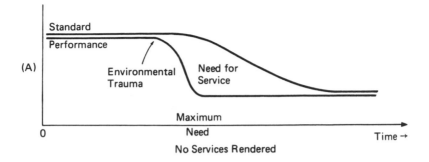

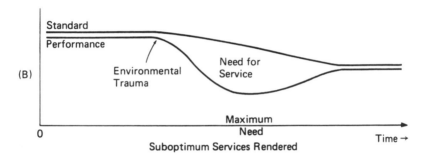

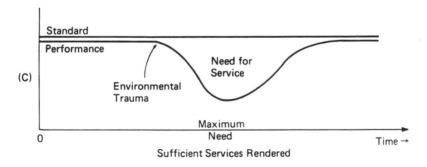

Figure 4. Results engendered by three levels of program response.

becomes psychotic) causes a sharp decline in functioning. In (A) the environmental force is unrelenting, and no effective services are secured. Equilibrium is restored by a gradual process of accommodation of standard of functioning to performance. After a time, the discrepancy vanishes, and the system is again stable. However, the person now functions at a reduced performance level. Note that the line representing performance drops precipitously following the environmental trauma, indicating that function is lost rapidly. The decline in standard does not fall so quickly; it lags behind and falls at a gentler rate. The size of the gap between the two lines indicates the client's need for service. Notice that this gap changes with time, small at first, then growing larger, and then shrinking again. The unfortunate outcome may be assumed to result from the fact that the gap between standard and performance of this scenario is at its maximum too small and of too brief duration to generate sufficient care-seeking behavior to secure active agency response. We do not assign blame to either party in this interaction but merely assert that the forces promoting their contact are not brought into play. Either a more demanding client or a more reactive agency might have produced more favorable results. The person in this scenario eventually adjusts to a new lowered level of stable operation.

In (B) we see that need is only great enough to elicit minimal response. In this case, reequilibration is achieved by a compromise between recovery of function and deterioration of standard. Of course, the most favorable outcome is the one described in (C). In this case sufficient need is generated to elicit a high level of demand upon the program. The program in turn renders sufficient services with the consequence of stable functioning at the pretraumatic level. A variety of other outcomes is possible, depending upon the strength and duration of the environmental trauma, the time required for changes in performance to be reflected in the standard, and the strength, rapidity, and duration of the program response.

Regulation of the intensity, frequency, and duration of a series of interactions between program and client is a central management objective in the delivery of any service. In various service systems, the optimum intensity, frequency, and duration differ depending upon the character of client need and of program service. For example, education systems usually require intense and frequent contacts extending in duration for years. In contrast, immunization programs may require only a single contract in the lifetime of the client.

We now examine the character of the program's response to the client's needs. Figure 5 depicts a more complete picture of our analysis of a service delivery system. Look at the program or agency side of the diagram. Services rendered depend upon a pool of resources apportioned. Although the agency policy may not be explicit and the staff may not be aware of it, the apportionment of resources is controlled by four factors: assessed need (4), reward for service (7), resources available (8), and services demanded (12). Two of these, (4) and (12), are affected by the client sector. The linkages of the remaining two, (7) and (8), are not shown in the diagram and are here assumed to derive from

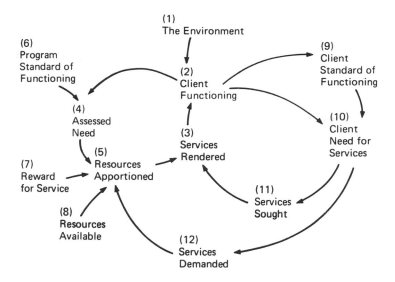

Figure 5. A more complete version of the general theory.

exogenous sources. In some types of relationship between client and program—for example, fee for service medical practice—rewards must be considered within the system boundary, which is affected directly by the services rendered within the system. Apportionment policy with respect to a given client or a population of clients is regulated in this fashion: The program compares the client's present level of functioning (2) with *its* standard (6), deriving an assessed client need (4). Thus, if the client is functioning at 80 and the program defines its standard of adequate functioning as 100, the client's need is 20. The resources deemed to be required to close this gap are then apportioned or reserved for the client. The actual intensity of treatment given each week or month also depends on the highly variable time period over which attempts are to be made to render the apportioned services. This time period is determined primarily by the total amount of discretionary resources resident in the agency at any moment (8), and secondarily by the size of the reward to the agency (7) and by the amount of pressure for services placed upon the agency by the client (12). If the combined strength of these forces is great and resources are available, the program attempts to satisfy the assessed need rapidly. If these forces are weak, the rate at which attempts are made to render the needed service is extremely slow. The difference between these extremes in an actual service situation may be very great. The dentist setting up private practice in a suburban community who has only a few patients is able to satisfy their dental needs nearly instantaneously; a public dental clinic in an urban area may be subject to such heavy demands that extremely long waits for service are encountered.

Note that in Figure 5 structural boundaries distinguishing the program sector from the client sector have been omitted. Variables (1), (2), and (3) are at the interface between the two sectors and conceived of as integral to both. From a functional perspective, such boundaries are cumbersome and misleading. Also note that the environment has been acknowledged explicitly as a pressure that influences client functioning.

On the client side of the diagram, a link has been added connecting client functioning to client standard of functioning. This expresses the premise that, in the long run, the client's standard of performance is a simple reflection of his or her actual past performances. In the short run, a vast gap may exist between performance and standard, such as when someone is stricken with acute illness. We have seen that precisely such a gap generates the need for services. The behavior of the system shifts from a steady state to a transient condition only when one or more of these gaps exist. The goal of this system, and of all negative feedback systems, is to eliminate the discrepancy, to reestablish a steady state. We have seen that the effectiveness of services rendered determines how close the final equilibrium comes to the original level of performance.

The greater the service, the greater the client functioning; the greater the functioning, the less the client need. Finally, the less the client need, the less the service. And, on the program side, the greater the service, the greater the functioning; the greater the functioning, the less the assessed need; the less the assessed need, the less the service. Thus, these two principal loops, as shown in Figure 6, operate to maintain the client's functioning at an optimum level, via their control of services rendered. These two interacting negative feedback loops constitute the causal structure of the human service delivery process and allow us to make sense of and to think rigorously about a wide range of phenomena of interest to those who study and manage such systems. The theory has been used to integrate and derive research priorities from a diverse literature on patient dropout (Levin & Roberts, 1976, Chapter 4), to study the life cycle of a typical human service agency (Levin & Roberts, Chapter 3), to conceptualize a national study of community mental health centers (Levin, Wilder, & Gilbert, 1978), to guide the development of a system for evaluating the management performance of mental health centers (Levin & Lewis, 1977a,b) and below is applied to a series of computer simulations aimed at developing strategies of reducing treatment dropout among chronically mentally ill patients.

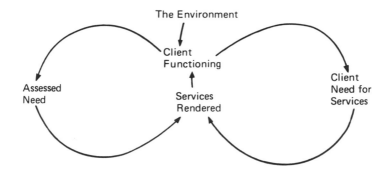

Figure 6. An overview of the general theory.

Computer Simulation of the Dropout Phenomenon: System Structure

This analysis focuses upon the interactions between an individual patient suffering from major recurrent chronic mental illness and a multi-modal community mental health program designed to maintain the patient in the community at his or her optimal level of functioning, to prevent hospitalization, and to reduce the duration of any needed hospital stays. In the patient of concern to this analysis, periods of relatively stable functioning alternate with periods of disrupted functioning. Mental health professionals believe that early identification of slippage in functioning and prompt intensive intervention reduce the duration and severity of an acute episode. They advocate that programs maintain contact with patients during periods of recovery and provide them with a broad range of rehabilitation services to reduce any residual disabilities. Dropout from a treatment program by such patients represents an important challenge.

The formulations that make up the system structure were devised by us and our colleagues, based upon our experience and supported by the available empirical literature whenever possible. A listing of the DYNAMO equations of the model are reported in Levin and Roberts (1976). The structure shown in Figure 7 parallels very closely that of the more general theory presented in Figure 5. A few of the names of variables have been changed to terms more common in mental health parlance; for example the *client* is here called the *patient*. In addition two variables in the patient sector (habit and benefit expected) have been added as key elements. Within the program sector, several variables have been added that relate specifically to the way in which

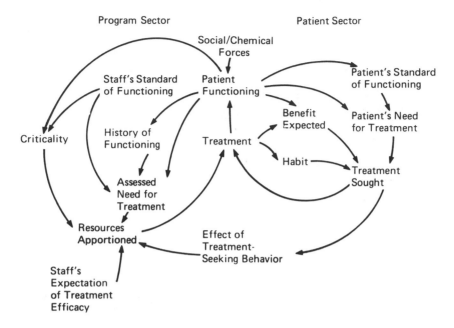

Figure 7. The causal structure of treatment dropout.

program staff interact with their patients—history of patient functioning, criticality, and staff expectation of treatment efficacy—all of which
influence professional assessment of the client's need for help when
the problem is one of serious mental illness. The import of these variables will be clarified in the following discussion.

The model is dominated by two by now familiar causal feedback
loops. On the patient side, a decline in functioning produces efforts to
secure treatment that, in proper intensity, restores functioning to the
preepisodic level. In the program sector, the staff perceives the drop in
the patient's functioning and allocates resources so as to produce treatment in the amount needed to effect the restoration. A return of functioning to the preepisodic level acts to shut off the system by eliminating the need for treatment until such time as another drop in the
patient's functioning of sufficient magnitude sets the system into motion again. Even this skeletal model depicts the dropout as a dynamic
rather than a haphazard occurrence. As did the general theory of program client interaction, it makes very clear that the rendering of treatment depends, not only on the staff's assessment of the patient's need
for treatment and the allocation of resources for this purpose, but also
on the patient's collaboration, that is, his or her willingness to seek or
receive service.

The Patient Sector

Patient functioning refers to the individual's ability to function independently, for example, socially and vocationally, according to general community norms for behavior. The model postulates that this level of functioning is influenced by two factors, by treatment and by social and/or chemical forces, the latter of which are exogenous to the treatment process and to the model (Figure 8). The basic model makes the simplifying assumption that treatment always has a positive effect on the patient's functioning and furthermore that the relationship is a linear one: The greater the amount of treatment given, the greater the resulting improvement in the patient. The model's structure, of course, permits analysis of alternative assumptions about treatment efficacy: for example, that small treatment doses have disproportionately great effects or that disproportionate gains are achieved under conditions of very high treatment intensity.

Social and chemical forces, as well as treatment, affect the patient's level of functioning. They may operate either to improve or to impair his or her performance at any given point in time. Although they may be central to the course of the illness, they are conceptualized here as aspects of the general theory's "environment" that, though very relevant to mental illness, reside outside of any system structure that our present state of mental health knowledge permits us to construct.

Patient's standard of functioning is the equivalent of the patient's expectations for his own level of functioning, that is, his present estimate of his maximum capacity to perform in the roles appropriate to his life situation. This estimate is based upon a long-term averaging of his actual functioning as is shown in Figure 9. Small variations in functioning are reflected to a small degree in a patient's standard, but large and sustained changes in his performance over time have a substantial effect on his expectations for himself and on the standard of functioning to which he aspires. Thus, if improvement does not occur

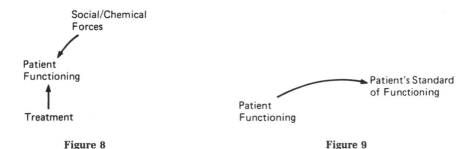

Social/Chemical
Forces

Patient
Functioning

Treatment

Figure 8

Patient
Functioning

Patient's Standard
of Functioning

Figure 9

after a decline in performance, the standard eventually declines to accommodate the new lower level of functioning. In our later discussion of system behavior, the term *eventually* will take on considerable significance, both as a source of variation among different kinds of patients and as a potential avenue for influencing their treatment behavior. The interaction between performance and standard is of particular relevance to understanding the dynamics of chronicity in patients in whom the cycle of impaired functioning and declining expectations seems to play such a large role.

Patient's need for treatment refers to the discomfort experienced by an individual and others in his environment when a discrepancy exists between his actual level of performance and the expectations or standards concerning how he should be functioning in his life roles. When the two are equal, no discomfort is felt and no need for treatment is perceived. When, however, actual performance falls below the standard, a need arises (Figure 10).

Benefit expected refers to an explicit or implicit forecast by the patient of the change in his level of functioning that he expects to experience if he continues to accept treatment from the program. The patient's initial set of anticipations about the value of treatment (which may range from very high to very low) is influenced, once in treatment, by a continuing appraisal of the impact on his level of performance of the treatment received thus far (Figure 11). Although changes in functioning during treatment may originate in social or chemical forces outside of the treatment relationship (see Figure 8), it is unrealistic to assume that the patient can precisely attribute them to the appropriate source. The model assumes therefore that the patient attributes the changes—regardless of their actual causes—to the treatment process *in proportion to its present intensity*. For example, if his performance happens to be improving while treatment intensity is high, more of the improvement is attributed to treatment than if treatment intensity happened to be low at that time. The result of such an attribution is an increase in the benefit expected by the patient from continued treatment. Conversely, should the patient experience a decline in perfor-

Figure 10   Figure 11

mance coincident with a period of intense treatment, he associates that decline with the treatment process and reduces his expectations of benefit from further treatment. Thus if a patient with an initially high level of conviction about the value of treatment finds his first contacts less beneficial than he expected, his level of conviction will diminish, and he may indeed drop out of treatment. Conversely, a patient with an initially low level of hopefulness may find his first therapeutic contacts more beneficial than he had anticipated, may increase his expectations of being helped by treatment, and may be encouraged to continue.

Habit is an inertial effect that asserts simply that a person with little or no recent treatment experience is not likely to seek it out in the absence of very pressing need, whereas a person presently involved in treatment is likely to continue in that relationship even after he has ceased to derive benefit from it. This momentum factor is a function solely of the prevailing rate of treatment, as shown in Figure 12. Furthermore, "habit" is conceptualized as a threshold phenomenon, that is, resistance is overcome (or habit is established) at a relatively low level of engagement in the treatment process. But once that threshold is crossed, more intensive participation in treatment has little impact upon the habit.

Treatment sought refers to all efforts exerted by the patient, or others in his social network acting on his behalf, with the aim of obtaining professional help. These efforts include both the behavior involved in the initial search for help and in ongoing cooperation with staff in order to enable the appropriate rate of treatment to be given. The magnitude of these efforts is a product of the interaction of three variables already discussed: the need for help, the estimate of the value of treatment, and the habit of contact with the program (Figure 13). These three variables parallel three tenets of traditional learning theory: Habit operates according to the principle of association or contiguity; benefit expected embodies a cognitive–economic view of learning; and the patient's need for treatment is formulated as a drive reduction mechanism.

Figure 12   Figure 13

The Program Sector

Parallel to the patient's personal standard of functioning, staff behavior in a treatment program is also influenced by standards of patient functioning. The model asserts that the staff is subject to not one, but two standards—one very general in its application ("staff standard of functioning"), the other specific to the characteristics of each particular patient ("history of functioning").

Staff standard of functioning, the first of these standards, refers to the level of functioning that the staff believes, based on its training and experience, is needed by any patient to maintain a minimal community adjustment. This standard is a general one and has its origin in sources apart from the individual patient. It tends to be highly stable for any single professional group, and any change occurs only over a very long period of time, and not as a consequence of any one particular staff–patient interaction.

A staff standard of functioning may be said to be operative whenever a "diagnosis" is being made, in the sense that the various aspects of a patient's condition are being measured against some generally held criterion of "normalcy" or "health." This standard originates outside the model and operates within the treatment process as one of the determinants of how the staff assesses the patient's need for treatment. By defining a "floor" under which patient functioning should not be allowed to fall, it triggers another mechanism called "criticality."

Criticality, diagrammed in Figure 14, refers to the discrepancy between the patient's actual performance at a given time and the staff standard of functioning, the fixed estimate of the level of performance required for a person to maintain a stable community adjustment. Thus, for example, if a patient threatens harm to himself or others, his behavior would produce a criticality factor that would impel staff to increase greatly the amount of treatment they judge to be needed. This extra impetus to intense treatment is eliminated quickly when the patient's performance returns to a level commensurate with the staff's standard of functioning.

History of functioning is the basis for the second standard of performance, alluded to earlier, against which the staff measures the patient's current status. Unlike the first fixed and general standard, however, this one is specific to the particular patient and originates in a knowledge of the course of his functioning over a long period of time. As a professional group, staff has access to the "history" of the patient's illness. On the basis of this long-term trend, it formulates a clinical prognosis of the patient's performance potential, against which his actual functioning can be evaluated at any given point. This staff stan-

Figure 14   Figure 15

dard, like the patient's own standard of functioning, is derived from observations of the patient's behavior. Because it is supported by a broader and more objective view of the situation, however, it is more durable—that is, less susceptible to fluctuations in response to short-term changes in the patient's functioning—than is the patient's standard for himself. This perspective is suggested by Figure 15.

Assessed need for treatment, shown in Figure 16, reflects the staff's assessment of the improvement in performance that may be reasonably expected in the patient's response to treatment. A knowledge of the history of his illness—its duration and its course—provides a standard for the patient's optimal level of functioning. The difference between that potential and his present level of function defines the magnitude of the individual's need for treatment at a given time.

Resources apportioned refers to the investment in treatment that staff is willing to make for a specific patient. Two of the determinants of the magnitude of this investment are indicated in Figure 17. The most potent of the two is the assessment of the patient's need for treatment, that is, the amount of help deemed necessary to restore his performance to its "average" level within a reasonable period of time. If the "criticality" factor becomes operative, staff will greatly increase the resources apportioned to treatment (as represented, for example, by hospitalization). Once the criticality dimension is overcome, however, the extra resources engendered by it will be quickly reduced. The resources allocated to a patient's care depend on more than a statement of his need. As seen next, they are also influenced by the beliefs of staff about

Figure 16   Figure 17

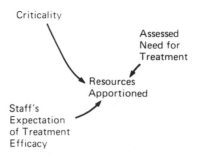

Figure 18

the efficacy of the treatment they are rendering, and by their perception of the patient's willingness to cooperate in the treatment process.

Staff's expectations of treatment efficacy, which is derived from its training and experience, is indicated in Figure 18. Although two professionals may judge a patient very similarly in terms of his present clinical condition and his consequent need for help, they will prescribe different treatment programs for him if one of them believes that each treatment unit is enormously effective and the other believes it is only minimally so. To achieve the same clinical goal within the same period of time, the former allocates fewer treatment resources to his case than the latter. At first encounter, this formulation appears contrary to our expectation that a therapist who believes treatment is more potent will prescribe more of it than the therapist who believes it is less so. The real process, however, is counterintuitive. Assume, for example, that the clinical assessment of a patient reveals a discrepancy between his actual and potential functioning to the magnitude of X, and a goal is set to eliminate that discrepancy within a fixed period of time. One therapist believes that each unit of treatment is moderately effective (let us say capable of restoring $1/X$ of the gap) and plans Y units of treatment to meet the goal; the other attributes twice the power to each unit of treatment (capable of restoring $2/X$ of the gap), and therefore plans only half the number of treatment units ($Y/2$) to achieve the same goal within the same period.

Effect of patient's treatment-seeking behavior is greatest in the initial phase of the treatment relationship when the staff is made aware of the patient's efforts to get care. Once the program has been engaged by the patient's efforts, additional effort achieves sharply diminished returns. Thus, although some additional effort to get care, for example, a telephone call, may result in an earlier appointment, once the patient is in treatment the staff is highly autonomous in regulating the intensity of care and does not permit the patient to dictate it. This "patient-

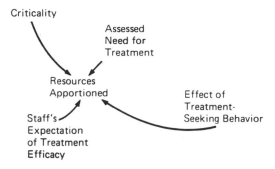

Figure 19

effect" variable builds into the model formal recognition of the fact that staff takes account of the patient's cooperation in treatment: The patient who refuses to cooperate ultimately has a discouraging effect on staff and causes it to diminish the resources apportioned to his treatment (Figure 19).

Treatment, shown in Figure 20, refers to any services rendered by program staff to the patient for the improvement or restoration of his level of functioning. Treatment is assumed to vary only in intensity. High-intensity treatment simulates such modalities as hospitalization. Lower intensity treatment simulates outpatient services. For treatment to occur, two conditions are necessary: The patient must seek help and the program staff must allocate resources to serving him. Treatment terminates if either of these conditions is absent. When it comes to a halt—because the patient ceases to seek or accept help while the staff is still offering resources to him on the basis of its assessment of his need—a situation exists that is commonly labeled a treatment dropout.

Computer Simulation of the Dropout Phenomenon: System Behavior

In our computer-aided exploration of the factors contributing to dropout, we first report the behavior of the model with a basic set of as-

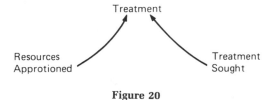

Figure 20

sumptions. Next we show the effects of a number of changes representing differences in patient characteristics and treatment responses. In some cases the changes have little effect on the model's behavior. In others, the changes have profound effect on behavior and help to identify factors that contribute importantly to treatment dropout. Clinically, dropout indicates leaving treatment while it is still being offered before full recovery occurs. Generally, dropout occurs early in treatment. The model takes into account not only when treatment stops but the level of the patient's recovery at that point. A return to almost a full level of functioning is not considered in the model as a serious dropout issue. Second, for simulation purposes, the model is run over a 5-year period. Although dropouts in the real world occur at much earlier points in treatment, the model assumes that the forces and counterforces at work in the model are similar to those that produce early dropout. For some dropout-inducing factors, remedial efforts have been simulated and their effects shown. Some changes produce uncertain results and suggest new areas for empirical research.

Consider first the case of a successful response by the system to a crisis. The behavior characteristics of such a successful pattern of interaction between program and patient is displayed (see the run shown in Figure 21). In this simulation, the patient experiences a 33% decrease in his level of function as the result of a sudden crisis suffered in the tenth month of the run. His level of functioning (F) drops immediately after the crisis from 60 to 40. The drop motivates the patient to seek, and the program to provide, care. Average rate of treatment (T) rises to a peak of about 12 treatment units per month soon after the crisis and subsides as health begins to return to its precrisis level. By the twenty-fourth month of the run, 14 months after the crisis, functioning has returned to its original level and rate of treatment has gone to zero. This represents a successful completion of treatment without dropout.

The full recovery was produced by the feedback structure of the model shown early in Figure 7. On the patient side, the treatment sought (S) is stimulated primarily by the patient's need for treatment, as measured by the discomfort gap between the current level of functioning (F) and the patient's standard of functioning (N). The large magnitude of this gap right after the crisis initiates a strong care-seeking response. Because treatment is so successful in a short period, patient anticipation of treatment benefit (B) remains high and supports the patient's efforts to get care. Habit (H) is quickly established by the rapid beginning of an intense treatment program, and any resistance to continuing treatment is rapidly overcome.

On the program side, the large initial gap between the patient's

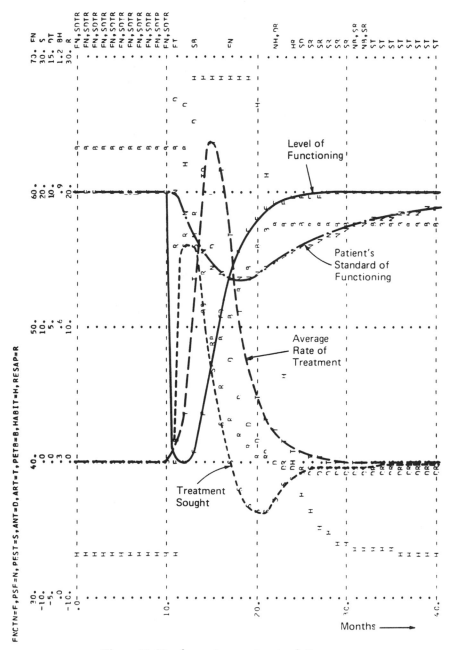

Figure 21. Nondropout computer simulation run.

functioning and the staff's historical record of the patient's functioning motivates the staff to allocate resources (R) so as to increase treatment provided by the staff. The program's responses are amplified by the criticality factor that intensifies treatment for lower absolute values of functioning. The rapid seeking and provision of treatment work together to quickly return the patient's functioning to its precrisis level. Restoration of this level acts to return the system to a steady-state condition until such time as another significant drop in the patient's functioning perturbs it once again.

Our baseline run is a success in that the patient was sufficiently well motivated and the program was responsive enough to prevent dropout or waste. The patient returned fully to his precrisis level of functioning. We are now in a position to examine less favorable circumstances. Using the basic run as a benchmark, many other factors can be examined, and their importance in causing dropout can be determined. On the patient side, the investigation will focus upon three elements: (1) patient's standard of functioning, (2) benefit expected (and a combination of patient's standard of functioning and benefit expected), and (3) habit. Three program variables will also be considered: (1) treatment timetable, (2) expectation of treatment efficacy, and (3) the role of criticality.

Patient Characteristics

Unstable Patient's Standard of Functioning

By the stability of a standard of functioning, we mean the readiness with which the standard changes in response to new levels of health or functioning. What is the effect of an unstable standard of functioning on the treatment system? A person with an unstable standard is little influenced by his previous level of functioning after a crisis and is unlikely to strive to regain that level. A person with a very stable standard is likely to maintain that standard long after he had suffered a loss of functioning, even though this persistence may require him to endure considerable discomfort.

The baseline run assumed that a lengthy time was required for the patient's standard of functioning to reflect accumulated changes in his actual functioning. Specifically we assume that a patient's standard would drift toward acceptance of the actual accumulated change in functioning over a period of about 1 year. This time constant was reduced to only 3 months in Run 2 (see the run shown in Figure 22). One would expect that a greatly reduced time required to adjust standard of

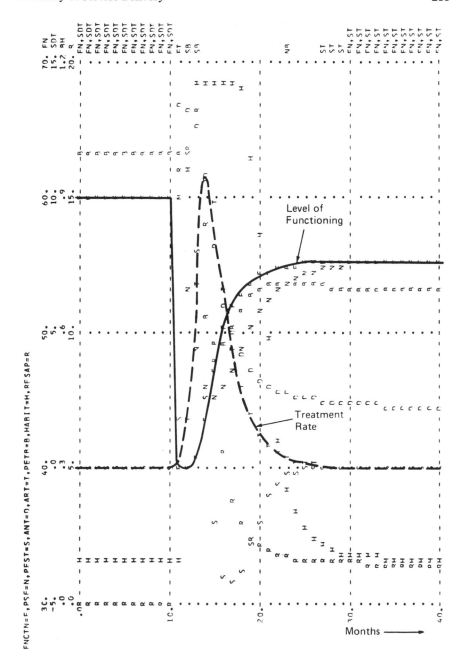

Figure 22. Unstable standard of functioning.

functioning would greatly increase the likelihood of dropout. This was not the case. Under this experimental condition, the patient's level of functioning (F) reaches 55, or 92% of the precrisis level of 60. No significant waste occurs, as the patient was able to maintain a standard of functioning sufficient to support an adequate degree of recovery. Recovery occurs rapidly and health returns to a higher value before the standard (N) has had much opportunity to deteriorate. Apparently, the patient's initial expectation of benefit was strong enough to sustain his efforts to get care. Because the treatment response was adequate, the patient was able to enter treatment and to sustain a belief that the process was beneficial.

Indifference toward Benefit Expected

The patient's anticipation of treatment benefit represents his estimate of the improvement in health that can be gained from treatment. The patient's initial anticipations result from previous personal experiences with the program, impressions of program effectiveness based on the experience of others, and knowledge, if any, of the field of mental health care. These anticipations change, depending on changes in health and treatment intensity. If the patient is undergoing an intense program of treatment, he is more likely to attribute changes in health to treatment rather than to social and chemical forces. As changes in level of functioning are attributed to treatment, patient anticipations change, and anticipation of benefit expected is reinforced. If the patient's expectation of benefit from treatment (or anticipation, or hope) is a critical variable, weakening its influence in the model would be expected to result in less complete recovery than in the baseline run. To test this hypothesis, the function that describes the influence of benefit expected upon patient's efforts to get care (or "treatment sought") was changed so as to make the patient's behavior indifferent to the benefit expected.

Run 3 shows (see the run shown in Figure 23) the behavior of the system when confronted with a patient whose tendency to seek care is impervious to his expectations of benefit. Again, the illness pattern simulated is of a crisis nature. The results show that the patient's indifference toward expected benefits has only a minor effect on the behavior of the system. The level of functioning recovers to a value of 55 rather than 60 as in the base run. Again, the rapid recovery due to treatment leaves little time for the patient's skepticism to outweigh his discomfort and cause him to drop out. These results are contrary to expectation because 92% recovery was achieved by the patient in this run. We can say that the patient took reasonably full advantage of the

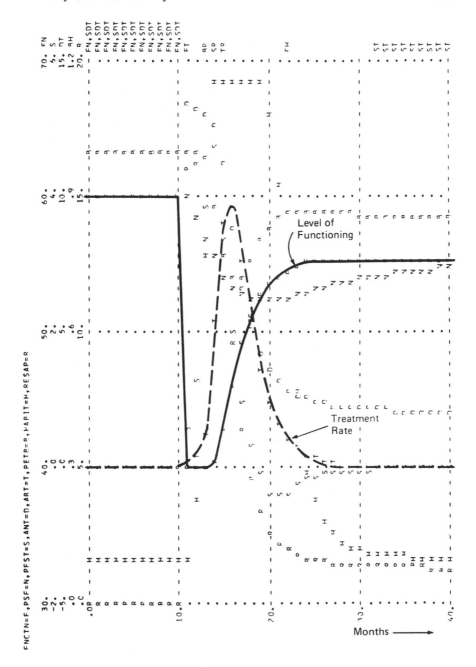

Figure 23. Skepticism in benefit expected.

program resources at his disposal. The *post hoc* explanation for this result is that the discomfort experienced by the patient as a result of the discrepancy between functioning and standard of functioning was sustained long enough to cause the patient and program to develop a self-perpetuating relationship (assisted by the habit effect), strong enough to sustain itself despite counterpressure created by declining patient belief in treatment efficacy. If this explanation is correct, then altered model behavior would stem from the weakened effect of "benefit expected" only if it were accompanied by instability of patient standard of functioning. This latter experimental manipulation would cause a more rapid reduction of the discomfort that is the primary force moving the patient to seek and use the program. The patient would become resigned to a lower level of performance and remove himself from an unpromising relationship before significant benefit could be derived from it.

Combination of Unstable Standard of Functioning and Indifference toward Benefit Expected

A simulation run was tried with a patient in crisis with both an unstable standard and indifference toward the benefit of treatment. The results are shown in Run 4 (see the run shown in Figure 24). Clearly, the combination of these detrimental characteristics is sufficient to reduce this patient's efforts to get care long before much of the lost functioning has been recovered. The final level of functioning (F) in this run is 44, as compared to 60 in the crisis baseline run. In this case, his attitude toward anticipation of benefit (B) causes the patient to seek a smaller amount of treatment (S) than is required by his discomfort. Treatment starts off at a lower intensity, and the functioning level begins to rise very slowly. In the run with unstable standard alone, the intense treatment rate permitted by the importance the patient attributes to the possibility of treatment success raised the functioning substantially before the falling standard eliminated the discomfort gap and caused the patient to stop seeking help. However, the slow growth in treatment intensity due to patient indifference keeps health at a low level for a long enough time to allow a large decrease in the (unstable) standard. The drop in the standard wipes out the discomfort gap, eliminating the patient's felt need for treatment and causing dropout when functioning is still far below its precrisis level. The health "deficit" (initial level of functioning minus final equilibrium level) left at the end of the run is greater than the sum of the deficits left in the runs with unstable standard or indifferent anticipation alone.

Although each of the detrimental factors alone was insufficient to

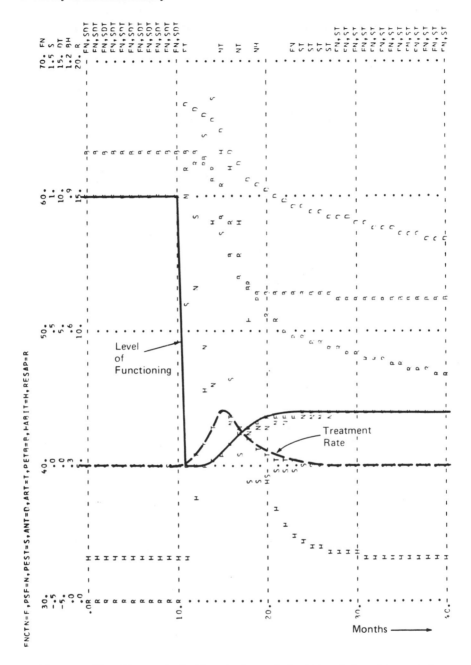

Figure 24. Unstable standard of functioning and skepticism in benefit expected.

produce dropout, the combination of the two crippled the system of staff–patient interaction and caused an early dropout. This is typical of nonlinear dynamic systems in which combinations of factors can act synergistically to produce disproportionate behavior changes.

Decreasing the Effect of Habit

Habit represents the degree to which a patient is involved in the treatment process. As a unitary factor, a high habitual commitment to the program implies that the patient has major involvement in treatment and openness to increasing the rate of treatment. Low level of habit implies little involvement in treatment and probably resistance to increasing the rate of treatment. The effect on the performance of the system of this element, however, doesn't seem to be substantial. In runs wherein the habit factor was made more unfavorable compared to the baseline run, few differences were generated in the performance of the system.

Program Characteristics

A Delay in Treatment Timetable

We assume that program staff are reasonably accurate in assessing the patient's deficit and in estimating and offering the total amount of care required to return him to his premorbid condition. The main avenue of fallibility open to staff is the rate at which the needed care is provided. In our model, the staff regulates this rate through several explicit or implicit assessments and decisions. First, an estimate is made based upon history taking and observation of ongoing behavior of the level of recovery likely to be achieved and sustained by this patient; second, by comparing his present level of functioning with the level of recovery expected, an accurate projection is made of the amount of treatment effort that will be required to achieve this recovery goal; third, a treatment plan is devised based upon these considerations, including as well a timetable for achievement of this result. In the baseline run, this time constant was set at 6 months: Staff makes treatment available for the patient at whatever intensity is required to attempt to close the gap between performance and expected recovery state in 6 months. It follows, then, that the shorter the timetable, the greater the treatment intensity required to close the gap within the time period; the longer the time schedule, the less the treatment intensity required. If patients were infinitely patient, this time constant would be unimportant. The appropriate amount of treatment spread over any period, however long, would result in eventual full recovery. In reality a patient's involvement in the treatment process is subject to the influence of a number of variables, including the discomfort he experiences, the benefit he feels he is deriving from it, the expectation he has of his own potential recovery, and the intensity of his relationship with program staff. At very low rates of treatment intensity, these variables would stabilize his performance at a lower-than-optimum recovery level, causing him to leave treatment long before the full plan is executed. Our simulation experiments show that the tendency for the interaction system to break down and produce dropout varies directly with the projected length of time for recovery.

An experimental run (not shown) was conducted using 1 full year rather than 6 months as the time constant regulating the rate at which treatment is made available by the program. This variation has only limited effect upon recovery. The patient still makes adequate use of the resources offered, recovering to 54 units or 90% of his premorbid condition. Extending the timetable goal to 24 months, however, leads to a further decrease in the performance of the program–patient interaction system, with a final health value of 50 units. An extremely long timetable goal of 60 months leads to a further (but not proportionate) drop in level of functioning to 46 units (Run 5; see the run, shown in Figure 25).

System performance for the crisis case undergoes a moderate decrease over that of the run with the 24-month goal. In this case, the patient is already taking advantage of most of the care that is offered. The decrease in system performance merely reflects the lower treatment intensity offered by the staff. This large a decrease came about because the treatment rate in the 60-month run takes longer to grow to the point where it can counteract the deterioration of standard of functioning. A delay in the growth of treatment rate is serious because the continual deterioration in health goes on for that much longer and results in a significantly lower final value of health level. The decrease in system performance merely reflects the weak program response by the staff.

What are the implications of this set of runs? One is that responding more slowly to a crisis will not have a great effect on system behavior as long as the delay is not too long. These shorter delays are realistic representations of the effect of overloads in operating programs. Recovery time as long as 60 months more realistically represents a situation in which treatment resources are extremely scarce and the patient must wait a very long time before receiving treatment at all.

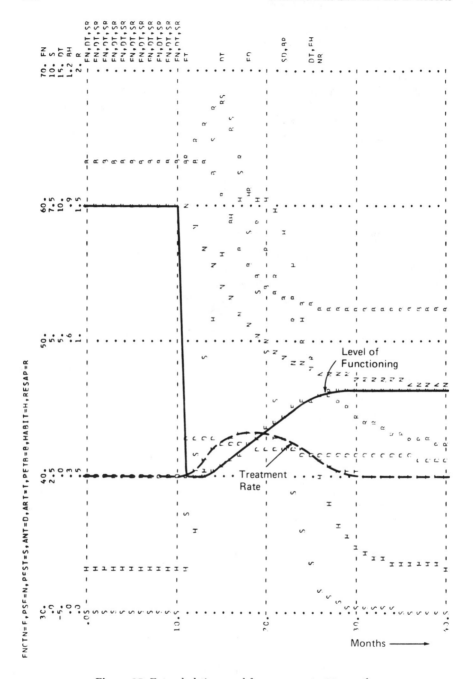

Figure 25. Extended time goal for recovery to 60 months.

Exaggerated Expectation of Treatment Effectiveness

Misjudgment of treatment efficacy can have an effect similar to that of a
long delay time. In Run 6 (see the run shown in Figure 26), the staff's
estimate of efficacy is assumed to be four times the actual efficacy of
treatment, prompting the staff to give too weak a schedule of treatment.
As a result, the level of functioning (F) reaches only about 51 units by
the end of the run. As in runs with long recovery timetables, the patient's standard of functioning (N) and anticipation of treatment benefit
(B) both decrease rapidly, whereas health (F) is rising slowly. This
result implies that research is needed on the degree of treatment
efficacy when dealing with various forms of mental illness. It also implies a need for careful monitoring of the progress of patients in treatment. This needs to be coupled with action to prevent dropout in
patients making slow progress.

Criticality Factor

Recall that the criticality factor served to amplify the treatment intensity offered by the staff when the patient's health level is at a low
enough value (on an absolute scale) to indicate potential danger to the
patient and those around him. For lower values of functioning, the
treatment intensity is multiplied by a factor of up to 3. As it turns out,
though, the criticality factor does not play so important a role. This is
shown in Run 7 (see the run shown in Figure 27), which is the result of
a simulation run with the effect of the criticality factor eliminated (i.e.,
set equal to 1 for all values of health). The final value of the health level
is about 58 rather than 60, as in the base run, representing an insignifi-
cant change. In the case of more severe crises, extending over a longer
period of time, the criticality factor might be more important. The results of these simulations are reported in Table 1.

Efforts to Correct the Dropout Problem

Having produced a simulation model to account for patient dropout
under a variety of conditions, our attention naturally turns to devising
changes in policy or system structure that will correct or alleviate the
dropout problem. The most promising corrective strategies would seem
to be those that would most directly influence the mechanism that
created the problem. Assume that the staff learns to plan realistic timetables for the recovery of patients and develops realistic estimates of
treatment effectiveness. The focus then turns to remedial efforts to

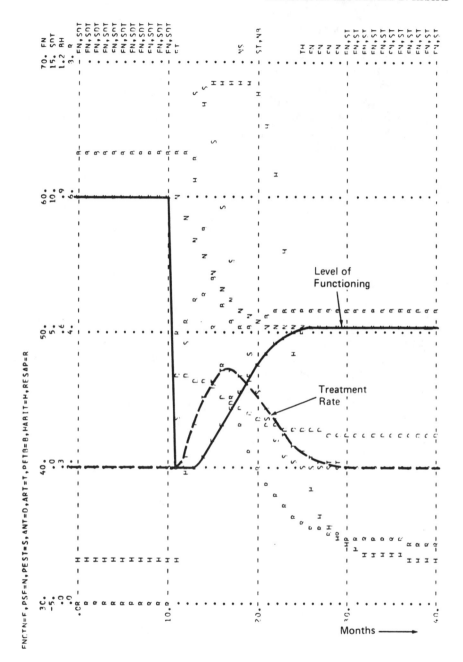

Figure 26. Exaggerated estimate of treatment efficacy.

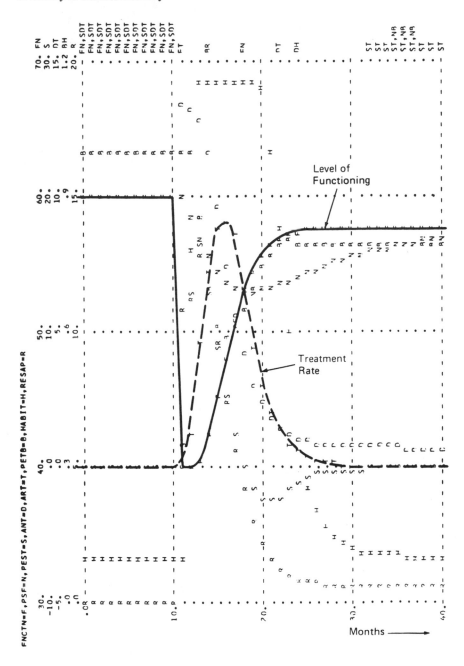

Figure 27. Elimination of criticality factor.

Table 1. Simulations of Patient Dropout

Condition	Final level of functioning	Percentage of recovery to preacute state
(a) Baseline run	60 units	100
(b) Doubling the instability of the standard of functioning	55 units	92
(c) Increasing indifference toward benefit expected	55 units	92
(d) Combination of (b) and (c)	44 units	73
(e) Extending the treatment timetable by factor of 10	46 units	77
(f) Quadrupling the estimate for treatment effectiveness	51 units	85
(g) Eliminating the criticality factor	58 units	97

counteract the detrimental effects of a patient with (1) an unstable standard of functioning and (2) an indifference toward benefit expected.

Efforts to Support Standard of Functioning

This run assumed that the program has some quantity of staff resources that were neither used nor accounted for in any of the previous runs. These resources are used to support the patient's standard of functioning in the two runs. The result of this experiment as shown in Run 8 (see Figure 28) are disappointing. The patient recovers to a level of 46, or 77% of his precrisis condition. This is only a 4% improvement over the symptomatic run (see Figure 24). Further examination of the model structure reveals that a supported standard has a significant effect only after the patient is firmly established in a treatment relationship. Moreover, once this treatment relationship has developed, it is so well supported by other mechanisms in the model that dropout is not likely anyway. We see that the support of standard of functioning is, by itself, not a promising strategy.

Efforts to Enhance Patient's Initial Level of Benefit Expected from Treatment

In this simulation, staff resources are allocated to rapidly double the level of benefit expected from treatment as compared to the symptomatic run. Otherwise the conditions are identical. The result depicted in Run 9 (see the run shown in Figure 29) is a recovery to 54 units, or 90% of the premorbid adjustment. Implementation of this strategy is effec-

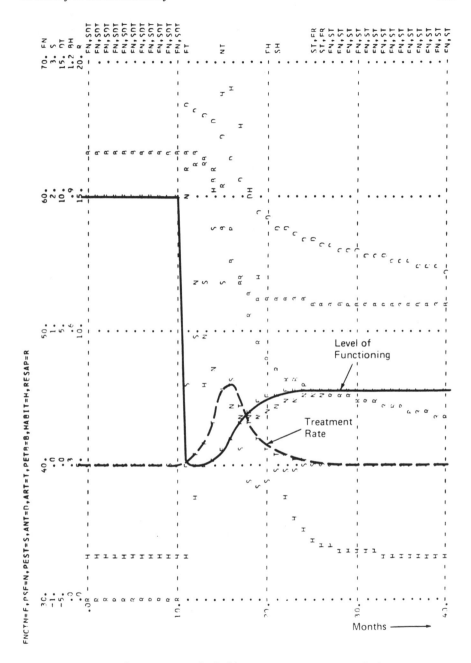

Figure 28. Support of patient's standard of functioning (in patient with characteristics of unstable standard of functioning and skepticism in benefit expected).

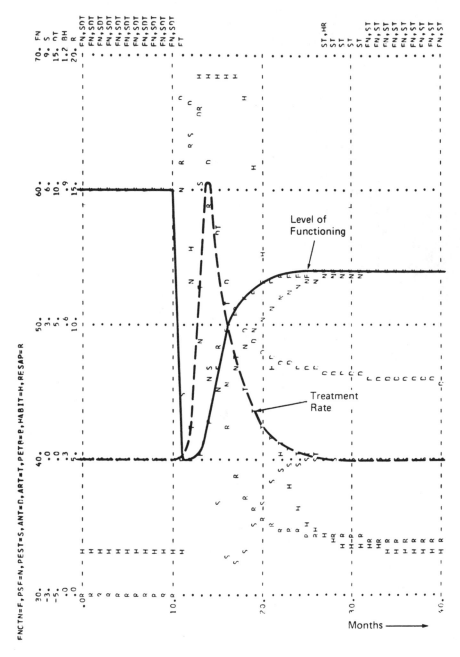

Figure 29. Support of initial benefit expected (in patient with characteristics of unstable standard of functioning and skepticism in benefit expected).

tive in coping with the problem of patient dropout. It should be noted that combining this strategy with ongoing effort to support standard of functioning achieves essentially full patient recovery, as shown in Run 10 (see the run shown in Figure 30). In terms of relative effectiveness, however, a program that would increase the initial level of benefit expected in a patient with both an unstable standard of functioning and skepticism in benefit expected would appear to be sufficient to reduce importantly the problem of dropout. The results of the corrective simulations are shown in Table 2.

Conclusions

When most patients with chronic mental illness spent long periods of their lives as inpatients in hospitals, dropout from treatment was not as pressing an issue as it is today, as the patients were captive in the treatment system. The current emphasis on the care of such patients in the community highlights the importance of maintaining continuity of the interaction between a mental health program and its patients. Customarily, studies of medical care (and, indeed, of human services in general) conclude that the remedy for less than optimal delivery of care lies in increasing the resources allocated for the purpose. Although we do not dispute that recommendation, the application of the system dynamics approach has uncovered a causal structure of the care–delivery system that suggests that treatment dropout by patients suffering chronic disability is not solely a function of scarcity of resources, but, in the case studied, of the interaction of an unstable standard of functioning and indifference toward the benefits expected from treatment. We suggest that a program use some of its resources to raise and maintain the level of anticipation of benefit or hope in such patients to stabilize the treatment system and reduce dropout.

The patient's forecast of the benefit he expects to experience in treatment stems from an act of anticipation present at the onset of treatment and developed during treatment by a continuing appraisal of the impact on his level of performance of the treatment received thus far. These formulations, whether correct or incorrect, appear signifi-

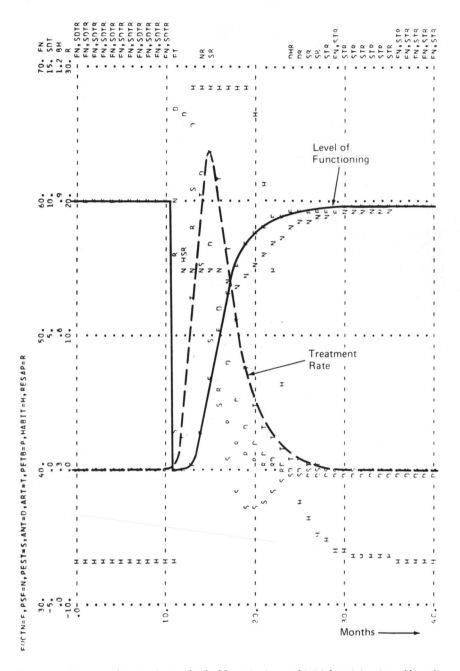

Figure 30. Support of patient's standard of functioning and initial anticipation of benefit expected (in patient with characteristics of unstable standard of functioning and skepticism in benefit expected).

Table 2. Efforts to Reduce Dropout

Condition	Final level of functioning	Percentage of recovery to preacute state
(a) Baseline run	60 units	100
(b) Combination of unstable standard of functioning and indifference toward benefit expected	44 units	73
(c) Support of standard of functioning	46 units	77
(d) Doubling initial benefit expected	54 units	90
(e) Combination of (c) and (d)	59 units	98

cantly to affect the patient's participation in the treatment system. The patient approaches the treatment process with an initial set of beliefs about the help he may receive from it. Some of these beliefs originate in his social class and cultural milieu and in the basic personality traits of the patient—his optimism and suggestibility—and, as such, are not readily subject to influence. Other beliefs, stemming from prevalent attitudes about the nature of psychiatric illness and its treatment and from prior experience he or others in his social network have had with care-giving systems, are more readily subject to influence. The patient's assessment of how his functioning has changed in response to treatment is also potentially subject to influence.

Let us look at a number of strategies that could be employed on an experimental basis and, if successful, could be incorporated into the standard procedures of a mental health program. First, a mental health program could attempt to educate the citizens and referral sources in its community in the nature of mental illness and in the availability and efficacy of its treatment programs. Second, the program could identify those patients with chronic psychiatric disability who do seek help but who are most vulnerable to dropout from treatment and, as such, require interventions beyond the usual measures of promoting and maintaining a therapeutic alliance. Following the findings of the simulation experiments, staff would administer interviews, attitude questionnaires, and other tests that could uncover those patients who enter treatment with both little hope of treatment benefit and with unstable standards of functioning or who develop these two conditions during the process of treatment. Special remedial efforts could be employed with this group. Third, the program could make sure it has hired staff who have a genuine interest and concern in treating patients with chronic disability, especially those patients from the lower social class who most often constitute the group vulnerable to treatment drop-out. The staff should be well trained in the methods most appropriate for treating patients with chronic disability, and should be encouraged to have confidence in their treatment theories and methods. Fourth, the early phase of treatment could be marked by an intake process that is especially responsive to the acute needs of the patients and their families. The therapist, patient, and his family, where appropriate, could discuss their expectations of treatment benefit in order to get this critical issue into the open, where it could be more readily subject to influence. The patient and his family, where appropriate, could participate maximally in discussing the treatment goals and the planned treatment procedure and the relationships between the two. The value of the planned treatment regime could be enhanced through educational measures; for example, orientation interviews and reading and audiovisual material. The patient and his family, where appropriate, could participate in group discussions about treatment benefit with other patients who are at various stages of recovery from illness, and their families, to gain group support for a new value system. And fifth, the staff could reinforce expressions of hope during the treatment process, highlight patient gains and their contiguity to treatment input and be sensitive to drops in the level of benefit expected that might require repeated "injections of hope," using the interventions noted.

ACKNOWLEDGMENT

Support for this work was provided by the National Institute of Health under Grant 16586.

References

Hertzman, M., & Levin, G. (1974a). The uses of a macro-systems modelling approach to mental hospitalization. *International Journal of Social Psychiatry, 20*, 225–231.

Hertzman, M., & Levin, G. (1974b). Empirical confirmation of a simulation model of mental hospital. *International Journal of Social Psychiatry, 20*, 218–224.

Hertzman, M., & Levin, G. (1975). An index of responsiveness to patient needs: The synchrony ratio. *Community Mental Health Journal, 11*, 44–47.

Hirsch, G. B., & Killingsworth, W. R. (1975). A new framework for projecting dental manpower requirements. *Inquiry, 12*, 126–142.

Hirsch, G. B., & Miller, S. (1974). Evaluation HMO policies with a computer simulation model. *Medical Care*, August.

Kligler, D. S., Levin, G., Hirsch, G., Roberts, E., & Wilder, J. F. (1971). System simulation of program-patient interaction. *Proceedings of the Summer Computer Simulation Conference*, Boston.

Levin, G. (1974). How treatment decisions are made: An analysis of the intake-deposition system. *International Journal of Mental Health, 3*, 93–103.

Levin, G. (1977). Poor quality is the solution, not the problem. *Health Care Management Review, 2*, 69–72.

Levin, G., & Lewis, K. (1977a). SNAPOR: Evaluation from the clinicians' point of view. *Evaluation, 4,* 77–78.

Levin, G., & Lewis, K. (1977b). A practical approach to need-based program evaluation. In W. Neigher *et al.* (Eds.), *Emerging developments in mental health program evaluation.* New York: Argold Press.

Levin, G., Hirsch, G., & Roberts, E. B. (1972). Narcotics and the community: A system simulation. *American Journal of Public Health, 62,* 861–873.

Levin, G., Roberts, E. B., & Hirsch, G. B. (1975). *The persistent poppy: A computer aided search for heroin policy.* Cambridge: Ballinger.

Levin, G., Roberts, E. B., with Hirsch, G. B., Kligler, D. S., Roberts, N., & Wilder, J. F. (1976). *The Dynamics of Human Service Delivery.* Cambridge: Ballinger.

Levin, G., Wilder, J. F., & Gilbert, J. (1978). Identifying and meeting clients' needs in six community mental health centers. *Hospital and Community Psychiatry, 29,* 185–188.

Levin, G., Roberts, E. B., & Hirsch, G. (1981). A system dynamic view of substance abuse. In J. H. Lowinson & P. Ruiz (Eds.), *Substance abuse: Clinical problems and perspectives.* Baltimore: William & Wilkins.

Roberts, E. B., & Hirsch, G. B. (1976). Strategic modelling for health care managers. *Health Care Management Review,* Winter.

Roberts, N. (1974). A computer simulation of student performance in the elementary school classroom. *Simulation and Games, 5,* 265–290.

12

Educational Policy Analysis

An Analysis of Effective and Ineffective Schools Using the System Dynamics Method

Karl H. Clauset, Jr. and Alan K. Gaynor

Introduction

The purpose of this chapter is to illustrate the application of the system dynamics method to an educational policy problem of wide current interest. Other applications of system dynamics to educational problems include the work of Andersen *et al.* (1980) and Weaver (1982) on school finance, Blakeslee (1984), Weaver (1983), and Gaynor and Clauset (1983) on teacher supply and demand, Gaynor and Clauset (1985) and Garet (1979) on innovation and institutionalization, and Roberts (1974) on teacher–student interactions.

What is presented here is a theory about a set of feedback dynamics that distinguish structurally effective from ineffective schools.[1] Concepts are explained with reference to relevant educational research leading up to the presentation of a dynamic model in the form of a

---

[1] The following material is excerpted from an earlier paper submitted to the American Educational Research Association (Clauset & Gaynor, 1982).

Karl H. Clauset, Jr. • Education Development Center, Inc., 55 Chapel Street, Newton, Massachusetts 02160.   **Alan K. Gaynor** • School of Education, Boston University, Boston, Massachusetts 02215.

Analysis of Dynamic Psychological Systems, Volume 2: Methods and Applications, edited by Ralph L. Levine and Hiram E. Fitzgerald. Plenum Press, New York, 1992.

causal-loop diagram. The emphasis here is on the presentation of the theory. Although findings are presented in general at the end of the chapter, detailed discussion of them, along with the full literature review, structural diagrams, and model equations, occurs elsewhere (Clauset, 1982).

The Problem

The current emphasis on reform and improvement in U.S. elementary and secondary schools underscores a belief in and a concern for public education. The belief is that all but clearly exceptional students should be able to acquire a repertoire of basic skills and minimal competencies during the course of their schooling. Both educators and the public are concerned that elementary and secondary schools are not achieving this goal. It is in the nation's urban schools where the problem seems to be the worst. The continued failure of most urban elementary schools to educate students effectively has been well documented (cf. Edmonds, 1979; Kozol, 1967; Silberman, 1970; Weber, 1971).

There is repeated reference in the literature to a widening gap in reading achievement between poor and middle-class children. By and large, poor children enter elementary school at a lower level of reading readiness than do middle-class and upper-middle-class children. As these children move through elementary school, the gap persists and widens. The percentage of poor children who drop out of school by the time they reach age 16 is far greater than that of middle-class children (Dearman & Plisko, 1981). The problem can be viewed, then, as a discrepancy between the actual patterns of reading achievement for initially low-achieving students in urban elementary schools and a desired level of achievement for all students.

This problem can be conceptualized from the perspective of cohorts of students moving through an elementary school or from the perspective of the school monitoring patterns in reading achievement for successive cohorts over a period of years. From the cohort perspective, if students entering the first grade in a given year are divided into initially low, average, and high achievers, based on indicators of reading readiness, the graph of cohort achievement in reading over the 6 years of elementary school would show a widening gap between the low and average achievers. The initially high achievers who began first grade above grade level maintain or increase their gap relative to the average achievers. Differences in achievement among groups of students increase, rather than disappear at later grade levels. As Bloom (1976) points out in his review of the few longitudinal research studies in this area, "There is a substantial relation between achievement differences among a group of students at one time and their achievement differences several years later" (Bloom, 1976).

From the school perspective, if one examined the reading achievement scores of sixth graders who were initially high-, average-, and low-achieving students, one would find that initially low-achieving sixth graders would maintain a pattern of achievement 1 to 3 years below the average achievers and the initially high achievers would be 1 to 2 years above the average achievers.

Research Objectives

The work described here has sought to achieve two goals: (1) to develop a causal theory for understanding the problem of a widening reading achievement gap for initially low-achieving students, and (2) to use the theory to evaluate the likely consequences of policies implemented by school administrators to ameliorate the problem.

To develop a theory about effective and ineffective schooling, we focused on the interactions among a mutually interdependent set of variables that, over time, produce the patterns of reading achievement already described. This set of variables represents a *causal* structure of the problem system. The growing evidence that schools do have an impact regardless of family and socioeconomic background factors implies a focus on the constellation of variables within the boundaries of the school setting. In addition, we have developed a computer simulation model of a typical urban elementary school—the School Effectiveness Model—that is designed to address the following question: What mix of strategies available to administrators at the school or district level is most effective in closing the achievement gap for initially low-achieving students? The policy options examined with the model cluster in four areas: (1) changes in the school's student population; (2) changes in the quantity, or intensity, of instruction delivered to initially low achievers; (3) changes in the quality or appropriateness of instruction; and (4) changes in the school climate.

The Dynamic Hypothesis

The problem behaviors described earlier in this chapter suggest the existence of a multiplier effect in the ineffective school that operates to reinforce the initial achievement differences among entering students. We argue that the fundamental difference between schools that are effective and ineffective for initially low-achieving children lies in the relationship between *observed achievement* and the *appropriateness and intensity of instruction* that the school delivers to different achievement groups. Considerable research on "direct instruction" (Rosenshine, 1979) and "mastery learning" (Bloom, 1976) suggests that in all schools, effective and ineffective, there is a direct causal relationship between the appropriateness and intensity of instruction and the rate at which children, especially poor children, learn to read (Benjamin, 1980; Medley, 1979; Rosenshine, 1979; Salganik, 1980). We hypothesize that effective schools provide instruction to *low-achieving* students that is appropriate and *more intense* in order to bring their reading achievement up to grade level. In these schools, grade-level performance is the norm for all but clearly exceptional children. By contrast in the ineffective school, instruction is *most intense and appropriate* for children whose achievement is *already* at grade level or above and increasingly *less intense and appropriate* for children who read further and further below grade level (Rist, 1973).

Thus effective schools are characterized by "negative feedback" between observed achievement and appropriateness and intensity of instruction (where *lower* achievers get *more* intense instruction), and ineffective schools are characterized by "positive feedback" (in which lower achievers get *less* appropriate and intense instruction). We are arguing that it is *differing expectations of teachers* for low-achieving students that determine different patterns of appropriateness and intensity of instruction in the effective and ineffective schools. This is our dynamic hypothesis.

It is central to our general thesis that differences in native ability and home environment, although seriously compounding the problem of differential achievement among students, constitute clearly separable aspects of the problem. Ineffective schools are characterized by institutional dynamics that operate *in addition to* genetic and environmental differences among students. Our argument is that there is a set of transcendent structural characteristics that determine how schools *respond* to student differences and that mark schools as structurally more or less effective.

According to this hypothesis, appropriateness and intensity of instruction constitutes an *institutional response* to a perceived learning gap. The perceived learning gap is the difference between expectations teachers and principals hold for students and their perceptions of how well students are actually achieving. It is the perceived learning gap that exerts pressure on teachers and principals to accept professional responsibility for low-achieving students and to work *institutionally* to increase the appropriateness and intensity of instruction for them.

In the *ineffective* school, teacher and principal expectations for students regress toward actual achievement, a dynamic that has the effect of "writing off" initially low-achieving children from the very beginning as students who cannot keep up. This dynamic obviates any perceived learning gap and any *institutional responsibility* for improving instruction. Thus ineffective schools are dominated by positive feedback that amplifies over time the initial differences in student achievement. This positive feedback centers around the interactions between student achievement and motivation. As achievement rises or falls below grade-level standards, student motivation to learn is increased or decreased. Changes in motivation affect the current learning rate that affects achievement.

In *effective* schools, the teachers and principals maintain their belief that all but clearly exceptional children can learn *at grade-level standards*. In these schools, teachers and the principal perceive a learning gap for low-achieving students, accept institutional responsibility for closing this gap, and work to do so by improving the appropriateness and intensity of instruction for low-achieving students. Effective schools exhibit goal-seeking behavior where grade-level standards are the performance goals for initially low achievers. Bloom (1976, p. 212) characterizes effective schools as "self-correcting" systems where feedback to teachers and students can reveal the errors in learning shortly after they occur and where appropriate corrections are introduced.

The Basic Feedback Structure

To understand the factors affecting achievement, one must focus on the student's learning rate. The student's learning rate for any subject is directly dependent on the amount of time the student is successfully engaged in instructional activities related to that subject. This focus on time and learning draws heavily on the work of the Beginning Teacher Evaluation Study (BTES) in California (see Fisher *et al.*, 1978; Denham & Lieberman, 1980).

Achievement, Learning Rates, and Engaged Time

Central to this concept is the notion of "engaged time" in instructional activities, or what the BTES study refers to as "academic learning time." Engaged time is a function of the amount of time available for instructional activities in a subject and the student's engagement rate in those activities. In this discussion, engaged time is *student* time. It is the amount of time students are engaged in learning activities.

The amount of time available for instruction in a particular subject area is a function of school policies, teacher effectiveness, and student behavior. The student's engagement rate in learning activities depends both on student motivation for learning in that particular subject area and on the appropriateness of the activities planned and presented by the teacher. Activities are appropriate if they are at the right level of reading comprehension, culturally and topically relevant, and properly sequenced with reference to prior learning (cf. Bloom, 1976).

Thus appropriateness of instruction depends heavily on teacher skills. It is also affected by class size because class size mediates the effects of skill. Another way to think about appropriateness of instruction is to think about a teacher's "instructional efficiency" or how efficient a teacher is in converting time available for instruction into student engaged time. In the theory and model developed here, we have used the concept of instructional efficiency.

A teacher's instructional efficiency *for a given achievement* group may vary considerably from his or her general instructional efficiency. These variations are a function of teacher expectations for the group and teacher emphasis on the group. The effect of expectations on appropriateness of instruction (and, therefore, on cohort instructional efficiency) is largely a function of the perceived learning gap for the cohort. This gap is dependent on the level of teacher expectations and current achievement.

There is also a relationship between the appropriateness and intensity of instruction and student motivation. The literature on achievement motivation indicates that more appropriate and more intense instruction has a positive impact on student motivation to learn, whereas less appropriate and less intense instruction has a negative impact (Atkinson et al., 1976; Kolesnik, 1978; Russell, 1971; Watson, 1963).

Teacher expectations also affect student motivation. High teacher expectations for presently low-achieving students reinforce learning and help to raise student motivation. Students pick up the verbal and nonverbal signals from a teacher that say "you can do it." There is no such positive reinforcement in an ineffective school.

Appropriateness and Intensity of Instruction

As the name suggests, appropriateness and intensity of instruction is a two-dimensional construct. Engaged time in reading is increased both by more time spent in reading activities and by more appropriate reading activities. These are the companions—quantity and quality of instruction.

Time Available for Instruction

The quantity of instruction depends on a number of variables. First, it is a function of school policies that determine how much time is to be spent in the classroom on instructional activities and how much time is to be spent in other activities in or outside the classroom. Several studies (Fisher et al., 1978; Harnischfeger & Wiley, 1980; Karweit, 1985) report wide variations in time allocations both between and within schools.

Second, quantity of instruction is a function of how classroom time is apportioned among the different subject areas. Often there are school- or district-level guidelines for the number of minutes per day or per week for instruction in each academic subject.

Third, the amount of time available for reading instruction can be eroded by the time teachers spend in transitional and classroom management activities. This time is dependent on classroom student behavior and on teacher effectiveness. The more highly skilled a teacher is in classroom management and in dealing with behavior problems, the greater the fraction of time allocated for instruction that can actually be spent in instruction.

The Appropriateness of Instruction

The quality or appropriateness of instruction can vary from achievement group to achievement group. Ethnographic studies such as Rist's (1973) show that the quality of instruction provided to students in different achievement groups *within the same classroom* can vary considerably. These variations stem from differences in the teacher's instructional efficiency for the different achievement groups. Although a given teacher has a *general* level of instructional efficiency, his or her efficiency for a cohort depends on the teacher's emphasis on that group. Teachers who place more emphasis on a group are more efficient in their use of time *for that group* and more effective in their instruction. Students who receive greater emphasis have more time on task and higher engaged time than students who receive less emphasis.

It is central to our theory of schooling that the perceived learning gap is a major determinant of teacher emphasis for a particular achievement group. A teacher will devote more emphasis to the group if the teacher perceives a gap in reading achievement for that group (i.e., the teacher perceives that reading achievement is below the teacher's expectations for the group). If, because of low expectations, there is no gap in achievement, the teacher will make no effort to increase the emphasis on the group.

Teacher expectations also have a direct impact on teacher emphasis. There is research to suggest a systematic bias against those students for whom the teacher has below-normal expectations (cf. Rist, 1973). The effect of the bias (to reduce instructional emphasis) increases as the gap between grade-level standards and teacher expectations widens. The teacher who has below-normal expectations for a particular achievement group will place less emphasis on them, and the teacher's instructional efficiency for that achievement group will be less. Therefore, the appropriateness and intensity of instruction and the engaged time for that achievement group will decline.

Motivation and Behavior

Student motivation and behavior are important contributors to achievement patterns. They are also affected by achievement patterns and the instructional process. They are embedded in a set of positive feedback relationships that increase or inhibit achievement (Bloom, 1976). Motivation to learn in a given content area directly affects a student's engagement rate in learning activities. The lower the level of motivation, the lower the student's interest in and involvement with instruction. Motivation also has a direct effect on behavior.

Motivation

Motivation is primarily a function of environmental, rather than hereditary, variables (Watson, 1963). Although scholars agree that home and preschool experiences are important determinants of a child's attitude to school and general achievement motivation, school experiences are also important (Bloom, 1976; Kolesnik, 1978; Russell, 1971; Watson, 1963). Children tend to become success or failure oriented, and these patterns tend to be set early in the schooling process (Russell, 1971).

What school factors directly affect motivation? Clearly, feedback on achievement is important. Success reinforces success; failure reinforces failure. Students measure their own work against standards of performance set for them. Bloom argues (1976) on the basis of macro and microlevel studies that "the student's perception of his competence in school learning" is the major factor influencing motivation to learn. Kolesnik (1978) sees the quantity and quality of instruction as particularly important, and both Russell (1971) and Singer and Donlan (1980) underscore the importance of teacher expectations. In addition, Kolesnik suggests that an unruly and disruptive atmosphere (where levels of student behavior are low) exerts a negative influence on motivation for learning.

Consistent with the literature, then, the theory of schooling described here includes the effects of achievement, teacher expectations, appropriateness and intensity of instruction and behavior on motivation. These factors exert pressure on students to change their level of motivation. The pressures may reinforce or neutralize each other. Changes in motivation do not occur instantaneously but rather emerge over the course of time as the pressures for change continue.

Behavior

The case study literature on effective schools emphasizes the importance of good behavior and an orderly, quiet atmosphere conducive to learning (Benjamin, 1980; Phi Delta Kappa, 1980; Salganik, 1980). Students who are motivated to learn and who are experiencing success in their work are less disruptive and unruly. Achievement motivation is one determinant of student behavior in school. In addition, a student's behavior is influenced by the behavior of peers. Classmates have the greatest influence because most of the student's time is spent in his or her company, but the general level of schoolwide behavior is also important (cf. Duke & Meckel, 1980; Kozol, 1967).

The size of a cohort in the classroom also affects its level of behavior. Teachers have long been aware of the effects of a "critical mass" of disruptive students. When the number of disruptive students is large enough, each student's unruly behavior is reinforced by peers, and the general level of cohort behavior is lower than it would ordinarily be (cf. Duke, 1980; Kozol, 1967; Ryan, 1970). The same phenomenon occurs for exemplary behavior, but the effects are not so noticeable, probably because the difference between good and average behavior is not as great as the differences between bad and average behavior.

This description of the influences of behavior highlights the difficulties in trying to maintain good student behavior when motivation is low. Efforts by teachers and administrators to enforce strict discipline in an environment where the instructional process feeds feelings of failure and low motivation will only be met by resistance on the part of students. Under such conditions, the level of behavior desired by the staff is inconsistent with the level of behavior indicated by motivation (which is a direct effect of the learning environment). As with changes in motivation, changes in behavior do not occur instantaneously but rather emerge in response to the net effect of pressures from motivation and peer groups.

Interactions among Cohort Groups

The depiction of the theory thus far implies that decisions about reading instruction for a particular achievement group are made in isolation without considering the presence of other achievement groups in the classroom. This is not so. In heterogeneous classrooms typical of elementary schools, other groups can have a profound impact on the group's behavior and learning.

Interactions Affecting Behavior

Each achievement group's behavior is affected by its motivation to learn in *each* content area *and* by the general level of classroom and schoolwide behavior. Each cohort's behavior also affects the general classroom behavior. Larger cohorts have a greater effect than smaller ones. The general level of classroom behavior directly affects the amount of time available for instruction to all students in the classroom and, through out-of-class interactions, affects the behavior of students at other grade levels throughout the school.

Competition for Teacher Emphasis

Teacher emphasis is a finite commodity. Students in various achievement groups compete for teacher emphasis. The desired teacher emphasis for a given achievement group and the actual teacher emphasis for the group may differ. Although a teacher may wish to devote a great deal of his/her time to a particular achievement group, realities dictate that time must also be spent with the other groups. The teacher's instructional efficiency for a given achievement group is affected by the presence of the other achievement groups in the classroom.

The Theory

Figure 1 is a comprehensive picture of the relationships between achievement and instruction. It represents our dynamic theory of schooling.

The diagram depicts four feedback loops—one negative loop and

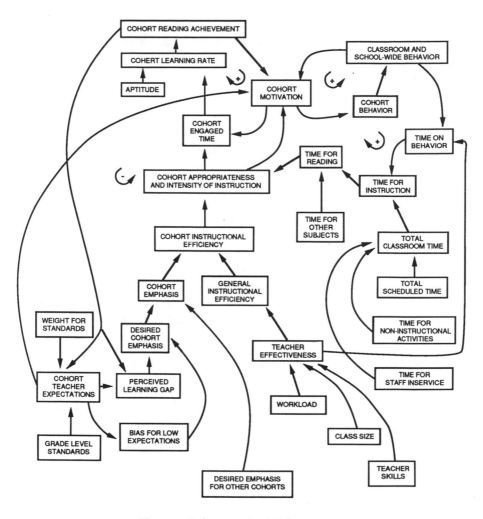

Figure 1. A dynamic theory of schooling.

three positive loops. The three positive loops highlight the self-reinforcing interactions among learning, motivation, behavior, and achievement. They operate to reinforce success in the effective school or failure in the ineffective school for initially low-achieving students.

The negative goal-seeking loop represents the mechanism by which teachers in an effective school strive to correct the learning gap for low-achieving children. A learning gap exists because of high teacher expectations for students. In the ineffective school, teacher expectations match current achievement, and there is no perceived learning

gap, even for underachieving children. As a consequence, the self-correcting feedback mechanism is disconnected, and failure is reinforced.

Several words of explanation are necessary about the figure. First,
in the lower left-hand part of the figure, the concept teacher "weight for
standards" is used to indicate where a faculty falls on the range of
choices for formulating expectations. A high weight for standards
means that teachers have expectations firmly rooted in grade-level standards. A low weight means teacher expectations vary with perceived
student achievement. This "weight for standards" concept is important,
even though it is not a familiar one, because of its role in setting teacher
expectations.

There are three ways one can form expectations:

1. Expectations can be based on a fixed set of standards as in an
   effective school.
2. Teachers can have no standards on which to base their expectations, as in an ineffective school, and base their expectations
   solely on a student's current level of achievement.
3. Teachers can be somewhere in between these two extremes,
   where they base their expectations partly on a set of fixed standards and partly on current achievement.

Second, the absence of school administrators from this causal theory of instruction is purposeful. School administrators are not directly
involved in the instructional process. In a McBer & Company study of
superior principals in Cleveland, Burruss (1978) found that the role of
the principal is that of *facilitator* and *maintainer* of the system. Principals work to reduce conflict by building norms and establishing consistent policies. As *leaders* of the educational process, superior principals maintain a strong institution in which education can take place.
They respect and support their teachers.

The thrust of our theory is that a crucial part of the administrator's
role in maintaining schools that are effective for all students is *to provide structure*. For effective schooling to take place, administrators
must work to create structures that *motivate* and *enable* teachers to
provide instruction that is appropriate for all students and that is so
skillfully managed that students achieve high proportions of engaged
learning time. In an important sense, then, the building principal does
not appear as a discrete element of the structure of effective schooling;
rather, *the structure, itself, is a manifest reflection of the skillful
administrator*.

Thus the extent to which teachers key their expectations for *all*
students to grade-level standards and are consciously aware of existing
learning gaps for particular students are critical outcomes of admin-

istrative performance. So, too, are teacher skills. Even when principals are able to help their faculties to monitor and accept professional responsibility for the learning problems of traditionally low achievers, how able are teachers to do anything about it. Have they developed a broad enough range of teaching styles to adapt to diverse learning needs of students? Do they possess the requisite skills to carry them out effectively? How accurately and quickly do they diagnose student learning needs and styles and match ways and contents of instruction to them? These questions focus us on the supervision and training functions of leadership.

In schools where high-level teaching skills are prevalent and where teachers are motivated to use them, even disadvantaged students learn. These are schools that manifest the effective schooling structure. It is this *structure* that stands as the essential embodiment of effective leadership in the principal's office: goal setting, planning, resource management, supervision, staff training and development, communications, and evaluation.

Intervention Analysis[2]

The goal of this research has been to try to understand the likely consequences and trade-offs associated with various efforts on the part of administrators to change a school that is ineffective for initially low achievers into a school that is effective for these students. Consistent with what appears to be the common reality, there is nothing in the model that will allow the ineffective school to self-generate the conditions required for the transition to greater effectiveness. In the model, at least for any reasonable time frame, the ineffective school maintains a dynamic equilibrium that is characterized by a persistent and widening gap from grade level to grade level in achievement between initially low achievers and other students in the school. Thus the ineffective school will not by itself begin the transformation process unless there is a *catalyst* for change.

That catalyst may be a strong, dynamic principal, a cadre of highly motivated teachers, or strong parental or school district pressure. However, the literature on effective schooling is very clear in stating that the principal plays a central role in the transition process (Benjamin, 1980; Brookover *et al.*, 1979; Edmonds, 1979; Phi Delta Kappa, 1980; Salganik, 1980).

---

[2]Findings have been highly summarized. Because of space limitations, data tables have been omitted.

Thus in the presentation that follows, interventions are ascribed as if initiated by someone in a leadership capacity. We believe that it would be a rare school in which such leadership could be manifested without the active collaboration of the school principal. It seems safe to say, at least in most schools, that without the active involvement of the principal in the intervention process, there is typically no concerted effort to bring about change, and no change occurs.

There are four classes of interventions that school administrators can make to try to improve reading achievement for initially low achievers. The first class has to do with *changing the size or demographics of the student population.* One tries to bring about an effective school through changing student inputs. Efforts to balance schools exemplify this strategy. The second class of interventions focuses on *improving the intensity or quantity of instruction for the initially low achievers.* This could be accomplished either by policies aimed specifically at increasing the intensity of reading instruction for low achievers or through grade-level or schoolwide policies aimed generally at improving the total time available for instruction. The third class of interventions aims to *increase the appropriateness of the instruction for initially low achievers.* One attempts to change the way teachers make decisions about instruction and to improve their general level of teaching skills. The fourth class of interventions focuses on the *school climate* and, in particular, on *student behavior.*

Some of these changes are focused specifically on the low achievers and have no direct impact on the other achievement groups. Other policies require diverting resources away from the average- and high-achieving groups to low achievers. Most of the policies, though, are ones that are schoolwide or grade-level policies affecting all three achievement groups.

The conclusions that follow are based on tests of the interventions described previously. The tests were performed on the School Effectiveness Model, a computer simulation model in which the theory described in this chapter is represented in more than 1,500 mathematical equations (Clauset, 1982).[3] The model simulates Grades 1 through 6 in a typical urban elementary school. Successive cohorts of low, average, and high achievers are identified entering the first grade and tracked through 6 years of schooling. Interventions were tested by making ap-

---

[3]The model was written in a system dynamics programming language called Dynamo II. The simulation runs were implemented on the IBM mainframe at the Boston University Academic Computing Center. Since this model was developed, several companies have developed software which enables modelers to build sophisticated models on personal computers.

propriate modifications in constants representing initial conditions and other model parameters.

With the exception of the demographic interventions, which were implemented as changes in initial conditions, all parameter changes were programmed to take effect in each test run *after the model had reached the ineffective school equilibrium*. Each policy was tried for a 5-year period, a 10-year period, and an indefinite period of time to assess its impact under various time constraints. Policies were also tested for different levels of intensity and for different target audiences (e.g., grades 1–6 vs. grades 1–3 vs. grades 4–6).

No attempt was made in the reported analyses to model the dynamics of the organizational response to policy interventions. Rather, it was implicitly assumed for the time being that conditions necessary to implement the test interventions were *unproblematic*. However, the focus of more current research is precisely on the complex of factors that typically interact in school districts to affect the success of efforts to implement and institutionalize school improvement programs (Gaynor & Clauset, 1985; Huberman & Miles, 1982; Huberman, 1983; Miles, 1983). It is anticipated that in the not too distant future it will be possible to formulate and test a composite theory of school improvement. Such a theory would incorporate the dynamics of both school effectiveness and the implementation and institutionalization of intervention strategies.

Changes in the Student Body

The results of interventions tested that examined alterations in the student body suggest that they have no effect on the *patterns* of appropriateness and intensity of instruction delivered to the different cohorts. Changes examined included alterations in school size, percentage of low achievers, and entry-level achievement characteristics of the low achievers. Across all of these interventions, the school remained an ineffective school. Achievement scores varied slightly from test to test, but initially low achievers still reached the end of the sixth grade with a larger achievement gap than when they entered school.

In fact, the test results illustrated how standard procedures for reporting average achievement scores may make a school that is *structurally ineffective* appear effective. Schools with a low fraction of low achievers were simulated. Model data showed that even where initially low-achieving students did poorly, *mean* achievement scores were above average. The research of Frederick (1984) and Silverman (1984) on suburban elementary schools illustrates precisely this type of masking by reports of average achievement scores.

Thus schools with large numbers of initially low achievers may simply *appear* less effective because (1) low motivation and negative behavior patterns have a deleterious impact on the entire school population and (2) purely mathematically, a high proportion of initially low achievers brings down the *average* score. The point is that structurally ineffective schools are bad for initially low-achieving children even when there are too few of them to damage the school's reputation.

Changes in the Intensity of Instruction

The class of interventions aimed at improving the intensity or *quantity* of instruction for initially low achievers produced mixed results. The only policy that resulted in clear and significant gains in achievement patterns was that aimed at reducing time spent on noninstructional activities. Policies aimed at varying class sizes among grade levels with a constant staff were not helpful. Policies aimed at increasing the staff and reducing class size had value only where the assumption was incorporated that it was possible to increase the proportion of highly skilled teachers for a long period of time.

Changes focused on adjusting the amount of time devoted to how reading instruction enhanced reading at the expense of instruction in other content areas. There did, however, appear to be a critical point at which small increases in the fraction of time for reading boosted both reading achievement *and* achievement in other subject areas. In summary, the two policies that seemed to be the most effective were also those that are probably the least costly to implement: decreasing the fraction of time for noninstructional activities and increasing the fraction of time devoted to reading instruction.

Changes in the Appropriateness of Instruction

Interventions aimed at improving the appropriateness of instruction simulated either general efforts to raise teacher expectations and skills or more specific efforts directed at the emphasis teachers place on low achievers. Focusing teacher expectations for low achievers more on grade-level standards and increasing the emphasis of teachers on low achievers both lead to significant gains for low achievers. However, these prove to be at the expense of other achievement groups. Increased teacher emphasis on low achievers lead to increased appropriateness and intensity of instruction for them but decreased emphasis on and instruction for average and above-average students. To what extent, in most schools, political and social realities can be expected to allow this diversion of resources is problematic (Weaver, 1982).

Staff development activities aimed at improving teacher skills benefitted all students. Simulation runs with the School Effectiveness Model suggest that initially low achievers can make dramatic gains if teachers skills rise to a sufficiently high level. In model runs, the improved instructional efficiency and reduced amount of time spent dealing with behavior problems lead to gains in achievement that, in turn, gave teachers more faith in the ability of initially low achievers to achieve grade-level standards. However, test results suggest that if *instructional time* is reduced to provide more staff development time, benefits to students will be significantly eroded. Furthermore, unless school administrators establish ongoing in-service programs to ensure that the level of teaching skills remains above average, normal staff turnover works to erode average faculty skills and recreate the dynamics of ineffectiveness.

Changes in School Climate

The results of interventions to improve student behavior suggest that efforts to improve schoolwide behavior in the halls, lunchroom, playground, and assemblies have only small effects on student achievement. Furthermore, our findings emphasize the negative effects on learning of time to improve student behavior that is taken away from instruction itself. Thus, we would argue that efforts to improve behavior directly should be considered only as a *supplement* to the main effort of improving instruction. However, common sense would dictate that if student behavior is too disruptive and chaotic, no learning can take place until some semblance of order prevails.

Summary and Implications

In this chapter, a theory was described that illustrates the application of system dynamics in the field of educational policy analysis. The theory was presented and explained in terms of the relevant educational research literature. It is founded on the hypothesis that effective schools distinguish themselves from ineffective schools in terms of important feedback dynamics. These dynamics describe how the appropriateness and intensity of instruction in schools works either, in the case of

effective schools, to bring initially low-achieving students toward
grade-level standards or, in the case of ineffective schools, to reinforce
poor academic performance. Following its substantive presentation,
the model was summarized visually in the form of a causal-loop diagram. The results of policy tests on the mathematical simulation model
were then described, and some recommendations were discussed.

This work has important implications for educators and psychologists. First, it examines problems of school effectiveness from the perspective of an underlying causal structure of mutually interdependent
variables. This is in sharp contrast to the typical approach of school
improvement programs that treat the 5, or 7, or 10 indicators of school
effectiveness (Edmonds, 1979) as separate, independent variables. Second, this approach to policy analysis allows both practitioners and
researchers to evaluate the likely consequences of different policies
before investing considerable time, money, and effort in actual interventions. Third, it is an example of modeling with "soft" variables
where the focus has been on understanding the relative impact of variables and the development of a unified theory without concentrating
on how to operationalize and measure individual variables.

Finally, we think of this work as an example of the use of computer
simulation modeling as a tool for further research—primarily qualitative research. The process of writing equations to translate our theory
into a computer model repeatedly forced us back to the literature and to
practitioners for more information on the relationships among variables. In a number of instances, the knowledge base was limited and
raised more questions than it answered. The model became our instruc-
tor and guide.

References

Andersen, D. F. (1977). *Mathematical models and decision making in bureaucracies: A
case study from three points of view.* Unpublished doctoral dissertation, Massachusetts Institute of Technology.
Andersen, D. F., Nguyen, T., & Chen, F. (1980). The dynamics of state aid to education:
Interactions between special education, regular education, and non-schooling education. *Proceedings of the International Conference on Cybernetics and Society.* New
York: Institute of Electrical and Electronics Engineers, 1980.
Atkinson, J. W., Lens, W., & O'Malley, P. M. (1976). Motivation and ability: Interactive
psychological determinants of intellectual performance, educational achievement,
and each other. In W. H. Sewell, R. M. Hauser, & D. L. Featherman (Eds.), *Schooling
and achievement in American society.* New York: Academic Press.
Averch, H. A., Carroll, S. J., Donaldson, T. S., Kiesling, H. J., & Pincus, J. (1974). *How
effective is schooling? A critical review of research.* Rand Educational Policy Study.
Englewood Cliffs, NJ: Educational Technology Publications.
Baldridge, J., & Deal, T. E. (Eds.). (1975). *Managing change in educational organizations:*

Sociological perspective, strategies and case studies. Berkeley, CA: McCutchan Publishing Corp.

Barr, R., & Dreeben, R. (1977). Instruction in classrooms. In L. Shulman (Ed.), *Review of research in education* (Vol. 5). Itasca, IL: F. E. Peacock.

Benjamin, R. (1980). Towards effective urban schools: A national study. In D. Brundage (Ed.), *The journalism research fellows report: What makes an effective school?* Washington, DC: Institute for Educational Leadership.

Berliner, D. C. (1979). Tempus educare. In P. L. Peterson & H. J. Walberg (Eds.), *Research on teaching: Concepts, findings and implications* (pp. 120–135). Berkeley: McCutchan Publishing Corp.

Blakeslee, G. E. (1984). *Science and mathematics teacher supply and demand in the competitive professional labor market: A system dynamics policy study.* Unpublished Ed.D. dissertation, Boston University.

Bloom, B. S. (1976). *Human characteristics and school learning.* New York: McGraw-Hill.

Bridge, G. R., Judd, C. M., & Moock, P. R. (1979). *The determinants of educational outcomes: The impact of families, peers, teachers, and schools.* Cambridge, MA: Ballinger Publishing Company.

Brookover, W. B., Beady, C., Flood, P., Schweitzer, J., & Wisenbaker, J. (1979). *School social systems and student achievement—Schools can make a difference.* New York: Praeger Publishers.

Brophy, J. E., & Good, T. L. (1970). Teacher's communication of differential expectations for children's classroom performance. *Journal of Educational Psychology, 61,* 365–374.

Brown, B. W., & Saks, D. H. (1980). Production technologies and resource allocations within classrooms and schools: Theory and measurement. In R. Dreeben & J. A. Thomas (Eds.), *The analysis of educational productivity Volume I: Issues in microanalysis.* Cambridge, MA: Ballinger Publishing Company.

Brundage, D. (Ed.). (1980). *The journalism research fellows report: What makes an effective school?* Washington, DC: Institute for Educational Leadership.

Burruss, J. A. (1978). *Characteristics of superior performing principals in the Cleveland Public Schools* (draft report). Boston: McBer & Co.

Centra, J. A., & Potter, D. A. (1980). School and teacher effects: An interrelational model. *Review of Educational Research, 50,* 273–291.

Chin, R., & Benne, K. D. (1976). General strategies for effecting change in human systems. In W. G. Bennis et al. (Eds.), *The planning of change* (3rd ed.; pp. 22–45). New York: Holt, Rinehart & Winston.

Clauset, K. H., Jr. (1982). *Effective schooling: A system dynamics policy study.* Unpublished doctoral dissertation, Boston University.

Clauset, K. H., Jr. & Gaynor, A. K. (1980, April). *The dynamics of effective and ineffective schooling: A preliminary report of a system dynamics policy study.* Paper presented at the Annual Meeting of the American Educational Research Association, Los Angeles.

Clauset, K. H., Jr. & Gaynor, A. K. (1982, March). *Improving schools for low achieving children: A systems dynamics policy study.* Paper presented at the Annual Meeting of the American Educational Research Association, New York.

Cohen, M. (1979, March). *Recent advances in our understanding of school effects research.* Paper presented at the Annual Meeting of the American Association of Colleges for Teacher Education, Chicago. (ERIC Document Reproduction Service No. ED 169 053.)

Coleman, J. S., et al. (1966). *Equality of educational opportunity.* Washington, DC: U.S. Government Printing Office.

Dabney, N., & Davis, A. (1982). *An ethnographic description of a successful inner city school and its community.* Washington, DC: National Institute of Education. (ERIC Document Reproduction Service No. 220 546.)

Davies, D. (1980). An afterword: Co-production as a model for home-school cooperation. In R. L. Sinclair (Ed.), *A two-way street: Home-school cooperation in curriculum decisionmaking.* Boston: Institute for Responsive Education.

Day, R. H. (1981, October). *Complex behavior in system dynamics models.* Paper presented at the 1981 System Dynamics Conference, Rensselaerville, NY.

Dearman, N. B., & Plisko, V. W. (1981). *The condition of education: Statistical report,* National Center for Educational Statistics. Washington, DC: U.S. Government Printing Office.

Denham, C., & Lieberman, A. (1980). *Time to learn.* Washington, DC: National Institute of Education.

Dreeben, R., & Thomas, J. A. (1980). *The analysis of educational productivity, Volume I, Issues in microanalysis.* Cambridge, MA: Ballinger Publishing Company.

Duke, D. L., & Meckel, A. M. (1980). The slow death of a public high school. *Phi Delta Kappan, 61,* 674–677.

Edmonds, R. (1979). Effective schools for the urban poor. *Educational Leadership, 37,* 15–18, 20–24.

Edmonds, R. (1982). Programs of school improvement: An overview. *Educational Leadership, 3,* 4–11.

Educational Research Service. (1980). Class size research: A critique of recent meta-analyses. *Phi Delta Kappan, 62,* 239–242.

Fisher, C. W., et al. (1978). *Teaching behaviors, academic learning time, and student achievement.* Final report of Phase III-B, beginning teacher evaluation study. Technical report V-1 (Summary). San Francisco: Far West Laboratory (ERIC Document Reproduction Service No. ED 183 525.)

Forrester, J. W. (1968). *Principles of systems.* Cambridge, MA: MIT Press.

Forrester, J. W. (1971). Counterintuitive behavior of social systems. *Technology Review, 73,* 52–68.

Forrester, J. W., & Senge, P. M. (1979). *Tests for building confidence in system dynamics models.* System Dynamics Group Working Paper D-2926-5, Alfred P. Sloan School of Management, Massachusetts Institute of Technology, Cambridge, MA.

Fowler, W. J., Jr. (1980, April). *Effects of school characteristics upon achievement test scores in New York State.* Paper prepared for presentation at the Annual Meeting of the American Educational Research Association, Boston.

Frederick, J. M. (1984). *A comparison of the major algorithms for measuring school effectiveness.* Unpublished doctoral dissertation, Boston University.

Garet, M. S. (1979). *The implementation of social policy: An assessment of organizational capability.* Unpublished doctoral dissertation, Massachusetts Institute of Technology.

Gaynor, A. K. (1977). The study of change in educational organizations. In L. L. Cunningham, W. G. Hack, & R. O. Nystrand (Eds.), *Educational administration: The developing decades.* Berkeley, CA: McCutchan Publishing Corp.

Gaynor, A. K. (1977, April). *Toward a dynamic theory of innovation in public schools.* Paper presented at the Annual Meeting of the American Educational Research Association.

Gaynor, A. K. (1980, April). *A dynamic model of mathematics curriculum change in an urban elementary school.* Paper presented at the Annual Meeting of the American Educational Research Association, Boston.

Gaynor, A. K. (1981, October). *The dynamics of stability and change in public schools.* Paper presented at the 1981 System Dynamics Research Conference, Rensselaerville NY.

Gaynor, A. K., & Clauset, K. H., Jr. (1981, April). *Theory of practice: A systems perspective*. Paper presented at the Annual Meeting of the American Educational Research Association, Los Angeles.

Gaynor, A. K., & Clauset, K. H., Jr. (1983, April). *Organizations and their environments: A system dynamics perspective*. Paper presented at the Annual Meeting of the American Educational Research Association, Montreal.

Gaynor, A. K., & Clauset, K. H., Jr. (1985, April). *Implementing and institutionalizing school improvement programs: A theoretical reformulation of the work of Huberman and Miles*. A paper presented at the Annual Meeting of the American Educational Research Association, Chicago.

Harnischfeger, A., & Wiley, D. E. (1977). Teaching–learning process in elementary schools: A synoptic view. In D. Erickson (Ed.), *Educational organization and administration*. Berkeley: McCutchan Publishing Corp.

Harnischfeger, A., & Wiley, D. E. (1980). Determinants of pupil opportunity. In R. Dreeben & J. A. Thomas (Eds.), *The analysis of educational productivity, Volume I, Issues in microanalysis*. Cambridge, MA: Ballinger Publishing Company.

Herriott, R. E., & Gross, N. C. (1979). *The dynamics of planned educational change: Case studies and analyses*. Berkeley, CA: McCutchan Publishing Corp.

Huberman, A. M. (1983). School improvement strategies that work: Some scenarios. *Educational Leadership*, November.

Huberman, A. M., & Miles, M. B. (1982). Innovation up close: A field study in twelve school settings. In *People, policies, and practices: Examining the chain of school improvement, Vol. IV*. Andover: The Network.

Karweit, N. (1985). Should we lengthen the school term? *Educational Researcher, 6*, 9–15.

Katzman, M. T. (1971). *The political economy of urban schools*. Cambridge, MA: Harvard University Press.

Kifer, E. (1973). *The effects of school achievement on the affective traits of the learner*. Unpublished doctoral dissertation. University of Chicago, 1973. Cited and discussed in B. S. Bloom. (1976). *Human characteristics and school learning*. New York: McGraw-Hill Book Company.

Klein, D. (1976). Some notes on the dynamics of resistance to change: The defender role. In W. G. Bennis, K. D. Benne, R. Chin, & K. G. Corey (Eds.), *The planning of change*. New York: Holt, Rinehart and Winston.

Kolesnik, W. B. (1978). *Motivation: Understanding and influencing human behavior*. Boston: Allyn & Bacon.

Kozol, J. (1967). *Death at an early age*. Boston: Houghton Mifflin Company.

Larkin, M. (1980, October). *The Milwaukee Teacher Expectation Project*. Abstract of a speech presented at the Conference on Urban Education, sponsored by the Council for Basic Education, Washington DC.

Leinhardt, G. (1980). Modeling and measuring educational treatment in evaluation. *Review of Educational Research, 50*, 393–420.

Mable, T. J. (1978). *Behavioral contracting with school discipline problems*. Unpublished doctoral dissertation, Boston University.

Mass, N. J., & Senge, P. M. (1978). Alternative tests for the selection of model variables. *IEEE Systems, Man and Cybernetics, 8*, 450–459.

Medley, D. M. (1979). The effectiveness of teachers. In P. L. Peterson & H. J. Walberg (Eds.), *Research on teaching: Concepts, findings and implications* (pp. 11–27). Berkeley: McCutchan Publishing Corp.

Miles, M. A., Farrar, E., & Neufield, B. (1983). *The extent of adoption of effective schools programs*. Washington, DC: Department of Education. (ERIC Document Reproduction Service No. ED 228 242.)

Miles, M. B. (1983). Unraveling the mystery of institutionalization. *Educational Leadership*, November.

National Education Association. (1977). *Class size*. Reference and Resource Series. Washington, DC: Author.

National Education Association. (1980). *Teacher supply and demand in public school, 1979*. Washington, DC: Author.

Peterson, P. J., & Walberg, H. J. (Eds.). (1979). *Research on teaching: Concepts, findings, and implications*. Berkeley, CA: McCutchan Publishing Corp.

Phi Delta Kappa. (1980). *Why do some urban schools succeed? The Phi Delta Kappa study of exceptional urban elementary schools*. Bloomington, IN: Author.

Pugh, A. L., III. (1977). *DYNAMO user's manual* (5th ed.). Cambridge, MA: The M.I.T. Press.

Purkey, S. C., & Smith, M. S. (1983). *Effective schools—A review*. Washington, DC: Department of Education. (ERIC Document Reproduction Service No. ED 221 534.)

Richardson, G. P., & Pugh, A. L., III. (1981). *Introduction to system dynamics modeling with DYNAMO*. Cambridge, MA: The M.I.T. Press.

Rist, R. C. (1973). *The urban school: A factory for failure*. Cambridge, MA: The M.I.T. Press.

Roberts, N. H. (1974). A computer system simulation of student performance in the elementary classroom. *Simulation & Games, 5*, 265–290.

Roberts, N. H. (1975). Parental influence in the elementary classroom: A computer simulation. *Educational Technology, 15*, 37–42.

Rosenshine, B. V. (1979). Content, time, and direct instruction. In P. L. Peterson & H. J. Walberg (Eds.), *Research on teaching: Concepts, findings and implications* (pp. 28–56). Berkeley: McCutchan Publishing Corp.

Rosenthal, R., & Jacobson, L. (1968). *Pygmalion in the classroom: Teacher expectations and pupils' intellectual development*. New York: Holt, Rinehart & Winston.

Rossell, C. (1977). *District council liaison committee monitoring report*. Boston, MA: Citywide Coordinating Council, Boston Public Schools.

Rowan, B., Bossert, S. T., & Dwyer, D. C. (1983). Research on effective schools: A cautionary note. *Educational Researcher, 4*, 24–31.

Russell, I. L. (1971). *Motivation*. Issues and innovations in education series. Dubuque, IA: Wm. C. Brown Company Publishers.

Rutter, M., et al. (1979). *Fifteen thousand hours: Secondary schools and their effects on children*. Cambridge, MA: Harvard University Press.

Ryan, K. (Ed.). (1970). *Don't smile until Christmas: Accounts of the first year of teaching*. Chicago: University of Chicago Press.

Salganik, M. W. (1980). Academic achievement in urban schools: What works in Baltimore. In D. Brundage (Ed.), *The journalism research fellows report: What makes an effective school?* Washington, DC: The Institute for Educational Leadership.

Sergiovanni, T. J., & Starratt, R. J. (1971). *Emerging patterns of supervision: Human perspectives*. New York: McGraw-Hill Book Company.

Silberman, C. E. (1970). *Crisis in the classroom: The remaking of American education*. New York: Random House.

Silverman, R. (1984). *Measures of school effectiveness: A comparison of the algorithms*. Unpublished doctoral dissertation, Boston University.

Singer, H., & Donlan, D. (1980). *Reading and learning from text*. Boston, MA: Little, Brown and Company.

Summers, A. A., & Wolfe, B. L. (1975). Which school resources help learning? Efficiency and equity in Philadelphia public schools. *Business Review*. Philadelphia, PA: Federal Reserve Bank of Philadelphia, February, 4–29.

Thomas, J. A. (1977). *Resource allocation in classrooms. Final Report.* Washington, DC: National Institute of Education. (ERIC Document Reproduction Service No. ED 152 729.)

Watson, G. (1963). Some differences between high-achievers and low-achievers. In G. Watson (Ed.), *No room at the bottom: Automation and the reluctant learner.* Washington, DC: National Education Association.

Weaver, W. T. (1982). *Contest for educational resources: A dynamic theory of equity.* Lexington, MA: Lexington Books/D.C. Heath.

Weaver, W. T. (1983). *America's teacher quality problem: Alternatives for reform.* New York: Praeger.

Weber, G. (1971). *Inner-city children can be taught to read: Four successful schools.* Washington, DC: Council for Basic Education.

Weiser, R. R. (1976). *The innovative process in a dynamic organization: An historical case study of Meadowbrook Junior High.* Unpublished doctoral dissertation, Boston University.

Wiley, D. E. (1976). Another hour, another day: Quantity of schooling, a potent path for policy. In W. H. Sewell, R. M. Hauser, & D. L. Featherman (Eds.), *Schooling and achievement in American society.* New York: Academic Press.

Wolcott, H. F. (1977). *Teachers versus technocrats: An educational innovation in anthropological perspective.* Eugene, OR: Center for Educational Policy and Management, University of Oregon.

Wynne, E. E. (1981). Looking at good schools. *Phi Delta Kappan, 62,* 377–381.

13

A Cognitive Control System

William T. Powers

Organisms create regular results by variable means. This phenomenon has been explained in the past as an artifact of observation (we class similar outcomes together), as a statistical effect (any set of variable acts has some average effect), or simply as an inconvenient property of organisms (behavior is inherently variable). Control theory explains it in a different way: The variations of means are simply what is required to keep some perception under control. When we understand what is being controlled by an organism's actions, the apparent variability of behavior all but vanishes. This chapter uses an experiment with control of a cognitive relationship to illustrate this property of behavior.

Control as a Phenomenon

As Marken (1988) has pointed out, we must distinguish between the phenomenon of control and the theory of control. The phenomenon can be studied without proposing any model of the internal system that brings it about. Other usages of the term aside, we can say that control exists when an organism acts on its immediate surroundings (or its physical relationship to them) in such a way as to bring about or maintain a particular state of some variable that an observer can measure, the organism's action protecting that variable against unpredictable in-

William T. Powers • 73 Ridge Place, Durango, Colorado 81301.

Analysis of Dynamic Psychological Systems, Volume 2: Methods and Applications, edited by Ralph E. Levine and Hiram E. Fitzgerald. Plenum Press, New York, 1992.

dependent disturbances. Another way to describe the same phenomenon is to say that the organism forces some variable to behave in a repeatable way that is different from the behavior that would be seen without the organism's actions. Variables acted upon by organisms in this way are called controlled variables.

The stabilization of variables (in fixed states or dynamic patterns) against disturbances requires that there be a specific relationship between the organism's effect on the variable and effects due to other simultaneously acting influences. The path of a car, for example, is maintained constant for long periods by a driver. For this constancy of path to happen, it is necessary on physical grounds alone that the forces created by the driver and transmitted through the steering linkage to the front wheels affect the car at all times in a direction opposite to the effects of crosswinds, soft tires, tilts in the roadbed, and any other forces that act on the car independently of the driver. The effect of the driver's actions must also be quantitatively related to the effect of the net disturbance: It must be equal as well as opposite to that effect, to a high degree of precision.

Because behaviors are normally carried out in a variable environment, most behaviors can be seen to involve control. It is easy, however, to overlook or take for granted the principal effect of this peculiar relationship of action to disturbance: the constancy of the controlled variable itself. There is a strong tendency to focus on the causal relationship or correlation between disturbance and action and overlook the resulting stabilization of some particular variable, as if the state of the variable itself were a side effect—beneficial to the organism, luckily. Behaviorists have tried to explain such beneficial outcomes by giving them special "reinforcing" traits.

By focusing on the controlled variable, it is possible to show (experimentally) that stabilization of such variables depends on the organism's ability to sense their states—to perceive them. The actions of organisms can affect physical variables, of course, without any sensory feedback about their states. Organisms can continue to produce organized movements when feedback is interrupted, but they cannot control the variables affected by these movements; that is, maintain the variables constant in the presence of unpredictable disturbances. When sensing of a controlled variable is prevented, control is immediately lost even if controllike actions continue to be produced: The controlled variable departs from its stable state under every disturbance, particularly so if the cause of the disturbance cannot be sensed. Clearly, perception of a variable is essential for accurate control of that variable and usually for any degree of control at all. Steering a car, threading a needle, and adjusting the contrast of a television set are impossible to do with the eyes closed. The deaf find speaking difficult and singing impossible. Fingers numb with cold can't pick up a pin because they can't feel it. A person with no sense of taste or smell would be a terrible cook.

Once the role of perception in control is seen, it is a short step to the realization that only the perception, not the objective situation it represents, is actually controlled. Distortions of perception do not prevent control: They simply redefine what is being controlled. A driver told to go 40 miles per hour without looking at the speedometer will maintain an apparent velocity of the car that is judged as 40 miles per hour, maintaining it constant uphill and down, even though the objective speed is really 30 or 50 miles per hour. This is the general basis for the statement that control involves actions that control inputs, not outputs or their objective conscquences.

Control and Conventional Concepts

Hershberger (1987, 1988) has summed up the relationship of control phenomena to conventional pictures of behavior in an elegant way. Motor actions, he points out, are "emitted behaviors" or "responses" in conventional terms. These emitted behaviors affect controlled variables. Disturbances act on these controlled variables as well: They result in changes of action that oppose the effect of the disturbance. If the resulting stabilization is overlooked, it then seems that the disturbance "elicits" the response, as indeed it does. The disturbance fits the definition of a stimulus. It has this relationship to the response, however, only in terms of the stabilized state of the controlled variable: If that variable were not being controlled, the disturbance would elicit no systematic behavior and would have no systematic stimulus value.

To this analysis we can add that "reinforcers" are simply controlled variables. We see such variables as having a reinforcing effect only because behavior regularly tends to restore those variables to particular states, when possible. Reinforcement theory is an attempt to explain this repeated outcome by attributing to the reinforcer a property that alters behavior in just the way that will generate more of the same reinforcement. This psychological property of matter has no physical existence; it is really an attribution of a property of the organism to the environment of the organism. Control theory explains the same outcome without giving any nonphysical property to the supposed reinforcer. Instead, it proposes that the organism compares the perceived state of the reinforcer with an internal reference criterion and acts to maintain the difference to be as small as possible. Artificial systems designed to imitate the ability of human beings to control external variables have shown us how such purposive systems must be internally organized to behave in this way. No nonphysical properties are required of the organism or its environment. What have been called reinforcers are just external variables under control by the organism. Figure 1 summarizes very briefly a control-system approach to analyzing operant conditioning.

The "internal criterion" determines the particular state in which

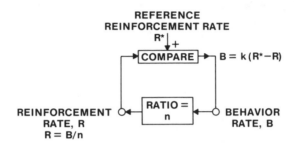

Figure 1. A minimal control-system model of operant conditioning. The rate of reinforcement, R, in number of reinforcements per unit time, is an input to the behaving organism above the dashed line. The organism responds, after learning is complete, with an output of B lever presses per unit time.

The variables R and B are related in two ways at the same time. R depends on B according to the schedule of reinforcements: On a fixed-ratio 10 (FR 10) schedule, one reward is produced for every 10 presses, regardless of the pattern of pressing. Simultaneously, B depends on R in a way that characterizes the organism. With two independent relationships and two unknowns it is possible to solve for both variables in terms of assumed constants.

A simple slope-intercept form for the organism's response is $B = k(R - R^*)$, where R^* is the value of reinforcement rate at which the behavior rate would just become zero. The apparatus equation is $R = B/n$ where n is the fixed ratio. Figure 1 shows the relationships involved in diagrammatic form. This is a standard diagram used in control theory, although here it does not provide for an independent disturbance as a complete analysis would require.

Note that under the usual rule of thumb that assumes food reinforcement to have a positive effect on behavior (k positive), we have a positive-feedback situation. If R^* is assumed zero, there is no solution to the equations other than $B = R = 0$.

However, under conditions where bar pressing produces life-sustaining amounts of food reward, the feedback is negative, for behavior then decreases with increasing reward. In that case, k is negative, and R^* represents the rate at which the organism would consume food if feeding freely. Staddon (1983, pp. 201, 212–214) has collected data from the literature showing that the negative feedback relationship applies under physiologically normal conditions. The organism normally controls the rate of reinforcement, maintaining it as near as possible to a specific reference rate (a full analysis will be given in Powers and Marken, in preparation). The commonly accepted rule of thumb applies only under conditions of extreme deprivation. If k is very large and negative, the steady-state conditions approach $R = R^*$ and $B = nR^*$.

any controlled variable is maintained. In principle, the driver's actions can maintain any path of the car, the rat's bar pressing can maintain any rate of reinforcement. Nothing in the environment prevents or encourages any particular target value of the variable about which control takes place. It is the driver who determines what path is to be maintained, the rat that determines what amount of food delivery is the right amount. Control theory offers simple explanations for the particular states being maintained, identifying them with physical signals inside the organism that specify the level of perception at which behavior ceases to alter the state of the perception. But it is not necessary to explain how this specification occurs (if at least one physical possibility is known, as it is). We need only recognize that there is some reference condition established inside the behaving organism. The state of the controlled variable affects behavior only with reference to this internal specification, generically termed the *reference signal* whether the specified condition be simple or complex.

Thus we can understand Hershberger's picture of control behavior in relation to the traditional concepts of behavior. The controlled variable is stabilized in a particular state specified inside the organism. A disturbance alters that variable, making it depart slightly from the reference condition. This departure immediately gives rise to an emitted behavior that keeps the controlled variable, as perceived, from differing more than a small amount from this reference condition. The result is to create the appearance that the disturbance elicits the error-correcting emitted behavior.

The internal reference signal, whether fixed or variable, therefore specifies the target state of the controlled variable, the state in which behavior normally maintains the variable, whereas external disturbances determine the amount and direction of action produced in the process of maintaining the controlled variable near its reference state. A closed-loop control system automatically produces the required action, through mechanisms that are well understood. The conventional picture is accurate as far as it goes: The control-system model completes the picture.

The process of control accounts for the observed relationships between disturbances or stimuli and the emitted actions they elicit and for the apparent reinforcing value of certain joint outcomes of action and disturbance. Once we see behavior as a process of controlling particular perceptions, we can understand how it can be that actions vary while their outcomes do not—how a driver fighting a crosswind can be turning the wheel apparently at random to right and left while the car proceeds down the center of its lane. The outcomes are under control by the organism; the actions adjust continually to prevent disturbances from altering the outcome. In fact, once we understand this control process, we no longer expect regular results to be achieved by regular actions: It is quite the opposite.

An Example: Cognitive Control

This reinterpretation of behavior leads to new understandings of what is going on in many kinds of behaviors. The behavior we will examine here is that of "answering questions." The normal way of interpreting this kind of behavior is to consider the question as an eliciting stimulus and the answer as a learned response to that stimulus. This interpretation carries the implication that the act of answering is produced by neural commands that cause the muscles of speech or writing to act in organized patterns that produce the sounds or marks recognized as answers: The answer to a question is basically a learned pattern of motor behavior. In the first of the two experimental conditions we will see, this interpretation appears satisfactory. In the second, however, there can be no regular connection between a motor response and the answer to a question: The best explanation appears to be the one offered by the control interpretation.

Experimental Conditions, General

In both experiments that we will examine, the physical situation has deliberately been set up to allow separation of the motor behavior involved in answering a question from the answer itself. A subject seated before a computer screen sees questions in the form of simple "sentence-completion" arithmetic problems and indicates the correct answer by moving a joystick (located at the side of the chair) that, by electronic means, causes a pointer on the screen to move up and down a displayed list of possible answers. The task is very simple: For example, a question such as "17 minus 12 = ?" appears on the left side of the display and the subject pushes or pulls on the handle to bring the pointer next to the numeral "5" in a list ranging from 0 to 22, in sequence down the right side of the display. The indication of the correct answer must be held for 3 seconds (which suffices to distinguish accidents from intended answers). There is, of course, no inherent connection between moving a joystick handle and answering an arithmetic question. Subjects, however, quickly learn the required actions. During one experiment, a new question is presented every 6 seconds for 5 minutes, a total of 50 questions. The range of handle position is scaled so that 23 distinct positions are resolved, numbered 0 to 22.

The questions are constructed automatically as follows. First, a list of 50 numbers is produced by a random-number subroutine, such that they all lie in the range from 0 to 22. These constitute the right answers to a series of problems. Then 50 more numbers are similarly selected to serve as the first term of the problem question, and finally the questions are completed by computing the second term of the problem question required to produce the desired answer (addition and subtraction are chosen as necessary). No attempt is made to be sure that the extremes are represented or that the distributions are uniform.

Each trial starts with a 50-millisecond tone and a new problem. The trial ends when the subject has held the screen pointer next to the correct answer for 3 seconds. A new trial immediately starts. If the subject indicates a wrong answer, the program simply waits until the right answer is indicated. Two subjects were tested: For reasons that may be obvious, they generated identical results, so only one set of data plots is shown.

Conditions, Experiment 1

In the first experiment, the answer pointer is positioned directly by the joystick handle. Moving the handle well forward positions the screen pointer at the "0" near the top of the screen, and pulling it well back positions the pointer next to "22" near the bottom of the screen. The handle positions numbered 0 through 22 in the figures correspond to pointer positions next to the answers 0 through 22.

Results, Experiment 1

Figure 2 plots obtained handle positions against correct answers for each trial. The right answer is indicated up the left side and the handle position along the bottom. The numbers in the plot represent the number of trials in which the same pair of positions was repeated. A moment's reflection will verify that all subjects necessarily produce this same plot, given the same list of problems, because each trial ends only when the pointer indicates the right answer; handle position corresponds to pointer position. That is why the plotted entries lie on a straight line. The point of this trivial exercise will be explained shortly.

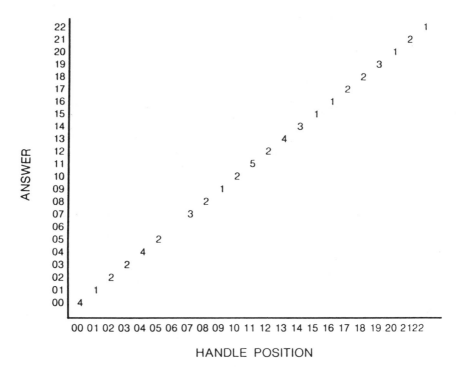

Figure 2. Plot of correct answers (vertical) against handle position that caused pointer to indicate correct answer (horizontal). Each pointer position depended directly on handle position, so the plotted points lie on a straight line. Numbers in plot indicate repetitions of same answer-handle combinations.

Conditions, Experiment 2

The second experiment is exactly the same except that a disturbance is introduced, acting directly on the answer pointer on the screen. The same list of answers and questions as before is used. Now the position of the answer pointer is determined by the sum of handle position and an independent disturbing number taken from a precalculated table. To make this table, first a table of desired handle positions is created, using a random-number algorithm restricted to the range from 0 to 22. Then for each problem, a disturbance is calculated such that for the pointer to indicate the right answer, the handle must be in the predetermined position. For example, if the answer to a given problem is 12 and if the random-number table yields a desired handle position of 1 for that problem, the disturbance entry is set at +11. Thus for the pointer to be in Position 12, the handle must be at position 1 to compensate for the disturbance of 11 position units and place the pointer at Position 12. The subject still positions the pointer on the screen opposite the correct answer: All that has changed is that the task can no longer be accomplished by moving the handle to a position corresponding to the right answer. Repeating a given handle position from one problem to the next will in general result in pointing to a different answer; repeating an answer will, in general, require putting the handle in a different position.

With the disturbance present, there is no longer any uniform relationship between final handle position and the position of the answer pointer when it indicates the right answer. To place the answer pointer next to one of the answer numbers, the handle might have to be in any position from fully forward (Position 0) to fully back (Position 22), depending on the value of the disturbance added to the pointer position for that problem.

Results, Experiment 2

Figure 3 shows the results. As in Figure 2, the right answer is plotted up the left side and the handle position along the bottom. The numbers in the plot again indicate the number of times a given combination of answer and handle position was repeated. Again, all subjects given the same list of problems and disturbances will produce exactly the same plot because each trial ends only when the pointer indicates the right answer continuously for three seconds.

Discussion

As a method for exploring arithmetic ability, these two experiments are useless, for all subjects always get the answers right for every problem. Correctness is not the point.

The first experiment is designed to fit the traditional explanation of question answering, in which a specific motor action leads to the production of a specific answer. The behavior we see could be explained by saying that the presentation of the problem (2 plus 3 = ?) constitutes a stimulus, and the handle movement to a new position constitutes a response to that stimulus. Reinforcement could even be imagined to operate here: The motor response that is reinforced is the one that brings the pointer to the correct answer.

The second experiment is designed to rule out the traditional explanation, as it does. The motor response is measured by the handle

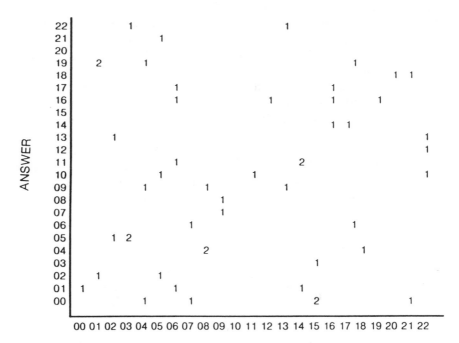

HANDLE POSITION

Figure 3. Plot of correct answers (vertical) against handle position that caused pointer to indicate correct answer (horizontal). Each pointer position depends on handle position plus a disturbing offset. The offsets were preselected to give a random distribution of handle positions. Numbers in plot indicate repetitions of same answer-handle combinations.

position number along the bottom of each figure. As can be seen in Figure 3, there is no regular relationship between the correct answer and the handle position by means of which it was indicated. In fact the handle positions required to produce the right answer were selected by a random-number algorithm, and if the experiment had been sufficiently protracted, the correlation of handle positions with the correct answers could have been made arbitrarily close to zero.

In one sense these experiments are trivial because they were arranged so that one trial could end only when the right answer was given. But there is another way to view them: If the traditional explanation had been correct, subjects should not have been able to complete the second experiment except perhaps through a long period of trial and error, which did not occur. Clearly the idea that an answer to a question is a motor response cannot explain the fact that subjects accomplished the task of Experiment 2 as easily as that of Experiment 1.

The handle positions required to indicate the right answers in Experiment 2 were randomly distributed: There was no possibility of producing the correct answer by responding to the question stimulus with any one "postural response." Yet the correct answers were quickly indicated every time.

Figure 4 is a model illustrating, if not how the real person works, at least how the author conceives the cognitive processes to relate to behavior during the experiments.

Conclusions

We have seen here a case in which a simple kind of high-level (as opposed to "sensorimotor") behavior seems to conform to the conventional interpretation until disturbances are introduced. Conventional interpretations generally omit the influence of independent variables on the very outputs that are usually called *behaviors* or *actions*. When such disturbances are introduced in real experiments, the result is not to make the outcomes correspondingly different but to produce a change in the action that leaves the outcome essentially undisturbed.

Only control theory can explain this phenomenon without resort to elaborate *ad-hoc* premises. The simplest explanation for what the subjects were doing in these experiments is that they were controlling a relationship: the relationship between a question and a number selected from a list. The overall reference condition was that the number seen next to the pointer should be recognized as an answer to the question. To meet this condition, the subject had to compute, internally, the correct answer and establish one position as a target for the pointer. Then a lower-level control system brought the pointer to that preselected position on the screen. Clearly, the answer had to be known before the target position could be selected, and the target position had to be selected before the right response could take place. The direction and amount of response depended both on where the correct answer was located and where the pointer was located before the response.

With the disturbance acting, it is not possible to conclude that producing an answer and emitting a motor response are the same thing. Neither can the answer be thought of as an intervening variable. If the internally computed answer were an intervening variable, it would correlate with the response, but there is no such correlation in the second experiment. Control experiments do not follow the logic of conventional conceptions of behavior.

Once seen this way, the experiment can easily be elaborated to demonstrate control of other relationships between the perceived ques-

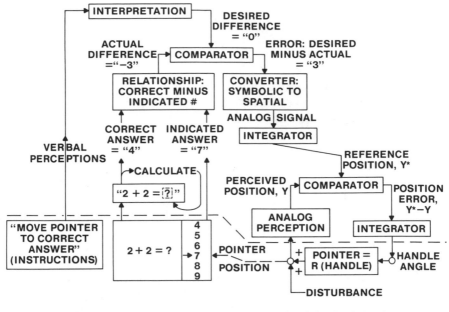

SYMBOLIC PROCESSES   **ANALOG TRACKING**

Figure 4. Model of behavior in these experiments. The lower right section is the same control model I have used elsewhere (unpublished) as a model for tracking behavior: a control system in which handle movement is proportional to the time integral of position error (the difference between intended and actual position of a pointer moved by the handle). The reference position, labeled y^*, is adjustable in this application. As y^* changes, the resulting error causes handle movement, which in turn causes the perceived y coordinate of the pointer to follow the changing reference signal. Disturbances applied directly to the pointer are met by equal and opposite effects from changes in handle position. This model can duplicate both pointer and handle movements, under random disturbance, with an RMS error of about 3% of the maximum movements (with y^* assumed constant). Perceptual signals in this diagram flow upward.

On the left, below, are three boxes showing aspects of the environment that relate to behavior. One is the "instruction" box. Verbal instructions are perceived by the subject (next box upward) and interpreted by unspecified cognitive functions to mean a specific target condition: the desired numerical difference between the right answer and the answer indicated on the screen. In this experiment the instructions imply that this difference should be zero, although other differences could be specified. This desired difference is the highest level reference condition considered here.

A second box at the lower left shows the question presented on the screen ("2 + 2 = ?"). This question is perceived as a problem, and through internal symbol manipulations the question mark is replaced by the correct answer, 4. The person then perceives that the correct answer is the symbol "4."

The third lower-left box shows the part of the screen where the pointer is: The box above is the perceptual function that interprets this appearance; the person perceives that the indicated answer is the symbol "7."

The perceptions of the calculated and indicated answers are passed upward to an-tion and the perceived answer. For example, the experimenter might ask the subject to indicate a number one greater than the correct answer (with suitable adjustment of the experiment to terminate each trial when that condition was met). Any number of other relationships could be invented, and unless they were very complex, all subjects would still show identical behavior.

Control theory, which originated in electrical engineering in the late 1930s, has been applied to various aspects of behavior by many workers. Conventional approaches can be found in Wiener (1948), Ashby (1952), Riggs (1970), McFarland (1971), Jones (1973), Gaarder (1975), Toates (1975), and for the engineering–cybernetics approach, Osafa-Charles et al. (1980). The particular approach used here, in which behavior as a whole is seen as a process of controlling perceptions, was introduced by Powers, Clark, and McFarland (1960) and elaborated further by Powers (1973, 1989). Many scientists, particularly during the past 4 or 5 years, have adopted this theory of behavior, two examples (Marken and Hershberger) having been cited here. More examples of this theoretical and experimental approach will be appearing in the literature of the social and biological sciences in the next several years.

---

other perceptual function that determines their relationship—their arithmetic difference—by subtracting the indicated answer from the right answer, to yield a perception of "-3," all in symbols. Because of the way this subtraction is done in the perceptual function, the resulting perception represents the magnitude of the relationship "correct greater than indicated." The right answer is 3 less than the indicated answer, so the perception carries a minus sign: "-3."

This perception of difference is compared with the desired difference, in the top box labeled comparator. The comparator computes the error as reference minus actual perception. The desired difference is "0," so the comparator reports a symbolic error of "3" units. This error information enters an output function box that translates this symbolic error into a quantitative positional error and integrates this error to produce an intended spatial position, y^*. If the symbolic error is positive, the intended position y^* increases with time (equivalent, because of the organization of the tracking control system, to an upward direction of movement of the target position on the screen). The control system first mentioned moves the handle so as to keep the perceived pointer position, y, matching the target pointer position y^*, automatically counteracting disturbances added externally to the pointer position.

Because two integrations are involved, this whole system would be unstable; some damping (in the form of a phase advance or "anticipation") must be added at one or both levels to make it behave correctly. Damping is not indicated in this semiqualitative model. Damping could be added, for example, by making y^* proportional to the integral of the symbolic error PLUS a constant times the symbolic error. The steady-state behavior of a computer simulation of the model would then, at the end of each trial, match that of the subject, although the dynamics would still probably be different.

References

Ashby, W. R. (1952). *Design for a brain*. New York: Wiley.

Gaarder, K. R. (1975). *Eye movements, vision, and behavior*. Washington, DC: Hemisphere.

Hershberger, W. A. (1987). Some overt behavior is neither elicited nor emitted. In A. C. Catania & S. Harnad (Eds.), *The selection of behavior* (pp. 107–109). New York: Cambridge University Press.

Hershberger, W. A. (1988). The synergy of voluntary and involuntary action. In W. A. Hershberger (Ed.), *Volitional action* (pp. 3–20). Amsterdam: North Holland.

Jones, R. W. (1973). *Principles of biological regulation*. New York: Academic Press.

Marken, R. S. (1988). The nature of behavior: Control as fact and theory. *Behavioral Science, 33*, 196–206.

McFarland, D. J. (1971). *Feedback mechanisms in animal behavior*. London: Academic Press.

Osafa-Charles, F., Agarwal, G. C., O'Neill, W. D., & Gottlieb, G. L. (1980). Application of time-series modeling to human operator dynamics. IEEE Transactions on systems, man, and cybernetics. *SMC-10*, No. 12, 849–860.

Powers, W. T. (1973). *Behavior: The control of perception*. Chicago: Aldine; now New York: Walter de Gruyter/Aldine.

Powers, W. T. (1978). Quantitative analysis of purposive systems: Some spadework at the foundations of scientific psychology. *Psychological Review, 85*, 417–435.

Powers, W. T. (1989). *Living control systems*. Gravel Switch, KY: The CSG Press.

Powers, W. T., McFarland, R. L., & Clark, R. K. (1960). A general feedback theory of human behavior, Parts I and II. *General Systems, V*, 63–83.

Riggs, D. S. (1970). *Control theory and physiological feedback mechanisms*. Baltimore: The Williams & Wilkins Co.

Staddon, J. E. R. (1983). *Adaptive behavior and learning*. Cambridge, MA: Cambridge University Press.

Toates, F. M. (1975). *Control theory in biology and experimental psychology*. London: Hutchinson Educational.

Wiener, N. (1948). *Cybernetics: Control and communication in the animal and the machine*. New York: Wiley.

14

Dynamics of Attitude Change

Stan A. Kaplowitz and Edward L. Fink

What should be the characteristics of a systems theory of attitudes and beliefs? First, a systems theory should be holistic. Any forces, variables, or behaviors considered within the systems framework should be seen as interdependent. In our chapter, you will see that (1) attitudes are structured so that changing one attitude has effects on other related attitudes, and that (2) the forces within the attitudinal system have effects that permeate the system.

Second, a systems theory should consider both the equilibrium states of a system and the process by which equilibrium is achieved. In this chapter, we show how trajectories for attitude change over time may be derived and how traditional attitude change research can be made relevant to understanding these trajectories.

Third, a systems theory should show how control mechanisms bring stability to a system. Typically, such stability comes about from negative feedback, and such systems may have conservation laws and the possibility of oscillation. In our model of attitude change, we use a mechanistic or physicalistic metaphor. According to Leatherdale (1974), models are generally built on metaphors. Mechanics not only

Portions of this chapter also appear in S. A. Kaplowitz & E. L. Fink (1988). *A spatial-linkage model of cognitive dynamics*. In G. A. Barnett & J. Woelfel (Eds.), *Readings in the Galileo system: Theory, methods and applications* (pp. 117–146). Dubuque, IA: Kendall-Hunt Publishing Co.

Stan A. Kaplowitz • Department of Sociology, Michigan State University, East Lansing, Michigan 48824.   **Edward L. Fink** • Department of Speech Communication, University of Maryland, College Park, Maryland 20742-1221.

Analysis of Dynamic Psychological Systems, Volume 2: Methods and Applications, edited by Ralph L. Levine and Hiram E. Fitzgerald. Plenum Press, New York, 1992.

has a great deal of richness and precision but is, in some respects, analogous to attitude change. Using our analogy, we derive trajectories for this system and show that oscillations of attitudes and beliefs are expected.

Fourth, we can derive, from our second and third principles, that *time* should play a significant role in any systems theory. In rhetoric, timing plays a significant role, and it makes sense that a theory used to explain the art of the rhetor should make time central as well. Timing of messages and the responses to them will therefore be discussed as crucial to the theory to be presented.

Finally, a system is a fiction in which we consider all components known; all sources of energy must be considered to be inputs from the environment or explained as transformations of the internal structure of the system. In our chapter, the external forces are created by messages received by the individual. Internal changes in the cognitive space include changes in linkages between concepts, changes in the mass of concepts, and changes in the location of concepts. One's thoughts, distractions from one's cognitive activity, and one's physiological arousal level all affect the way external messages work through the cognitive system to bring about changes in attitudes and beliefs. We briefly sketch the implications of our framework for understanding the role that internal activities play in the belief change process.

We will now review the status of research on attitudes and persuasion. After doing so, we will present our model of attitudes and beliefs, a dynamic model that integrates much existing literature.

The Current Status of Attitude and Persuasion Research

> The study of attitudes and persuasion began as the central focus of social psychology. However, . . . , after accumulating a vast quantity of data and an impressive number of theories—perhaps more data and theory than in any other single topic in the social sciences, . . . there was surprisingly little agreement concerning when, if, and how, the traditional source, message, recipient, and channel variables affected attitude change. Existing literature supported the view that nearly every independent variable studied increased persuasion in some situations, had no effect in others, and decreased persuasion in still other contexts. (Petty & Cacioppo, 1986, pp. 124–125)

We see several reasons for these weaknesses in the development of theory in the study of attitudes. First, much of the research on attitudes and persuasion considers the effect of source, message, or receiver characteristics with little integration among the various theories and propositions tested by different studies (see, e.g., Fishbein & Ajzen, 1975,

p. 52; McGuire, 1969, p. 271). Hence, each theory has a limited and imprecisely known range of applicability, and when different processes lead to seemingly contradictory predictions, such conflicts generally cannot be resolved by reference to a higher-level theory.

Second, the theoretical principles in this area are often stated in a very imprecise form. What is typically predicted, and tested for, is the *presence* and *direction* of a relationship—not its *strength* or *functional form*. The development of precision is hindered by the unfortunate tendency to operationalize variables as if they were dichotomous or trichotomous (e.g., distinguishing low-, medium-, and high-credibility sources) rather than assuming that these variables can be measured on continuous scales and attempting to do so.

Third, there is evidence and theory to support the idea that attitude change can take place in the absence of new external messages. The work of Tesser and his associates (Tesser & Conlee, 1975; Tesser, 1978) shows that merely thinking about a topic can cause an attitude to change over time.

Even more interesting is the observation that over time, an attitude that started moving in one direction can reverse its direction of motion. Such reversals have been demonstrated by Poole and Hunter (1979), who found that when attitude change is induced in a subordinate concept (such as "dog") but not in its superordinate one (here, e.g., "animal"), the change decays over time. By measuring attitudes at four points in time, Walster (1964) found evidence of *two* changes of direction. Each of Walster's subjects first chose one of two possible jobs. This was the first measurement of their attitudes toward these jobs. She then found that 4 minutes later, subjects regretted their decisions. After 15 minutes, however, she found the usual dissonance reduction effect (subject's being very happy with their decisions) and after 90 minutes, neither effect was evident. Walster's evidence, coupled with the experience many people report, of changing their minds, suggests the possibility of cognitive oscillations, a possibility empirically supported by Kaplowitz, Fink, and Bauer (1983), described later.

In short, evidence suggests that when researchers measure attitudes at more than one time point, they often find interesting time courses. Further, as we shall argue later, understanding the time course of attitude change is vital for understanding any cognitive forces involved in the change. Such time courses have, however, rarely been examined. On those few occasions where they have been, the attitude has not been measured at a sufficient number of time points, and/or the measurement of time or attitude has been too crude, and/or there has been little attempt to provide a theory that makes sense of the observed time course.

McGuire (1985) states that current research on attitudes tends increasingly to see attitudes and beliefs as an interacting dynamic system. Among such systems approaches, he includes information integration theory (e.g., Anderson, 1981) and the cognitive response approach (e.g., Petty, Ostrom, & Brock, 1981). We believe that although both of these approaches are important advances, neither of them fully combines a sense of process with generality (i.e., the capability of integrating many seemingly diverse phenomena) and precision. In this chapter, we shall sketch out a systems model that, we believe, moves us further towards a precise, general, and dynamic theory.

Some Key Theoretical Variables

A message has a *position advocated* by the source. The receiver has an *initial view* on the issue. The difference between the position advocated and the initial view is called *message discrepancy*, and various studies (e.g., Aronson, Turner, & Carlsmith, 1963; Bochner & Insko, 1966; Jaccard, 1981) have shown that attitude change is often maximized at moderate levels of discrepancy.

Another set of variables involves the source's characteristics, as perceived by the receiver. In one of the classic investigations in this area, Hovland and Weiss (1951) showed that sources who are presumed high in *credibility* are more effective than less credible sources. One aspect of credibility is *expertise* (see, e.g., Aronson, Turner, & Carlsmith, 1963; Bochner & Insko, 1966). Another aspect of credibility is *trustworthiness*, which, in turn, is related to whether the source is perceived as *unbiased* (see, for example, Eagly, Wood, & Chaiken, 1978). As Eagly *et al.* (1978) show, a source is especially likely to be perceived as unbiased when he/she takes a position that is substantially different from the position he/she was *expected* to take (i.e., when the message *disconfirms* expectations).

A third set of variables involves the relationship between the attitude being changed and other attitudes. Sherif, Sherif, and Nebergall (1965) write of *ego involvement* (the extent to which the individual is committed to an attitude) as creating resistance to change. Others have shown that such resistance can be increased by linking a belief to other beliefs (e.g., Holt, 1970; McGuire, 1964; Nelson, 1968; Watts & Holt, 1970) or to positively valenced groups or individuals (Kelly, 1955; Tannenbaum, 1967).

Links to other beliefs can promote, as well as inhibit, attitude change. A number of authors have shown that one can change beliefs about one concept, or a set of concepts, by providing messages or inducing thoughts about other concepts (e.g., Leippe, Greenwald, &

Baumgardner, 1982; McGuire, 1981; Rokeach, 1975). These findings are consistent with the various theories of cognitive consistency, such as balance (Heider, 1946), source-object congruity (Osgood & Tannenbaum, 1955), and dissonance (Festinger, 1957). Such linkages between beliefs are consistent with the idea that people's attitudes and beliefs are organized into *belief systems* (see McGuire, 1981) or into *schemata* (Bartlett, 1932).

Finally, there are relevant processes that may take place a good deal after the receipt of an external message. First, the message may be *forgotten*. As certain aspects of a message are forgotten, the attitude change that it induces may decay (see Cook & Flay, 1978, p. 31), or there may be a sleeper effect (see Hovland & Weiss, 1951). Second, receivers generate subvocal (cognitive) responses to messages. Those working in the cognitive response tradition (see, e.g., Petty, Ostrom, & Brock, 1981; Petty & Cacioppo, 1986) argue that whether these cognitive responses are favorable or unfavorable to the view expressed in the external message has an important effect on attitude change. The number of these cognitive responses appears to be reduced by distraction (Petty, Wells, & Brock, 1976), increased by involvement in the issue (see, e.g., Petty & Cacioppo, 1979), increased by exogenous factors that increase arousal (see, e.g., Cacioppo, 1979), and decreased by exogenous factors that decrease arousal. In short, memory and thought processes induced by a message may affect an attitude some time after receipt of the message.

The Dynamic System Model

Some models of attitude and belief systems emphasize *connections* between concepts (e.g., Heider, 1946; Phillips & Thompson, 1977), which may be of various strengths (J. R. Anderson, 1983). Other models represent systems of attitudes and beliefs as a configuration of concepts in a multidimensional space (see Woelfel, Cody, Gillham, & Holmes, 1980; Woelfel & Fink, 1980; Woelfel & Saltiel, 1978). For such spatial models, it is assumed that the distance between concepts corresponds to their dissimilarity and that concepts may be represented in a space whose dimensionality may be empirically determined. The model we have developed involves *both* a spatial structure and a set of linkages.

Spatial Models

One reason for using a spatial structure is that, as Tversky and Gati (1978, p. 79) have stated, "the notion of similarity—that appears under such different names as proximity, resemblance, communality, representativeness and psychological distance—is fundamental to theories of perception, learning, and judgment." Because the degree to which concepts are regarded as similar may involve considering many attributes, a spatial model is able to represent the complexity of thinking about relationships. Because the relations that Heider (1946) calls *unit* relations (i.e., similarity, causality, ownership) are closely related, they can all be represented in such a model. *Affect* (sentiment), the other relationship of Heider's concern, can also be represented within a spatial model—as the proximity of a concept to an evaluative concept such as "My preference" or "Things I like" (see Fink, Monahan, & Kaplowitz, 1989; Neuendorf, Kaplowitz, Fink, & Armstrong, 1986).

The spatial model has an important resemblance to balance theory, in that perceptual, cognitive, and affective judgments may all be represented within the same framework. It has, however, an important advantage in that degrees of similarity or affect are conceived of as distances, which may be measured on an unbounded, continuous magnitude scale rather than treated categorically or ordinally.

Although Tversky (1977) believes in the importance of similarity, he has regarded spatial models of attitudes and beliefs as inappropriate because the distances used in such models may violate the axioms for a space with a Euclidean metric (axioms satisfied by ordinary physical space). In particular, Tversky and Gati (1982) report violations of the axioms of symmetry (the axiom requiring that the direct distance between two points is the same, regardless of which is the starting point and which is the endpoint) and the triangular inequality (which requires that the sum of any two sides of a triangle be greater than or equal to the third side). For example, Tversky (1977, p. 325) suggests that Jamaica and Cuba will be regarded as quite similar (i.e., close) on the basis of geography, that Cuba and the USSR will be regarded as quite similar, based on political affinity, but that Jamaica and the USSR will be perceived as quite distant.

We see the violations of the distance metric more as problems of measurement than of theory. The observed violations of the Euclidean axioms are based on assuming that each concept is a *point* in the space. We believe that a concept is better regarded as a *region*, whose size is roughly proportional to the number of distinct attributes or meanings the concept contains.[1] The observed violations of the Euclidean ax-

---

[1]The assumption that the size of a concept is proportional to the number of meanings or attributes it "contains" makes intuitive sense. When the meaning of a concept contains other attributes or meanings, its geometric representation should contain the representations of those other concepts. Violations of Euclidean axioms have been found most likely to occur when dealing with concepts that contain such multiple meanings (e.g., "orange"; see Woelfel & Fink, 1980; Marron, 1985) or multiple attributes (see Tversky,

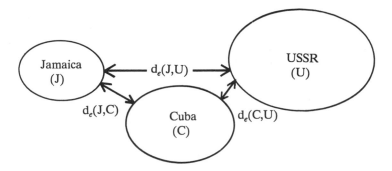

Figure 1. Hypothetical configuration of concepts, *Cuba* (C), *Jamaica* (J), and *USSR* (U), and interconcept distances reported by subjects. The concepts are assumed to be regions in a Euclidean space, but the *reported* distances are assumed to be the smallest edge-to-edge distances (e.g., $d_{e(J,C)}$ is the edge-to-edge distance between *Jamaica* and *Cuba*) rather than center-to-center distances. Hence, the reported distances violate the triangular inequality.

ioms, come, we believe, from respondents reporting distances that are not "center" to "center" distances. Returning to the previous example, we propose that Cuba, Jamaica, and the USSR might be regions in an ordinary Euclidean space, in which the triangular inequality is satisfied. The apparent violation of that axiom, predicted by Tversky (1977), would occur if respondents are reporting the shortest possible "edge"-to-"edge" distances between pairs of concepts (see Figure 1). To avoid such violations, our model assumes that we are able to obtain center to center distances between concepts.[2]

Dynamics

Not only are distances convenient for representing both similarity and affect, they are also very convenient for a dynamic model. Within a

---

[1] 1977). The larger the size of an object, the greater is the potential for a reported distance, to or from it, to deviate substantially from the distance to or from its center, thereby causing violations of the Euclidean axioms. Hence, the fact that violations are most common when concepts with multiple meanings or multiple attributes are involved is consistent with our assumption that such concepts have greater size.

[2] The actual method used to obtain distance judgments and to derive the spatial configuration of the concepts is discussed elsewhere (see Woelfel & Fink, 1980) and will not be presented here. Suffice it to say that subjects are asked magnitude estimation questions regarding the psychological distance between concepts, and this information can be used to generate the spatial coordinates of a multidimensional space (see Torgerson, 1958, pp. 247–297).

spatial model, attitude or belief *change* becomes analogous to motion. In keeping with out metaphor, if there is motion (other than at a constant velocity), there must be forces operating. Similarly, the absence of motion implies that forces are balanced (i.e., in equilibrium). We can borrow, from physics, equations relating force and motion, and construct, by analogy, a relationship between cognitive forces and cognitive motion.

For both parsimony and explanatory effectiveness, our model should posit a cognitive force that (1) is created by a message and therefore capable of causing motion and (2) whose effects continue after the message, so that any opposing force, generated by a later message, may be resisted. Such a force should also be able to account for the cognitive oscillations, or changes of mind, we have all experienced and that have been experimentally demonstrated by Walster (1964) and by Kaplowitz, Fink, and Bauer (1983).

We assume that a message linking two concepts, A and B, establishes a linkage between them, which creates a force satisfying the following equation:

$$F_{A,B} = K_{A,B}(d_M(A,B) - d(A,B)), \tag{1}$$

where $F_{A,B}$ is the force between the concepts, $d_M(A,B)$ is the distance (dissimilarity) specified in the message, $d(A,B)$ is the distance between those concepts in the receiver's cognitive space, and $K_{A,B}$ is the restoring coefficient of the linkage.

Several things should be noted about this equation: (1) The force generated by this linkage is zero whenever $d_M(A,B) = d(A,B)$ (i.e., when the message agrees with the recipient's current view). Hence, if this were the only linkage, the system would be in equilibrium whenever this condition held. (2) If the message stated a value that is *greater* than the subject's view (i.e., $d_M(A,B) > d(A,B)$), the force, $F_{A,B}$, is positive, which causes the distance between A and B to *increase*. If, on the other hand, the message value is *less* than the receiver's view, the force is negative, causing a *decrease* in the value of $d(A,B)$. In other words, the force is a restoring force in that it is always pushing $d(A,B)$ toward the equilibrium value (the message value). (3) Because this is a dynamic system, $d(A,B)$ is not constant. (4) As discussed later, the restoring coefficient is related to various source and message characteristics. In our initial discussion, we assume that $K_{A,B}$ is constant over time. Later, we shall relax this assumption. (5) This equation, which is analogous to the dynamics of a spring (see, e.g., Ingard & Kraushaar, 1960) and to an inductance–capacitance circuit (see, e.g., Halliday & Resnik, 1960, p. 799) is based on the assumption that the force inducing change is proportional to the departure of the system from equilibrium. Using the spring analogy, a large value of $K_{A,B}$ is analogous to a stronger or stiffer spring.

Thus far, we have discussed the creation of message linkages. We further assume that as long as the creating message is remembered, the linkage will continue to exist and to exert a force toward making the A–B distance the value specified in the creating message. Hence, a person's attitudes and beliefs are assumed to be a result, not only of the most recent message he/she has received, but of the entire system of linkages, resulting from the various messages that have been received over time.

Equilibrium Predictions of Attitude Change

Let us now derive the equilibrium value of attitude change resulting from a single persuasive message stating that the distance between A and B is $d_M(A,B)$. Assume that prior to receiving the message, the receiver, P, places concepts A and B at locations A_0 and B_0, respectively. Following Equation (1), the message should create a force, $F_{A,B}$. Hence, if $d_M(A,B)$ is less than $d(A_0,B_0)$ (P's initial view of the A–B distance), concepts A and B will be pulled toward each other.

Let us further assume that A and B are anchored to their initial locations by linkages from prior messages. Each set of anchoring linkages may be the result of a number of distinct linkages (i.e., linkages to conceptions of various significant others, to conceptions of a religious or ethnic community, or to an ideology). To simplify the algebra, however, we treat the combined effect of each set of anchoring linkages as a single anchoring linkage, linking A to its anchor $R(A)$ and B to its anchor $R(B)$. The anchoring linkages have their own restoring coefficients, $K_{A,R(A)}$ and $K_{B,R(B)}$, respectively.

Let us now illustrate this with an example. Let A be *President Reagan* and B be *trading arms for hostages*. Let us further assume that A is anchored to, and located at, $R(A)$, which is *Strong Foreign Policy*. B, on the other hand, is anchored to $R(B)$, which is *Weak Foreign Policy*. We further assume that the receiver views $R(A)$ and $R(B)$ as rather far apart (say, 100 units). Because A is initially located at $R(A)$ and B at $R(B)$, this means the receiver initially views *Reagan* as 100 units from *Trading Arms for Hostages*. In other words, $d(A_0,B_0) = 100$.

Suppose P, the receiver, now receives a message indicating that Reagan *has* traded arms for hostages. This message, therefore, claims that the distance between A and B is less than P had originally thought (say, 20 units). In other words, $d_M(A,B) = 20$. The linkage created by the

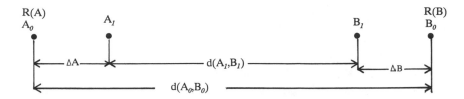

Figure 2. Geometric configuration of concepts A and B, and the concepts $R(A)$ and $R(B)$ to which they are respectively anchored. A_0 and B_0 are the equilibrium locations of A and B before they were linked by the message, and A_1 and B_1 are the postmessage equilibria. Note that prior to the message, each concept is considered to be at the same location as its anchoring concept.

$A–B$ message is now pulling A and B together. This motion, however, is being resisted by the linkages of A and B to their respective anchors. In other words, P's view of the $A–B$ distance is influenced not only by the $A–B$ message but also by P's prior beliefs, which were established by earlier messages.

The combined effect of these linkages is to establish new equilibrium locations, A_1 and B_1, of A and B (see Figure 2.) Although we have drawn Figure 2 for the case in which the message value, $d_M(A,B)$ is less than the original distance perceived by P, the derivation does not depend on this assumption. We have also simplified things somewhat by assuming that the anchoring concepts, $R(A)$ and $R(B)$, are so well anchored that their movement can be ignored.

To determine these new equilibrium locations, we must consider three linkages and hence, three forces. They are the force between A and B, the force between $R(A)$ and A, and the force between $R(B)$ and B. Following Equation 1, in order to calculate $F_{A,R(A)}$ (the force between $R(A)$ and A), we must not only know the distance between those concepts and the restoring coefficient in that linkage but also the implied distance between these concepts in the message that created that linkage. Although this linkage may be a result of many distinct messages, its net effect is such that prior to the new message, A was at equilibrium when its distance from $R(A)$ was zero. Consequently, we can say that $d_M(A,R(A)) = 0$. Analogously, $d_M(B,R(B)) = 0$.

After receipt of the message, $d_M(A,B)$, concept A is pulled in opposite directions by the linkages $K_{A,R(A)}$ and $K_{A,B}$. At the new equilibrium, A_1, the magnitudes of the opposing forces pulling on A must be equal. (The restoring forces in each individual linkage need not be zero, however.) Hence, we have

$$F_{A,R(A)} = F_{A,B}, \tag{2}$$

whereas from Equation 1, above, we have

$$F_{A,R(A)} = K_{A,R(A)}(d_M(A,R(A)) - d(A_1,A_0))$$ (3)

and

$$F_{A,B} = K_{AB}(d(A_1,B_1) - d_M(A,B)).$$ (4)

We can simplify these expressions as follows. First, recall that the message value from the anchor, $d_M(A,R(A))$, is zero. Second, we define $d(A_0,A_1)$ to be ΔA (ΔB is defined analogously) (see Figure 2). By simple geometry, the *new* distance between A and B, $d(A_1,B_1)$, must equal the *original* distance between them, minus the combined change in their locations, along the vector linking them.

Further, by definition, the message discrepancy D, is

$$D = d(A_0,B_0) - d_M(A,B).$$ (5)

Hence, in our example, $D = 100 - 20 = 80$.

Making use of all of this information and substituting into Equation 2 gives us

$$K_{A,R(A)}(\Delta A) = K_{A,B}(D - (\Delta A + \Delta B)).$$ (6)

Because concept B is also presumed to be at its equilibrium location, the two linkages at that point must also be generating equal and opposite forces. Hence,

$$K_{B,R(B)}(\Delta B) = K_{A,B}(D - (\Delta A + \Delta B)).$$ (7)

Because the left sides of Equations 6 and 7 are equal to identical expressions, they must equal each other, which gives us the relationship between ΔA and ΔB:

$$\Delta B = \Delta A(K_{A,R(A)}/K_{B,R(B)}).$$ (8)

Substituting from Equation 8 into Equation 7 and solving for ΔA, gives us

$$\Delta A = \frac{K_{A,B}D}{K_{A,R(A)} + K_{A,B} + \dfrac{K_{A,R(A)}K_{A,B}}{K_{B,R(B)}}}$$ (9)

Let us now examine the implications of Equations 8 and 9. First, because $K_{A,R(A)}$ appears only in the denominator, it is clear that the greater it is, the smaller is ΔA. In other words, the greater the anchoring of A, the less A will move. It is also clear that the stronger is the linkage between A and B generated by the new message ($K_{A,B}$), the greater is ΔA. From Equation 8, however, we see that the relative motion of concepts A and B is inversely proportional to the strength of their anchoring linkages. To return to our example, whether P's view of *President Reagan* or P's view of *Trading Arms for Hostages* will change more will depend on which of these concepts is more strongly anchored.

It is also clear that, other things being equal, the greater is D, the greater is attitude change. However, according to this model, the *total* attitude change will never exceed D and will only be equal to D if the restoring coefficients of the anchoring linkages are zero. (In other words, the recipient's view generally does not move all the way to the view advocated by the A–B message.)

To see this, we use Equation 8 to create an equation for ΔB that is analogous to Equation 9. We then add these together to get

$$\Delta A + \Delta B = \frac{(K_{A,R(A)} + K_{B,R(B)})K_{A,B}D}{(K_{A,R(A)}K_{B,R(B)})^2 + K_{A,B}(K_{A,R(A)} + K_{B,R(B)})} \qquad (10)$$

From Equation 10 we see that the total attitude change is D multiplied by a quantity whose denominator is larger than its numerator as long as neither $K_{A,R(A)}$ nor $K_{B,R(B)}$ equals zero. Hence, if neither of those anchoring linkages is zero, the attitude change will be less than D. If, however, at least one of the anchoring linkages approaches zero, the aforementioned numerator and denominator approach equality, and the total attitude change approaches D.

Relation of Source, Message, and Receiver Variables to Restoring Coefficients

The degree to which a concept is anchored depends, we assume, on the degree to which it is strongly linked to other well-anchored concepts. This, in turn, reflects *commitment* or *ego involvement*. Let us now relate the message linkage, $K_{A,B}$, to the variables that are known to influence attitude change. Because attitude change increases with increased source credibility, $K_{A,B}$ must be an increasing function of credibility. In order to account, however, for the finding that attitude change is generally maximized at moderate levels of discrepancy, $K_{A,B}$ must be

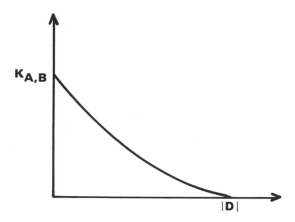

Figure 3. The restoring coefficient of the message linkage $K_{A,B}$ as a negative exponential function of the absolute value of the message discrepancy, D.

a decreasing function of the absolute value of D. In other words, highly discrepant messages are discounted. One function that would give us the desired result is an exponential decay function, which is shown in Figure 3.

Such an equation has been found satisfactory by Fink, Kaplowitz, and Bauer (1983). Further, it is consistent with the idea that the "energy" in the message undergoes an exponential loss as the linkage is stretched or compressed by a distance D, so that it can be "attached" to concepts A and B in the receiver's cognitive space.

Relationship to Information Integration Models

As indicated, McGuire (1985) cites information integration theory as an example of a dynamic systems theory. There is a mathematical model that is related to information integration theory (N. H. Anderson, 1974, 1981) and that predicts attitude change as a function of message discrepancy, source, credibility, and receiver's prior commitment. This model has been developed by Saltiel and Woelfel (1975) and Himmelfarb (1974) and elaborated by us (Fink, Kaplowitz, & Bauer, 1983; Kaplowitz, Fink, Armstrong, & Bauer, 1986). It is closely related to the work of Birnbaum and Stegner (1979). In this model, the equation for attitude change after one persuasive message is

$$\Delta A = w_1 D/(w_0 + w_1), \tag{11}$$

where w_0 represents the weight of the receiver's original attitude (i.e., his/her commitment to it) and where w_1 is the weight of the new message. If we return to equation (9) and assume that B is so tightly anchored that it does not move at all (i.e., that $K_{B,R(B)} = \infty$) then we see that the ratio $K_{A,R(A)}K_{A,B}/K_{B,R(B)}$ goes to zero as $K_{B,R(B)}$ approaches ∞. We then find that equation (9) is identical to equation (11) if we simply replace w_0 by $K_{A,R(A)}$ and w_1 by $K_{A,B}$. Therefore, the information integration model (called the "lever model" by Birnbaum & Stegner, 1979) can be regarded as a simplification of our multidimensional linkage model; in the information integration model, we assume that only one concept moves and all motion is in one dimension.

Source Bias

A number of studies (e.g., Eagly, Wood, & Chaiken, 1978; Walster, Aronson, & Abrahams, 1966) have found that the expected bias of the source has an effect on attitude change. Birnbaum and Stegner (1979) concluded that if a source who is a friend of the seller of a car specifies a certain value for the car, the receiver is likely to treat that message as the equivalent of a message from an *unbiased* source, which specifies a somewhat lower value. Furthermore, Eagly and Chaiken (1976) found that if a message departed too greatly from the position expected from the source (i.e., had too great a *disconfirmation*), the message was not believed.

Source bias may be incorporated into our model by assuming that for each concept, A, and for each source, S, there may also be, within P's cognitive space, the concept $A(S)$ (S's view of A). If P regards S as having a different view of A from P's own view, then A and $A(S)$ will occupy different locations in P's cognitive space. We further assume that when P receives a message linking A and B, the message also serves to link $A(S)$ and $B(S)$ in P's cognitive space. If the source's bias is well known and very salient, the $A(S)-B(S)$ linkage may be stronger than the $A-B$ linkage. Just as the strength of the $A-B$ linkage is assumed to be a decreasing function of *discrepancy*, the strength of the $A(S)-B(S)$ linkage is assumed to be a decreasing function of *disconfirmation*. In order for these linkages to cause motion of A and/or B, $A(S)$ and $B(S)$ must be linked to A and B, respectively. The strength of each such linkage should depend on S's expertise about the concept and on the degree of confidence P has in his/her prior judgment of S's bias or position.

An important consequence of assuming that P's cognitive space contains concepts such as $A(S)$ is that it permits the model to explain both the successes and failures of Osgood and Tannenbaum's (1955) source–object congruity theory. A message endorsing concept A is a message linking concept A with some positively evaluative concept, such as *Good*. If the source's viewpoint is salient, this should have the effect of moving $A(S)$ and $Good(S)$ (P's view of S's view of *Good*) toward each other. If A is a consensually understood concept, then A and $A(S)$ are very close together and tightly linked. But the concepts *Good* (P's own view) and $Good(S)$ need *not* have the same location in the space. If *Good* and $Good(S)$ are located on the same side of A (i.e., P views S's view of Good as similar to P's own view), then as $A(S)$ and A move toward $Good(S)$, they will also move toward *Good*. Hence, S's endorsement will be effective (See Figure 4a). If, however, *Good* and $Good(S)$ are in the opposite direction from A and $A(S)$, then as the latter two concepts move *toward* $Good(S)$, they will move *away* from *Good*. In this case, the endorsement from S is a "kiss of death" (see Figure 4b).

Thus far, we have shown how a theory that deals with discrepancy, credibility, and ego involvement can also explain the predictions of source–object congruity theory. We now demonstrate that our theory can also deal with findings that are within the scope of congruity theory but that it cannot handle. Suppose a negatively evaluated source (e.g., a convicted mobster) gives a message that greatly disconfirms

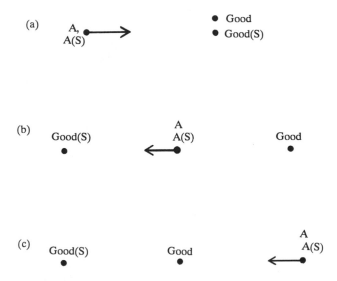

Figure 4. Configurations of object concept, A, source's view of *Good* ($Good(S)$), and recipient's own view of *Good*, resulting in (a) successful endorsement of A by Source S; (b) kiss-of-death endorsement of A by S; (c) successful endorsement of A by negatively evaluated source. Arrows indicate the predicted motion of A.

expectations (e.g., he advocates tougher law enforcement). According to congruity theory, such an endorsement should cause the subjects to be *less* favorable toward the object endorsed. As Walster, Aronson, and Abrahams (1966) found, however, such a message was effective in moving subjects in the direction advocated. This can be explained by assuming that Good(S), (where S is the mobster) and *Powerful Law Enforcement* (A) are in opposite directions from *Good*. As PA moves toward *Good(S)*, it moves toward *Good* (see Figure 4c). This movement makes P more favorable to PL.

The Internal Structure of a Compound Message

As Petty and Cacioppo (1986, p. 132) have noted:

> [L]iterally thousands of studies and scores of theories have addressed the questions of how some extra-message factor (e.g., source credibility, repetition) affects acceptance of a particular argument, but little is known about what makes a particular argument (or message) persuasive in isolation.

We believe our theory makes a contribution to this area as well.

An argument is a *compound message*. According to Woelfel and Fink (1980, pp. 150–159), such a compound message is composed of *simple* messages, each of which links two concepts. Each simple message will be most successful if not overly discrepant. The entire compound message will be most effective if each simple message creates sufficient attitude or belief change to make any simple message that comes later less discrepant and hence capable of producing a tighter message linkage.

Let us consider an example. Suppose a simple message asserted the desirability of a substantial tuition increase. Such a message would be found highly discrepant by most students and would be quite ineffective. But suppose, instead, that the following compound message was sent:

> A good education requires good faculty and good libraries. These things cost money—to pay higher salaries to attract good faculty, and to buy books for the library. The only way to get the necessary revenue is to substantially increase tuition.

Figure 5 shows a hypothetical space containing the relevant concepts. *Higher Tuition* is quite far from the concept *Good* in the recipient's cognitive space. Hence, a message simply asserting its desirability would be quite ineffective. *Higher Tuition*, however, should be quite close to *More Revenue for the University* because such a causal

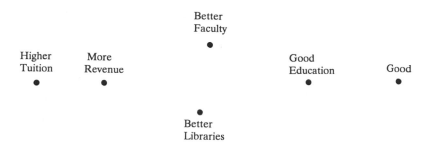

Figure 5. Hypothetical configuration of concepts invoked in an effective compound message advocating increased tuition.

relationship is quite plausible. *More Revenue* should, in turn, be close to *Better Faculty* and *Better Libraries*. Both of the latter concepts should be close to *Good Education*, which is presumably close to (P's view of) *Good*. In each case, a strong linkage should be established between a concept and one close to it. The effect of this set of linkages should be to pull *Higher Tuition* toward *Good Education* and *Good*.

Forgetting and Delayed Messages

Thus far, we have said that the strength of the linkage created by a message linking *A* and *B* is a function of message discrepancy and source credibility. We have also stated that this linkage persists after the delivery of the message. But does its strength remain the same? To answer that, we note that (1) over time, messages are often forgotten and (2) at least some aspects of memory of a message are related to the persistence of the attitude change induced by a message (see Cook & Flay, 1978, p. 31). Taking this into account, it is appropriate to regard forgetting as involving the weakening of one or more of the links established by a message. In view of Ebbinghaus's (1964) finding that forgetting follows a decelerating curve, we predict that the strength of a message-induced linkage is a negative exponential function of the time elapsed since the receipt of the message.

Although many messages are forgotten over time, some are not. In particular, messages that have been frequently repeated by external sources tend to be remembered, as do messages that are frequently thought about (i.e., subject to elaborative rehearsal; see Craik & Lockhart, 1972). Hence, forgetting (weakening of a linkage) can be counteracted through receipt of new messages on a topic. These messages may be either externally generated or self-generated. This suggests that cognitive responses may be treated as functionally equivalent to externally generated messages.[3] The causes of such cognitive responses will be dealt with below.

The Time Course of Attitude Change and Its Consequences

Thus far, we have discussed the equilibrium or final value of the attitude change caused by a persuasive message. However, systems may have dynamic equilibria, and it is a common practice to describe the trajectory (time course) predicted by a dynamic model. We shall do so in two ways. One is to present the position of an attitude as a unidimensional location, as a function of time. When graphing a trajectory in this way, we let the position, which we call x, occupy the vertical axis, whereas time, t is on the horizontal axis. A second way to describe dynamic systems is by their phase planes. Although phase planes often plot the interrelations among *distinct* state variables, the two variables we will represent are the position of the attitude, x, and its velocity, v. Our phase planes shall, to be consistent with other chapters in this volume, move position, x, to the horizontal axis and place velocity, v, on the vertical axis. We will now discuss the trajectories implied by this theory (see Figure 6).

The Simplest Variant of the Model

Recall that we derived, from Equation 1, that the restoring force in a linkage is proportional to the displacement of that linkage from its equilibrium. We combine this with Newton's law, that the magnitude of a force is equal to the mass being moved multiplied by its acceleration. This leads to the following equation of motion:

$$m(d^2x^*/dt^2) = kx^*, \tag{12}$$

where x^* is the distance of a concept from its equilibrium location (which may have changed as a result of a message) and m is the mass of a concept, which may, for simplicity, be assumed proportional to its size (number of meanings or attributes). If the restoring coefficient is constant over time, this leads to a trajectory of *undamped oscillations* with constant period (see Figure 6a for the trajectory as a function of time and 6b for the phase plane).

---

[3]It is also possible to treat such self-generated messages as mere epiphenomena, with no independent effects. This possibility is considered later.

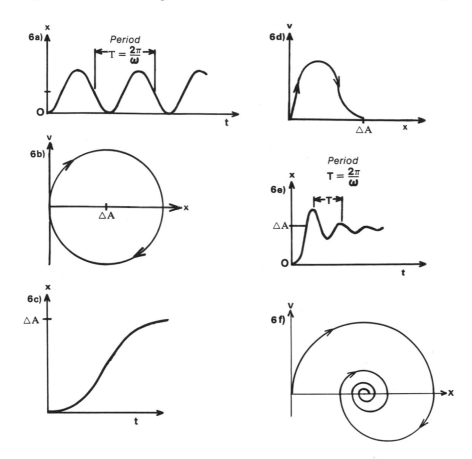

Figure 6. Various attitude change trajectories, all of which have x (attitude) = 0 at t = 0 and x = △A at equilibrium: (a) no damping, x as function of t; (b) phase plane of (a); (c) overdamped or critically damped, x as function of t; (d) phase plane of (c); (e) under-damped, x as function of t; and (f) phase plane of (e).

Both everyday experience and the behavior of many other systems suggest, however, that such cognitive oscillations should die out. This implies that there may be a *damping process*, which dissipates the cognitive "energy" on any one topic. Distraction, drugs that reduce arousal, and fatigue may contribute to such a process. Arousal, whether from being involved in an issue, or from exogenous sources (such as from drugs increasing central nervous system activity) should decrease such damping. If we let C reflect the degree of damping and assume that the damping force is proportional to and in the opposite direction from the velocity of a concept in motion, we get the following equation:

$$m(d^2x^*/dt^2) + C(dx^*/dt) + Kx^* = 0, \qquad (13)$$

where x^* is the displacement of concept A from its new equilibrium location. If x is the location of concept A, on a coordinate system such that $x = 0$ at $t = 0$ and $x = \Delta A$ at the new equilibrium, then

$$x = x^* + \Delta A. \tag{14}$$

If we assume that C and K are constant over time, then substituting from Equation 14, we can solve Equation 13 in terms of x. Although we shall not present the precise solutions to this equation,[4] we shall discuss the qualitative characteristics of these solutions.

If $C^2 > 4Km$, the system is overdamped (or critically damped, if $C^2 = 4Km$) and the motion of A describes the S-shaped trajectory shown in Figure 6c. The phase plane of this trajectory is shown in Figure 6d. As can be seen, v is zero at both the pre- and postmessage equilibria and reaches a maximum value when the attitude is approximately halfway between those points.

If $C^2 < 4Km$, we have an underdamped system and an oscillatory trajectory (see Figure 6e, if $C > 0$ and Figure 6a, if $C = 0$), with initial amplitude ΔA and period $2\pi/\omega$ (where $\omega = (4Km - C^2)^{1/2}/2m$). See Figures 6f and 6b for the phase planes corresponding to Figures 6e and 6a, respectively. In the underdamped system, with nonzero damping, shown in Figures 6e and 6f, the phase plane spirals in toward the equilibrium location, ΔA.

Certain key variables that have been discussed previously (discrepancy, credibility, and ego involvement) are not explicitly included in any of these equations. They are, however, implicitly included. First of all, as discussed for Equation 9, ΔA is a function of all of those variables. Second, K is the sum of $K_{A,B}$ (the message linkage) and $K_{A,R(A)}$, the anchoring linkage. As indicated, $K_{A,B}$ is assumed to be an increasing function of source credibility and a decreasing function of message discrepancy. $K_{A,R(A)}$ is assumed to be an increasing function of ego involvement or prior commitment. This implies that in an underdamped system, the period of oscillation will be an increasing function of discrepancy but a decreasing function of credibility and ego involvement.

Time Varying Restoring Coefficients

Thus far, we have derived solutions on the assumption that K, the restoring coefficient, is constant over time. If, however, as stated, forget-

---

[4]See Kaplowitz and Fink (1982) for the solutions to these equations and a book on differential equations (e.g., Haberman, 1977) for a more general discussion of equations of this type.

ting involves weakening of linkages, K gets smaller over time. Hence, the assumption that K is constant is a reasonable approximation only over small time intervals. If we relax this assumption, Equation 13 would then become a linear differential equation with *variable* coefficients. Although we no longer have a simple solution to Equation 13, we can still make some qualitative predictions. If linkages created by the stimulus message and the preexisting anchoring linkages all decrease proportionally, the equilibrium location of concept A will be unaffected. If, however, K decreases while C and m remain constant, and if the solution is initially undamped, ω will decrease, thereby increasing the period. After K has decreased sufficiently, the motion will become overdamped.

Dichotomous Decisions and the Gravitylike Force

Thus far we have been discussing attitude change in situations in which one's attitude can be placed on a continuum. But there are many situations in which one must decide which of two alternatives is preferred. In such cases, the equilibrium location resulting from the linkage forces may be somewhere between the cognitive locations of the two alternatives. Hence, a force other than the springlike linkage force is necessary to explain how people move cognitively to fully endorse one of the alternatives.

One possibility is that in addition to the springlike linkage force, there is also a gravitylike attracting force, which gets stronger as one moves toward an alternative (such a force is posited by Lewin, 1951). Examples are forces that are inversely proportional to the distance (or square of the distance) from each alternative.

If, however, such an additional force is posited and added to Equation (13), that equation would become a nonlinear differential equation and may exhibit chaotic behavior (see, e.g., Moon, 1987).

Attitude Trajectories and Cognitive Responses

Let us now consider the relationship between the trajectories predicted by the model and the cognitive responses (i.e., self-generated messages, or thoughts) that occur at a given time. There are three possibilities for this relationship. One possibility is that there is *no causal connection* between these phenomena (i.e., attitude change neither influences nor is influenced by cognitive responses). If this is the case, then, over small time intervals, the attitude trajectories should approximately follow the preceding solutions. If there are oscillations, they should have an approximately constant period and a monotonically decreasing amplitude. Although those employing the cognitive response perspective (e.g., Petty & Cacioppo, 1986) assume that such responses are a *cause* of attitude change, others (Eagly & Chaiken, 1984) regard this idea as unproven and believe that cognitive responses may be spuriously correlated with attitude change. The possibility that attitude "motion" may proceed in ways that are not accessible to conscious awareness is consistent with studies of creativity (see Perkins, 1981) and social cognition (see Nisbett & Wilson, 1977; Lewicki, 1986).

A second possibility is that cognitive responses *affect* the attitudinal trajectory. In other words, cognitive responses are messages, which create linkages like any other messages. Indeed, some of the cognitive response literature (e.g., Greenwald & Albert, 1968) has found that cognitive responses are even more effective than externally generated messages. If cognitive responses do, in fact, create new linkages or strengthen old ones, they can cause abrupt changes in K. This implies that the amplitude and frequency of oscillation and the equilibrium position can all abruptly increase or decrease. In short, this possibility suggests that the attitudinal trajectories can be quite irregular.

A third possibility is that cognitive responses not only *affect* the attitude trajectory but *are also caused by it*. In particular, cognitive motion should cause the anchoring linkages to be stretched and/or compressed. This, in turn, should change the restoring force (tension) in those linkages (even if K, the restoring coefficient, is constant). We assume that this change of tension constitutes a disturbance that may attract attention and generate cognitive responses.

In sum, the role of cognitive responses is an important question with which any complete theory of attitude change must deal. We are proposing several alternative submodels, each of which incorporates cognitive responses into the more general model. We hope this will guide further research in this area.

Implications of the Dynamic Model for the Timing of Messages

The notion that attitude and cognitive change is a continuous function of time has some important implications for the timing of messages. It is well known that an important part of being successful as a comedian (e.g., Wilson, 1979) or a persuader (see·McGuire, 1985, pp. 270–271) involves proper timing of message delivery. Our dynamic model is relevant here. In the case of persuasion, because we have assumed that the strength of the message linkage is greatest with small discrepancies, the best time to deliver a new persuasive message is at the moment at which the recipient's viewpoint is closest to the viewpoint advocated; the worst time is when the recipient's view is momentarily furthest from the position advocated.

In the case of humor, we have proposed a number of alternative models, based on the multidimensional linkage model, for the relationship between humor and timing (see Maase, Fink, & Kaplowitz, 1984). Many of these models assume that the joke body causes the focal concept in the joke (a concept that typically has at least two distinct meanings) to move in a direction that the punchline ultimately shows to be the "wrong" direction. One such model assumes that humor is maximized if the punchline is delivered just when the focal concept has reached the "wrong" location. Other variants assume that humor is maximized if the punchline occurs when the velocity (or acceleration) in the "wrong" direction is at a maximum when the punchline is delivered. Still another model assumes that humor is maximized when the punchline is delivered so as to resonate with the oscillations of the focal concept. (See Maase et al., 1984, for a more detailed description of these models.)

Thus, unlike static or nonsystem views of the attitude change process, our model makes clear that the *timing* of messages should have an effect. To explain this effect, we need not invoke any new theoretical concepts.

Testing the Model

The model's predictions can be divided into two classes—the equilibrium predictions and those involving the trajectories. Some of the equilibrium predictions, such as those involving the functional relationship between discrepancy, credibility, ego involvement, and attitude change, can be readily tested, provided we assume that most concepts are well anchored and stable. Indeed, some aspects of those functional relationships are currently being tested. All that is needed is to manipulate variables at many different levels and measure the relevant variables precisely.

Although testing the time course predictions is more difficult, it can be done given sufficient resources. There are two major ways of collecting appropriate data. One is to perform *repeated measurements* on each subject. In one such design, subjects could be placed in a language-laboratory-type setting. They would hear the instructions and persuasive message via headphones, and, at predetermined time intervals, they would be asked their opinions, which would be recorded. It would be emphasized that subjects should regard each response as tentative and not feel obligated to commit themselves to a final view. This technique also permits us to measure cognitive responses. Some subjects could be asked to verbalize all of their thoughts on the issue. These thoughts would be recorded as a running record of such responses and could be compared with attitude trajectories. (However, it is well known that verbalized thoughts are slower than nonverbalized thoughts, so this would have to be taken into account.)

Another form of repeated measure design has subjects continuously recording their responses at a computer, using a "mouse." Subjects would move the mouse as their view changed.

A very different design is the *surrogate repeated measures* design. This involves measuring each subject's view just once, but having many subjects measured at many different time intervals. The key theoretical variables in our model (discrepancy and various source and receiver characteristics) would be controlled, and we would randomly assign a relatively homogeneous sample to different time intervals. We can then assume that the trajectory we obtain, from combining subjects who have been measured at different times, approximates the trajectory followed by a single subject.

We have already used this technique to test a related model (Kaplowitz et al., 1983). In that study, student experimenters approached other students in the library, who became subjects. Each subject was given a message advocating an increase in the student health service fee and was asked to think about the issue prior to giving his/her own view. After a preassigned time interval, which differed for different subjects, the experimenter returned and asked the subject's view. Although our resources did not permit us to have either the control over independent variables, or the precise measurement of dependent variables we would have liked, with a sample of over 1,200 subjects we were able to find significant, though modest, support for a trajectory with undamped oscillations. A more precise test would again involve a large sample but would have each student interacting with a microcomputer. This would permit more complete control of message variables as well as more precise control over the time interval at which the subject responds.

Data Analysis

For the *repeated measures design*, each subject's time series data will be used to create an individual attitudinal trajectory. After possibly smoothing the data, we plan to use two data-analytic approaches. As an initial, exploratory approach, we will use spectral analysis. This will tell us if the trajectory is composed of one underlying frequency or if it is built on several superimposed periodic functions (see, e.g., Mayer & Arney, 1974).

The trajectories, or their phase portraits, may appear chaotic or random. To assess such a possibility, the Lyapunov exponent or other characteristics of the trajectory may be computed in addition to the spectral frequencies. We could then determine whether some of our manipulations affect the regularity of the observed attitudinal motion.

After identifying the relevant periodic functions, we will perform nonlinear regressions on each trajectory. This will enable us to estimate the parameters of each trajectory and the goodness of fit of the overall model. For the *surrogate repeated measures* design, a trajectory would be determined for each group of subjects in the same condition. Because the trajectories predicted by the model are nonlinear functions of time (see Figure 6), they must be estimated by means of nonlinear regression techniques.

If the trajectories are as shown in Figure 6, then they should be a function of the restoring coefficient, damping coefficient, and mass of the relevant concepts and linkages. One of the variables that is assumed to determine the restoring coefficient is message discrepancy. If we vary discrepancy across subjects (and hold constant other factors), we can see if the trajectories vary in a way that is consistent with the predicted effects of discrepancy on the restoring coefficients.

In order to determine the effect of cognitive responses on the trajectory, we may distract some subjects (to prevent such responses), whereas others would be encouraged to think about the attitudinal issue. If cognitive responses create or strengthen linkages, thereby causing abrupt changes in a trajectory, testing the model for such conditions may require identifying a number of the relevant linkages for each subject and comparing the data with results from computer simulations.

Conclusion

Clearly, testing this model is no trivial task. It requires significant resources. It also requires a control over many relevant variables. Lest this cause despair, we should consider the following thoughts of Einstein (1954, pp. 225–226):

> The physicist has to limit himself very severely; he must content himself with describing the most simple events which can be brought within the domain of our experience; all events of a more complex order are beyond

the power of human intellect to reconstruct with the subtle accuracy and
logical perfection which the theoretical physicist demands.

In short, the laws of physics can also be precisely confirmed only under
highly controlled, idealized conditions.

Although a good model must ultimately be testable, another function of a model is to integrate and synthesize existing knowledge. At
the beginning of this chapter, we argued that this field of research has a
wealth of theories and empirical findings but little integration among
them. We see the theory we are proposing as a step toward such integration. It integrates various source, message, and receiver variables into a
single framework and includes, within this framework, results from
research on cognitive consistency, cognitive responses, forgetting,
humor, and the time course of attitudes (also see Kaplowitz & Fink,
1982, 1988a,b). Because we propose precise functional relationships
among variables, we should be better able to explain why, as Petty and
Cacioppo (1986) have stated, "nearly every independent variable studied increased persuasion in some situations, had no effect in others and
decreased persuasion in still other contexts."

References

Anderson, J. R. (1983). *The architecture of cognition.* Cambridge, MA: Harvard University
    Press.
Anderson, N. H. (1974). Cognitive algebra: Integration theory applied to social attribu-
    tion. In L. Berkowitz (Ed.), *Advances in experimental social psychology* (Vol. 7, pp.
    1–101). New York: Academic Press.
Anderson, N. H. (1981). Integration theory applied to cognitive responses and attitudes.
    In R. E. Petty, T. M. Ostrom, & T. C. Brock (Eds.), *Cognitive responses in persuasion*
    (pp. 361–397). Hillsdale, NJ: Erlbaum.
Aronson, E., Turner, J. A., & Carlsmith, J. M. (1963). Communicator credibility and com-
    munication discrepancy as determinants of opinion change. *Journal of Abnormal
    and Social Psychology, 67,* 31–36.
Bartlett, F. C. (1932). *Remembering.* Cambridge: Cambridge University Press.
Birnbaum, M. H., & Stegner, S. E. (1979). Source credibility in social judgment: Bias,
    expertise, and the judge's point of view. *Journal of Personality and Social Psycholo-
    gy, 37,* 48–74.
Bochner, S., & Insko, C. A. (1966). Communicator discrepancy, source credibility and
    opinion change. *Journal of Personality and Social Psychology, 4,* 614–621.
Cacioppo, J. T. (1979). Effects of exogenous changes in heart rate on facilitation of thought
    and resistance to persuasion. *Journal of Personality and Social Psychology, 37,* 778–
    789.
Cook, T. D., & Flay, B. R. (1978). The persistence of experimentally induced attitude
    change. In L. Berkowitz (Ed.), *Advances in experimental social psychology* (Vol. 11,
    pp. 1–57). New York: Academic Press.
Craik, F. I. M., & Lockhart, R. S. (1972). Levels of processing: A framework for memory
    research. *Journal of Verbal Learning and Verbal Behavior, 11,* 671–684.

Eagly, A. H., & Chaiken, S. (1976). Why would anyone say that? Causal attributions of statements about the Watergate scandal. *Sociometry, 39*, 236–243.

Eagly, A. H., & Chaiken, S. (1984). Cognitive theories of persuasion. In L. Berkowitz (Ed.), *Advances in experimental social psychology* (Vol. 17, pp. 267–359). New York: Academic Press.

Eagly, A. H., Wood, W., & Chaiken, S. (1978). Causal inferences about communicators, and their effect on opinion change. *Journal of Personality and Social Psychology, 36*, 424–435.

Ebbinghaus, H. E. (1964). *Memory: A contribution to experimental psychology.* New York: Dover.

Einstein, A. (1954). *Ideas and opinions.* New York: Crown Publishers.

Festinger, L. (1957). *A theory of cognitive dissonance.* Stanford: Stanford University Press.

Fink, E. L., Kaplowitz, S. A., & Bauer, C. L. (1983). Positional discrepancy, psychological discrepancy, and attitude change: Experimental tests of some mathematical models. *Communication Monographs, 50*, 413–430.

Fink, E. L., Monahan, J., & Kaplowitz, S. A. (1989). A spatial model of the mere exposure effect. *Communication Research, 16*, 746–759.

Fishbein, J., & Ajzen, I. (1975). *Belief, attitude, intention, and behavior: An introduction to theory and research.* Reading, MA: Addison-Wesley.

Greenwald, A. C., & Albert, B. D. (1968). Acceptance and recall of improvised arguments. *Journal of Personality and Social Psychology, 8*, 31–34.

Haberman, R. (1977). *Mathematical models: Mechanical vibrations, population dynamics and traffic flow.* Englewood Cliffs, NJ: Prentice-Hall.

Halliday, D., & Resnik, R. (1960). *Physics for students of science and engineering* (Part II). New York: Wiley.

Heider, F. (1946). Attitudes and cognitive organization. *Journal of Psychology, 21*, 107–112.

Himmelfarb, S. (1974). "Resistance" to persuasion induced by information integration. In S. Himmelfarb & A. Eagly (Eds.), *Readings in attitude change* (pp. 413–419). New York: Wiley.

Holt, L. E. (1970). Resistance to persuasion on explicit beliefs as a function of commitment to and desirability of logically related beliefs. *Journal of Personality and Social Psychology, 16*, 583–591.

Hovland, C. I., & Weiss, W. (1951). The influence of source credibility on communication effectiveness. *Public Opinion Quarterly 15*, 635–650.

Ingard, U., & Kraushaar, W. L. (1960). *Introduction to mechanics, matter, and waves.* Reading, MA: Addison-Wesley.

Jaccard, J. (1981). Toward theories of persuasion and belief change. *Journal of Personality and Social Psychology, 40*, 260–269.

Kaplowitz, S. A., & Fink, E. L. (1982). Attitude change and attitudinal trajectories: A dynamic multidimensional theory. In M. Burgoon (Ed.), *Communication yearbook 6* (pp. 364–394). Beverly Hills: Sage.

Kaplowitz, S. A., & Fink, E. L. (1988a). *A dynamic spatial model of cognition and attitude change.* Paper presented at the Annual Meetings of the American Sociological Association, Atlanta.

Kaplowitz, S. A., & Fink, E. L. (1988b). A spatial-linkage model of cognitive dynamics. In G. A. Barnett & J. Woelfel (Eds.), *Readings in the Galileo system: Theory, methods, and applications* (pp. 117–146). Dubuque, IA: Kendall-Hunt.

Kaplowitz, S. A., Fink, E. L., Armstrong, G. B., & Bauer, C. L. (1986). Message discrepancy and the persistence of attitude change: Implications of an information integration model. *Journal of Experimental Social Psychology 22*, 507–530.

Kaplowitz, S. A., Fink, E. L., & Bauer, C. L. (1983). A dynamic model of the effect of discrepant information on unidimensional attitude change. *Behavioral Science, 28*, 233–250.

Kelley, H. H. (1955). Salience of membership and resistance to change of group-anchored attitudes. *Human Relations, 8*, 275–290.

Leatherdale, W. H. (1974). *The role of analogy, model, and metaphor in science.* Amsterdam: North Holland.

Leippe, M. R., Greenwald, A. G., & Baumgardner, M. H. (1982). Delayed persuasion as a consequence of associative interference: A context confusion effect. *Personality and Social Psychology Bulletin, 8*, 644–650.

Lewicki, P. (1986). *Non-conscious social information processing.* Orlando, FL: Academic Press.

Lewin, K. (1951). *Field theory in social science.* New York: Harper & Bros.

Maase, S. W., Fink, E. L., & Kaplowitz, S. A. (1984). Incongruity in humor: The cognitive dynamics. In R. N. Bostrom (Ed.), *Communication yearbook 8* (pp. 80–105). Beverly Hills: Sage.

Marron, T. A. (1985). *Ambiguity and the predictability of the violation of the triangle inequality.* Unpublished doctoral dissertation, University of Maryland at College Park, Maryland.

Mayer, T. F., & Arney, W. R. (1974). Spectral analysis and the study of social change. In H. L. Costner (Ed.), *Sociological methodology 1973–1974.* San Francisco: Jossey-Bass.

McGuire, W. J. (1964). Inducing resistance to persuasion: Some contemporary approaches. In L. Berkowitz (Ed.), *Advances in experimental social psychology* (Vol. 1, pp. 192–229). New York: Academic Press.

McGuire, W. J. (1969). The nature of attitudes and attitude change. In G. Lindzey & E. Aronson (Eds.), *The handbook of social psychology* (Vol. 3, pp. 136–314). Reading, MA: Addison-Wesley.

McGuire, W. J. (1981). The probabilogical model of cognitive structure and attitude change. In R. E. Petty, T. M. Ostrom, & T. C. Brock (Eds.), *Cognitive responses in persuasion* (pp. 291–308). Hillsdale, NJ: Erlbaum.

McGuire, W. J. (1985). Attitudes and attitude change. In G. Lindzey & E. Aronson (Eds.), *The handbook of social psychology* (3rd ed., Vol. 2, pp. 233–346). New York: Random House.

Moon, F. C. (1987). *Chaotic vibrations: An introduction for applied scientists and engineers.* New York: Wiley.

Nelson, C. E. (1968). Anchoring to accepted values as a technique for immunizing beliefs against persuasion. *Journal of Personality and Social Psychology, 9*, 329–334.

Neuendorf, K., Kaplowitz, S. A., Fink, E. L., & Armstrong, G. B. (1986). Assessment of the use of self-referential concepts for the measurement of cognition and affect. In M. McLaughlin (Ed.), *Communication yearbook 10* (pp. 746–759). Beverly Hills, CA: Sage.

Nisbett, R. E., & Wilson, T. D. (1977). Telling more than we can know: Verbal reports on mental processes. *Psychological Review, 84*, 231–259.

Osgood, C. E., & Tannenbaum, P. H. (1955). The principle of congruity in the prediction of attitude change. *Psychological Review, 62*, 42–55.

Perkins, D. N. (1981). *The mind's best work.* Cambridge, MA: Harvard University Press.

Petty, R. E., & Cacioppo, J. T. (1979). Issue involvement can increase or decrease persuasion by enhancing message-relevant cognitive responses. *Journal of Personality and Social Psychology, 37*, 1915–1926.

Petty, R. E., & Cacioppo, J. T. (1986). The elaboration likelihood model of persuasion. In L. Berkowitz (Ed.), *Advances in experimental social psychology* (Vol. 19, pp. 123–205). New York: Academic Press.

Petty, R. E., Ostrom, T. M., & Brock, T. C. (1981). *Cognitive responses in persuasion.* Hillsdale, NJ: Erlbaum.

Petty, R. E., Wells, G. L., & Brock, T. C. (1976). Distraction can enhance or reduce yielding to propaganda: Thought disruption vs. effort justification. *Journal of Personality and Social Psychology, 34,* 874–884.

Phillips, J. L., & Thompson, E. G. (1977). An analysis of the conceptual representation of relations: Components in a network model of cognitive organization. *Journal for the Theory of Social Behaviour, 7,* 161–184.

Poole, M. S., & Hunter, J. E. (1979). Change in hierarchical systems of attitudes. In D. Nimmo (Ed.), *Communication yearbook 3* (pp. 157–176). New Brunswick, NJ: Transaction.

Rokeach, M. (1975). Long term value change initiated by computer feedback. *Journal of Personality and Social Psychology, 32,* 467–476.

Saltiel, J., & Woelfel, J. (1975). Inertia in cognitive processes: The role of accumulated information in attitude change. *Human Communication Research, 1,* 333 -344.

Sherif, C. W., Sherif, M., & Nebergall, R. E. (1965). *Attitude and attitude change: The social judgment-involvement approach.* Philadelphia: W. B. Saunders Co.

Tannenbaum, P. H. (1967). The congruity principle revisited: Studies in the induction, reduction, and generalization of persuasion. In L. Berkowitz (Ed.), *Advances in experimental social psychology* (Vol. 3, pp. 271–320). New York: Academic Press.

Tesser, A. (1978). Self-generated attitude change. In L. Berkowitz (Ed.), *Advances in experimental social psychology* (Vol. 11, pp. 289–338). New York: Academic Press.

Tesser, A., & Conlee, M. C. (1975). Some effects of time and budget on attitude polarization. *Journal of Personality and Social Psychology, 31,* 262–270.

Torgerson, W. S. (1958). *Theory and methods of scaling.* New York: Wiley.

Tversky, A. (1977). Features of similarity. *Psychological Review, 84,* 327–350.

Tversky, A., & Gati, I. (1978). Studies of similarity. In E. Rosch & B. B. Lloyd (Eds.), *Cognition and categorization* (pp. 79–98). Hillsdale, NJ: Erlbaum.

Tversky, A., & Gati, I. (1982). Similarity, separability, and the triangle inequality. *Psychological Review, 89,* 123–154.

Walster, E. (1964). The temporal sequence of post-decision processes. In L. Festinger (Ed.), *Conflict, decision, and dissonance* (pp. 112–128). Stanford: Stanford University Press.

Walster, E., Aronson, E., & Abrahams, D. (1966). On increasing the persuasiveness of a low prestige communicator. *Journal of Experimental Social Psychology, 2,* 325–342.

Watts, W. A., & Holt, L. E. (1970). Logical relationships among beliefs and timing as factors in persuasion. *Journal of Personality and Social Psychology, 16,* 571–582.

Wilson, C. P. (1979). *Jokes: Form, content, use and function.* New York: Academic Press.

Woelfel, J., & Fink, E. L. (1980). *The measurement of communication processes: Galileo theory and method.* New York: Academic Press.

Woelfel, J., & Saltiel, J. (1978). Cognitive processes as motions in a multidimensional space. In F. Casimir (Ed.), *International and intercultural communication* (pp. 105–130). New York: University Press.

Woelfel, J., Cody, M. J., Gillham, J. R., & Holmes, R. (1980). Basic premises of multidimensional attitude change theory. *Human Communication Research, 6,* 153–167.

15

Toward Building Psychoanalytically Based Mathematical Models of Psychotherapeutic Paradigms

Robert Langs

Psychoanalysis has long been the stepchild of science. Though its origins include Freud's (1895) valiant efforts to establish a neurophysiological substrate for both conscious and unconscious mental phenomena, psychoanalysts—including Freud himself—have long since abandoned their own psychophysiological roots. Nonetheless, as psychoanalysts struggle to justify their labors and to provide their efforts with some foundation in research, there is a small but perceptible movement toward developing a truly significant foundation for psychoanalytic observations, clinical practice, and theory. Indeed, these efforts now include some crude but vital first attempts to generate mathematical models in realms pertinent to psychoanalytic investigation. Sashin's (1985) work with the application of catastrophe theory in understanding the vicissitudes of affect is pioneering in this respect.

The present chapter will offer what is perhaps the first effort to develop blueprints for psychoanalytically oriented mathematical models of the therapeutic interaction and process. Although the models suggested here are designed to account for both conscious and unconscious processes and transactions, their application includes psycho-

Robert Langs • Nathan S. Kline Institute for Psychiatric Research, Building 37, Orangeburg Road, Orangeburg, New York 10962.

Analysis of Dynamic Psychological Systems, Volume 2: Methods and Applications, edited by Ralph L. Levine and Hiram E. Fitzgerald. Plenum Press, New York, 1992.

therapeutic paradigms of all types—whether classical psychoanalysis or family therapy. To arrive at these models, I will briefly define a special realm of psychoanalytic observation, indicate likely reasons why psychoanalysts have attempted to sidestep scientific standards in measuring and researching their approach, and explain why the communicative approach to classical psychoanalysis (Langs, 1982, 1985, 1987), among the many variations in psychoanalytic thinking, is well suited to the challenge of creating mathematical models for the psychotherapeutic process (and for the nature of psychopathology as well, though this topic will not receive extended consideration here). Finally, with the history of these developments in place, several specific mathematical considerations will be presented and discussed.

Background Factors

The special purview of psychoanalysis is that of *unconscious psychodynamic* functioning. The realm of psychoanalysis is the processing of emotionally charged information and meaning. As such, it is an area excluded from research exploration by most cognitive scientists (Gardner, 1985). It is also a realm of observation that involves anxiety-provoking and extremely disturbing data. Sometimes the enormous quota of anxiety generated by such data is itself outside the conscious awareness of the individuals involved—patient, therapist, or researcher.

Perhaps the most controversial and disarming issue in psychoanalysis is the very definition of *unconscious mental processes*—variously referred to as *unconscious* thinking, mentation, imaging, perception, or experience. The confusion in this area is one of the major reasons that psychoanalysis lacks a firm database for its clinical theory and its theory of the structure and function of the mind—and in consequence, for a scientific approach to these areas. Virtually anything outside of awareness is labeled *unconscious*. This may include body posture, implied meanings, the functioning of internal physical organs such as the lungs or the kidneys, and a host of other phenomena. However, the relevance of these phenomena to the vicissitudes and nonmanifest basis of *madness* (a term defined here as referring to all forms of psychopathology, ranging from emotional to physical and behavioral dysfunctions) and of the therapeutic process has not been established. Furthermore, psychoanalysts have not been able to define the domain of their observations and theory, nor have they been able to suggest the limits of this domain—clinical phenomena for which their theory cannot sufficiently account. It is perhaps this last failing above all that makes it difficult for psychoanalysis to qualify as a form of science at this juncture.

The communicative approach to psychoanalysis is founded on a basic definition of those unconscious phenomena that are relevant to madness and to dynamic interactions. It does so in terms of encoded, or transformed (i.e., displaced and disguised; see below) expressions. It has found clinically that these transformed messages are vital to human efforts to adapt to emotionally charged issues. It has shown, too, in some general fashion, that these same transformed meanings are vital to the nature of madness or psychopathology. By maintaining a consistent focus on this level of expression, certain seemingly lawful interactions have been discovered in the general clinical situation. It is the regularity of these processes that have invited mathematical applications and that have led communicative therapists to turn to research investigations of both communicative style and the therapeutic interaction.

In general, the communicative approach has been shown to have considerable clinical predictive capacity. The approach has also been able to define the range and limits of its domain. It has discovered clinical situations in which transformed or encoded messages from patients run counter to the understanding of the therapist. Furthermore, these encoded and disguised images are also discordant with the patient's own behaviors. Thus a patient may on an unconscious and disguised level support a therapist's efforts to maintain the ideal conditions of treatment and to generate interpretations of the patient's unconscious encoded messages. Nonetheless, the patient's encoded images also reflect a belief within the patient that these confirmed conditions of treatment or validated understanding will ultimately lead to their psychic disintegration. The apparent contradictions between the patient's encoded support of these interventions and their profound sense of dread has led to the search for new levels of understanding— in substance, for new psychoanalytic paradigms. It is this quest that has also suggested the use of mathematical models.

To briefly clarify these issues, we may recognize that the concept of "transference" is basic to classical psychoanalytic thinking. Transference has been variously defined as the total relationship that a patient has with the analyst, or as its essentially unconscious and mainly distorted component. As such, transference is viewed almost entirely as an intrapsychic process within the patient and as a fantasy formation. For some analysts, transference is seen to be based on both conscious and unconscious fantasies, the latter having been subjected to mechanisms of disguise such as displacement, symbolization, and condensation—the means by which dreams are created.

This intrapsychic focus, which only vaguely acknowledges the actual interaction with the analyst, has created many problems for psychoanalysis. Among these is an extraordinary focus on issues of psychodynamics and genetics (i.e., developmental origins—early childhood factors in both madness and "transference experiences") and more recently on issues of narcissistic equilibrium and identity. As a result, such basic matters as the nature of the therapeutic *interaction*, the implications of *communications* from both patient and analyst, and the role of the *conditions of treatment*—the therapeutic *space*—are almost entirely ignored. Without a sense of communication and information of two human "systems" engaged in a spiraling interaction, the psychoanalytic approach lacks a sense of movement, system, transaction, feedback and feedforward, and similar conceptions that would foster both the development of measures that could be validated by scientific standards and the use of dynamical mathematical models.

In brief, the communicative approach to psychoanalysis is founded on the basic proposition that patients and therapists are engaged in a spiraling, mutually influential communicative interaction in which both conscious and unconscious communication play a significant role. Patients and therapists are seen as human systems engaged in efforts at adaptation. The conscious and unconscious meanings of the communications from a patient are fundamentally constituted as adaptive reactions to the therapeutic setting and to the interventions of the psychotherapist. (Outside influences play a secondary role.) It is understood that the patient may be aware of some of the meanings contained in his or her associations and in the therapist's efforts and that others are beyond his or her awareness. Similarly, the therapist may be aware of a range of comparable meanings and unaware of others. The psychoanalytic investigator is especially concerned with all meanings communicated by patient and therapist whether outside of the awareness of one or the other. However, the overriding focus is on *unconscious* expressions, those that are communicated in a special, *encoded* form— that is, through messages that have been disguised or transformed by the mental mechanisms of *displacement* and *symbolization*.

The communicative approach has shown that the associations, symptoms, and behaviors of a patient are adaptive responses to the stimuli or triggers (adaptation-evoking contexts or *adaptive contexts*) constituted in the main by the interventions of the therapist. In particular, the information and meanings that a patient encodes through displacement and disguise are virtually always initially structured as transformed perceptions of the *implications* of the therapist's efforts. Such perceptions are entirely unconscious for the patient and often for the therapist as well. The patient then responds to these unconscious perceptions, selected from a set of universal implications primarily in keeping with the patient's own madness and inner issues, in a variety of ways. Chief among these are unconscious efforts to correct and rectify the errant ways of the psychotherapist when he or she is in error.

In practical terms, the communicative approach first studies the nature of communication within the therapeutic interaction. Material from patients may be formulated in terms of their manifest or direct meanings. In addition, a therapist may *directly extract* a variety of *implications* from these same surface meanings. Every message is constituted with manifest or direct meaning and directly available implications. And even though Freud (1900), in *The Interpretation of Dreams,* offered a rather different definition of unconscious thinking and therefore of unconscious meaning and information processing, most psychoanalysts limit themselves to these directly available implications from messages from patients when formulating so-called unconscious meanings.

It is distinctive of the communicative approach to focus on a very different level of unconscious communication—that which is *encoded* or *transformed.* Messages with transformed meanings are called *derivatives.* Such messages have three rather than two levels of meaning: manifest and directly implied, and in addition, encoded or transformed (see Figure 1). The encoded meaning is carried by the same message but

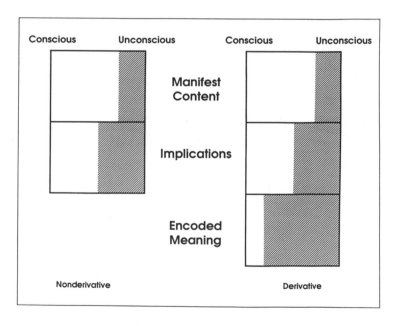

Figure 1. A derivative and nonderivative message.

can be arrived at only through a *decoding process* that undoes the basic mechanisms of disguise—displacement and symbolization. It is this level of unconscious meaning and processing that is most critical to both the transactions of the therapeutic interaction and to the nature of madness in regard to their unconscious foundations. Without a full appreciation of this particular level of meaning, it is highly unlikely that psychoanalytic investigations will meet the standards of scientific investigation or generate nontrivial, creatively revealing models of the therapeutic experience (see Figure 2).

To cite a brief clinical example, we may consider a patient whose therapist inadvertently overcharged her $10.00 on her monthly statement. In beginning the next hour, the patient pointed out the therapist's error and indicated that she (consciously) felt it was quite unimportant. Her thoughts quickly *shifted* to an incident with her butcher, who had mistakenly overcharged her $10.00 for some meat she had bought. The patient ranted and raged against this man, stating that he was an idiot, a charlatan, greedy, and out to exploit her.

In this relatively simply structured incident, the *manifest* material excuses the therapist's error on the bill and recounts an incident and a set of feelings directed by the patient toward her butcher. On the level

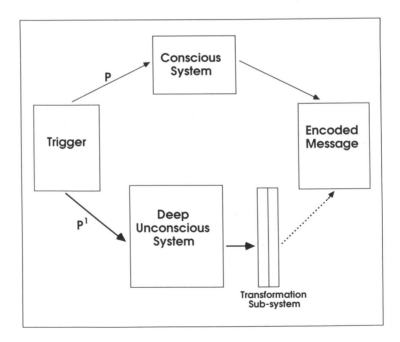

Figure 2. The creation of an encoded message.

of *direct implications*, the patient's material implies that she is forgiving of the analyst, quite understanding of his foibles, and yet in a murderous rage toward her butcher. There are as well many other implications to this surface material that can be *directly extracted* by patient, therapist, and reader.

But this material reveals another level of meaning, one that is both unconscious and encoded. We may safely propose that the story of the butcher has been *displaced from* the treatment situation. It would appear that the patient is entirely unaware of this displacement, and we can see as well that her use of symbolism or disguise is minimal—an overcharge by the butcher is used to represent an overcharge by the therapist. It appears that the patient's *conscious* view of the *butcher* is her *unconscious* (i.e., displaced) view of her *therapist*. Frightened by these perceptions of the therapist, the patient unconsciously invokes displacement in order to safely express her view and to vent her feelings. In order to arrive at this third level of meaning, a level that involves transformation through displacement, it is necessary to *undo the use of this mechanism* and to restore the manifest images to their latent or rightful scene—to shift the material of the butchershop to the therapist's office. It is this undoing of a shift in scene that characterizes the psychoanalytic decoding process and a sound interpretation as well. In essence, then, decoding involves the reversal, or undoing, of displacement and disguise (symbolization), using the trigger, or stimulus, as the guide for these efforts. Trigger decoding requires knowledge of the attributes of the trigger, both manifest and implied. It is this stimulus that has *shaped* the encoded message; the same stimulus shapes the therapist's efforts to decipher the disguised image.

There are many ramifications that follow from distinguishing between manifest contents and direct implications, on the one hand, and encoded or displaced (indirect) meaning, on the other. Perhaps most important among these is the realization that in terms of unconscious perception and sensitivity, it is the therapist's creation and management of the *basic ground rules* of the treatment experience that are most crucial. To state this in terms of a communicative model of the mind (Langs, 1986), processing in the direct realm (termed the *conscious system*, with its directly accessible unconscious component—the realm of all prior cognitive research) of the ground rules of treatment is variable and unpredictable, whereas sensitivity in the *deep unconscious system* (a system of silent and unconscious processing whose efforts are revealed only through transformed or displaced images), there is an enormous sensitivity to the explication of the conditions of treatment. It is for this reason, as we will see, that the models described in this chapter pay careful attention to the multiple variables pertaining to the therapeutic space as constituted by the ground rules of therapy. Indeed, it is this new model of the mind (see Figure 3) that has guided the directions of the research to be described in this chapter.

In brief, information is processed initially without awareness in all cases. Those meanings that are tolerable to consciousness are forwarded *directly* to the conscious system and to conscious awareness, which is itself a seeking and somewhat selective system. Once registered consciously, this information is stored in *direct* form in the super-

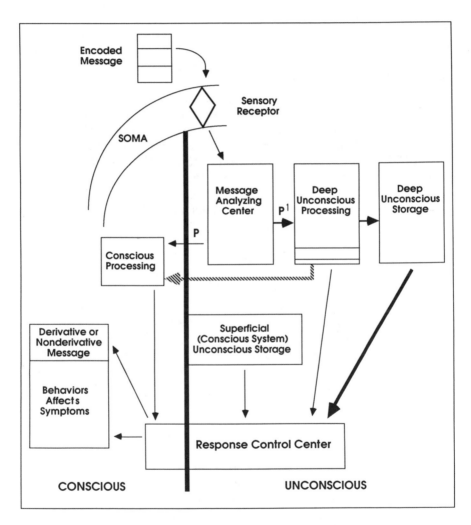

Figure 3. The fundamental system for processing emotionally charged information (feedback loops not shown; modified from Langs, 1986).

ficial unconscious system—a system with its own defensive or repressive barrier that maintains contents more or less outside of awareness depending on the degree of anxiety created by recall. It is to be stressed, then, that the superficial unconscious system contains information that is laid down in a memory bank in direct form, and that it is, when required, also retrievable in direct form—whether as a thought or as an image.

On the other hand, anxiety-provoking images and contents are not allowed access to the conscious system in direct form. They are processed in a different system, termed the *deep unconscious system*, whose operations are entirely outside of awareness. Upon entering this system, information and meanings are subjected to a set of unique premises, such as *parts* equal *whole; possessions* equal *the possessor*; and the like. This system also processes this information in a relatively raw manner, in which instinctual drive attributes are stressed—for example, bald considerations of sexuality and aggressiveness.

Given these premises, it is to be noted that the information is then processed quite logically and at a high level of intelligence. The central aspect of this system, however, is that none of this processing is revealed directly to awareness. A transformation subsystem safeguards its processing and alters all outgoing information by means of displacement and disguise. Thus all outputs from this system are registered in awareness in *indirect* form, and as such, are unavailable to conscious intelligence and adaptation.

The model of the mind accounts not only for conscious awareness but also for behavioral, symptomatic, and other forms of output or human expression. Through a series of direct and indirect feedback and feedforward loops, all incoming and internally generated information is processed in a response-control system that determines the form of outgoing expression. In the emotional realm, the influence of the deep unconscious system is pervasive, whereas in regard to immediate adaptation to external threats the conscious system holds sway (see Langs, 1986, for details). Overall, the model has proven to be a strong theoretical basis for the development of systems-type thinking and for ideas that call for mathematical treatment.

Some Fundamental Considerations

Inherent in the communicative view is the understanding that patient and therapist are each psychobiological entities and that the therapeutic interaction involves the interplay between two psychobiological systems. The mind is understood to be a function of the brain, though the psychoanalytic realm of observation is confined to *mental phenomena*. Although the issue of brain correlates is, of course, of great importance, it is necessary at present to define the microstructure of human communication and interaction so that the psychobiological laws that such structures and functions inevitably follow can be properly studied and defined before taking into account the substrate on which they are founded. It is in this quest for a molecular and systems analysis of the conscious and unconscious transactions within the therapeutic interaction that has led communicative therapists to seek out meaningful interaction with mathematicians and mathematical models.

As described earlier in this volume, there is a strong and highly creative trend toward the use of dynamical models in both the physical and biological sciences, including that of neuroscience. Because of the biological substrata of mental events, there is a strong reason to believe that there are important *parallels* (with whatever distinctions as well) between physical and biological phenomena and those in the mental realm. Abraham (1983), Garfinkel (1983), Prigogine and Stengers (1984), and others have offered convincing evidence of the value of mathematical models in understanding both physical and biological processes. These writings point to metaphors, analogies, and isomorphisms that suggest that these models will find considerable applicability to mental processes.

For example, Garfinkel (1983) and Rapp (1987) stress the importance of limit cycles—that is, of periodicity—in physiological systems, suggesting that homeostasis or equilibrium involves cyclical states that come to rest at maxima and minima points rather than involving simple point attractors or single steady states. Clinically, the communicative approach has observed cyclical variations in the communication of derivative or encoded messages, thereby creating an important area of observation that lends itself well to mathematical modeling (see below).

Priogogine and Stengers (1984) discuss many issues in physics that appear to be relevant to the modeling of mental transactions. For example, they characterize the Newtonian model of science as one that involves the passive observation of nature by a detached observer outside of the sphere of influence of the phenomena under observation. The model of the world derived in this way stresses stability, order, uniformity, linearity, universal laws, and the reversibility of time. These attributes are characteristic of the thinking of Freud and most modern-day psychoanalysts.

There is, however, another view of the universe that developed under the influence of thermodynamics (and later, quantum physics) and its second law, that of entropy (i.e., that relatively closed systems move toward disequilibrium and disorder through the loss of structure). With this law, the notion of irreversibility and the arrow of time are introduced, and with them, concepts of information and innovation, disorder, instability, diversity, disequilibrium, and especially nonlinear relationships. It becomes important to know the state of both the observer and the observed, because both interact. Feedback, chance, and probability become important considerations. Differences in how systems behave near and far from equilibrium are discovered, and the existence of bifurcation or catastrophe points is discovered. Chaos is seen to characterize many systems far from their equilibrium states. Yet surprisingly, such chaos is enormously creative in providing the system with many innovative options. Often, new order arises out of surface chaos through the development of dissipative structures. Through both cyclical and chaotic activities, physical and biological systems prove to be capable of developing surprising levels of innovation and order. Furthermore, beneath such chaos a measure of order prevails—as shown by several mathematicians in recent years.

Many of these conceptions are likely to apply to the therapeutic interaction. The therapeutic process appears to have a configuration characterized more by turbulence and chaos than by simple equilibria and steady states. The appearance of symptoms or their alleviation may well be conceived of as bifurcation points whose prediction requires extensive knowledge of the conscious and unconscious state of the two interacting systems—patient and therapist. Furthermore, bifurcation points involve qualitative changes based on quantitative factors. It therefore seems likely that some form of catastrophe theory (Thom, 1975; Zeeman, 1977) should prove to be a useful tool in identifying control factors and state variables that will account for the vicissitudes of emotionally charged systems and behaviors. In this regard, the dynamical systems approach utilized by Abraham and Shaw (1982) should prove especially facilitating.

One final analogy, however crude, deserves mention. For many years physicists investigated particles and considered space either as a mathematical convenience or a hypothetical construct (Feinberg, 1986). In recent years, largely through the development of Einstein's theory of relativity, space was established as a distinctive entity and afforded major dynamic importance—so much so, that there are physicists who feel that space is primary and particles are secondary, and some who even believe that space is everything and particles are hypothetical. In the main, however, it is now recognized that the configuration of space has definitive shape and form and that the very nature of particles, once thought to be quite fixed, depends on the conditions of space and environment.

By analogy, it could be proposed that the therapeutic space, once all but neglected, is in some ways more crucial to the therapeutic interaction and experience than the two individuals who populate that space. Certainly, it is clear that the spatial configuration of the therapeutic environment and the conditions of treatment as constituted by the ground rules and their explication are a highly critical and changeable configuration whose basic structure greatly determines the specific nature of the communications, as well as the conscious and especially unconscious experience, of both participants to treatment. This proposition appears clinically to be of such great importance that it seems fair to state that models of the therapeutic interaction, whatever the paradigm, that fail to take into account the therapeutic space are likely to be either crude or trivial, and without predictive value.

In all, there are ample reasons to believe that mentally founded processes and transactions follow biological or psychobiological laws and that dynamical mathematical modeling will be of great help in determining the specific nature of these laws to the point where effective prediction and profound new understanding proves feasible.

Prior Psychodynamic Efforts at Modeling

Zeeman (1977) made the first attempt to create a dynamical model of a psychotherapeutic process, one that involved the amelioration of anorexia nervosa. He used catastrophe theory to develop a series of predictions as to the inner mental state necessary within a patient to facilitate the process of cure. Although the clinical data on which the model was based, as well as its predictive powers, are open to considerable criticism, Zeeman's effort nonetheless inspired at least two further endeavors in this direction. The first of these was that of Galatzer-Levy (1978) who transferred Zeeman's dynamical model of rage and fear in animals to issues of love and hate (ambivalence) in psychoanalytic patients. Galatzer-Levy suggested that catastrophe theory, designed to account for bifurcation points and for hysteresis (the appearance of love at one quantitative juncture and of hate at a rather different quantitative juncture), should lend itself well to investigating a variety of qualitative changes that take place in the psychoanalytic experience on the basis of quantitative change. It is this search for a mathematical technique that could account for qualitative forms of motion and make use of topological approaches (Garfinkel, 1983) that was of special interest to Galatzer-Levy. Nonetheless, this author has not made actual attempts to apply catastrophe theory to psychodynamic mental phenomena or to the therapeutic experience.

A more recent attempt along these lines was reported by Sashin (1985) in which catastrophe theory was used to model the regulation of affects or feelings states (emotions). Sashin suggested that responses to affect-evoking stimuli possess the characteristics of the butterfly catastrophe described by Thom (1975), including the existence of multiple states, sudden jumps in response, inaccessibility of aspects of the curve, hysteresis, and divergence (points initially close together in the model are subsequently widely separated). The model views the mental apparatus as a dynamical system of forces in search of equilibrium, and catastrophe theory is used to model the discontinuance phenomena of the dynamical systems that account for affect responses.

Sashin's model involves four factors as determinants of affective response. The model traces the vicissitudes of these complex factors and predicts ways in which they influence affective behavior; the model also predicts certain kinds of therapeutic influence and techniques.

Sashin's effort is sophisticated and complex. Nonetheless, the variables chosen to account for affect vicissitudes, the question of how each variable can be measured, and the problems inherent in attempting to apply this model to the actual clinical situation are considerable. Particularly problematic is the generation of appropriate quantitative measures of the control factors that can be applied to clinical data. As a result, a definitive clinical test of the model and clear research investigations have eluded the author. Sashin's effort is both pioneering and a reminder of the enormous difficulties inherent in efforts to model psychodynamic transactions.

Psychoanalytic Mathematical Models of the Therapeutic Interaction

The psychotherapeutic experience, whether constituted as individual therapy or as a three-or-more-person interplay, is dynamical in many senses of the term. There is a sense of dynamic movement and conflict within those designated as patients and as therapists as well. There is a sense of movement to the course of a treatment experience and a powerful sense of motion to the unfolding interaction itself. These movements tend to be complex and nonlinear, and they suggest the use of topological models and dynamical systems approaches. The calculus of movement should permit studies of the quantitative aspects of these actions, and the mathematics of catastrophe theory and of the factors in the development of bifurcation points should prove helpful.

The therapeutic interaction presents many issues and problems in need of mathematical consideration. Both the development of symptoms in the course of a psychotherapy and their alleviation may be thought of as bifurcation points. Dynamical models based on the careful development of molecular-level scoring systems that take into account the fine details of both conscious and unconscious communication should prove extremely helpful in determining the as-yet unknown factors that create these bifurcation moments or catastrophes.

Similar considerations apply to the often striking alterations in the type of interventions offered by therapists in the course of their work with patients. Furthermore, as we study the interaction between the patient's system and that of the therapist, we will require mathematical assistance in sorting out the influence of the state or control variables and in defining the fascinating interplay of two or more systems that may be harmonious or disharmonious from moment to moment.

Nonlinear systems are drawn to attractors. These include the point attractor, which draws the system toward a single equilibrium point or stead state; the limit cycle or cyclical attractor, which characterizes systems prone to cyclical activities; and the chaotic or strange attractor, which characterizes systems that are seemingly random or turbulent. It is only through the use of mathematical models that the major attractors within patients, therapists, and their interaction can be determined. With such knowledge in hand, the prediction of future transactions and of the vicissitudes of symptoms may well be feasible.

In all such work, it is essential to establish a dialogue between the research psychotherapist and the mathematician. Computer simulations may also play a role. It is the responsibility of the research therapist to provide reliable and valid and especially meaningful quantitative and qualitative data for mathematical modeling. Prigogine and Stengers (1984) stress the importance of the knowledge of the initial state of any system in order to predict the nonlinear course of the system, the points of bifurcation, and the nature of its irreversible states. For this reason, the communicative psychoanalytic research that will contribute to model making should include a variety of measures that pertain to the state of the therapist, as well as to that of the patient. In addition, it will be essential to provide data as to the state of the therapeutic interaction. Finally, the configuration or state of the therapeutic space within which the therapeutic interaction takes place, a factor that also helps to shape the specific state of both patient and therapist, must also be known in great detail. Communicative researchers have already developed an extensive scoring system with which to quantify and qualify each of these important dimensions.

Initial Efforts at Modeling

The present research was initiated several years ago via long and arduous efforts to generate a meaningful means of quantifying critical dimensions of the therapeutic interaction. However, the definitive research on which mathematical models will be developed is less than a year old. Dialogue between this author and his colleagues and a number of mathematicians (in particular, Ralph Abraham, Richard Albanese, Paul Rapp, and Michael Maller) have proven enormously evocative and rewarding, despite the fact that present efforts are still in a preliminary and pilot-study phase. Nonetheless, through these endeavors it has been possible to generate the 35-item scoring system involving descriptive items that are theory free and quantifiable. In this initial phase, a single broad hypothesis is dominant, namely that there are a finite number of communicative interactional patterns between patients and therapists and that these patterns occur in predictive, though probabalistic form. It is the present goal of this research to identify these patterns, to discover factors within patients and therapists that contribute to them, and to identify the broad psychobiological/mental laws that they follow.

These initial measures are designed to quantify the traumatic and helpful power of the therapist's interventions, the degree to which they are in keeping with the patient's material or idiosyncratic, and the points at which a therapist shows some type of bifurcation process—a sudden alteration in the nature and mode of intervening.

The measures are also designed to reveal the influence of the conditions of treatment on the communications and state of both patient and therapist in the interaction of their respective mental systems. Points of bifurcation in the patient's communications, as well as in symptomatic manifestations, are also being measured. The mutual influence of patient and therapist on their respective communications is being considered in great detail.

In this initial phase, it has become clear that mathematical thinking based on applications to biological and physical systems has an innovative influence on psychodynamic measures. Thus in helping to fashion the quantitative scoring measures, Abraham demonstrated the mathematical advantages of scoring all items for both patient and therapist. In this way, both could be placed in the same phase space, enhancing the use of dynamical models. In the psychological realm, this innovation has proven to be remarkably evocative and insightful. By this means, scoring measures will indicate moments at which a therapist becomes a functional patient and behaves in a manner that reflects the typical measures of patient-oriented scores. Similarly, those moments in which a patient functions as therapist can also be identified. Initial studies indicate a lawful relationship between those moments in which the therapist functions in patient fashion—for example, by introducing new themes, by becoming self-revealing, and by seeking help of the other—and the patient shows therapistlike responses, such as direct efforts at inquiry and interpretation. Fundamentally, the realization that concepts that make sense mathematically enhance psychological understanding is to be of utmost importance.

Along different lines, Rapp suggested that errant and traumatic interventions by therapists may have calming effects on a disturbed patient. This formulation is derived from the observation of biological systems in states of disequilibrium or chaos but that are nonetheless brought to a relatively steady state by a chaotic input. This mathematically derived proposition is in keeping certain psychological measures that indicate the appearance of symptom alleviation in response to highly traumatic and deviant interventions by the therapist. Here, too, there is a notable blending of mathematical concepts and psychodynamic observation.

The final example along these lines arose when Rapp raised the question as to whether there is any evidence of sudden shifts in the functioning of the conscious or deep unconscious systems so that one system takes over the usual functioning of the other. The idea of a threshold for such jumps in functioning is derived from the mathematical study of biological systems.

This particular question has led to an entirely unprecedented set of observations that are captured and reflected in the quantitative measures that are being applied to the therapeutic interaction. In essence, these measures indicate that there are moments in which the traumatic power of a therapist's interventions exceeds a certain definable quantity; at such junctures, functions usually assigned to the patient's deep unconscious system are taken over by the conscious system. Such a leap makes clinical sense, in that interventions beyond this threshold create a shift in the traumatic power of the therapist from a level that is relatively subtle and implied to one that is gross and manifest—in substance, the shift from unconscious to conscious and immediate threat.

To clarify this point, we may consider a session that has been scored quantitatively, in which a noncommunicative therapist indicated several times to the patient that he wished to strike or otherwise assault her because much of what she was saying was so disturbing to him. With an accumulation of this type of intervention, the patient shifted from encoded or derivative response (e.g., she spoke of a stabbing pain in the stomach—an encoded image of the stabbing qualities of the therapist's comments; these and other types of images and themes pertain to scores postulated to reflect derivative expressions) to a direct response in which she asked the therapist why he had become so assaultive and was attempting to drive her mad. This communicative shift was reflected in the quantitative scores utilized in these mathematically oriented investigations.

Using systems theory, we have carried out a preliminary sampling of segments of a therapeutic consultation carried out by a therapist using the communicative approach. An important therapist variable lies in the extent to which he or she understands and responds to the patient in terms of encoded communication or fails to do so. Indeed, there are several complex components that characterize the communicative tendencies of patient and therapist and that generate specific constants that reflect this influence on their respective communications and on the nature of their interaction. Still, given the communicative understanding of this therapist, coupled with a cameraman who was recording this particular consultation session, it is possible to define in simple fashion a system of interaction between the interventions of the therapist (I) and the derivative communications of the patient (D) as each was effected by the background conditions of the interview (B; see Figure 4).

In Figure 4, the horizontal axis indicates the positive and negative power valence of the therapist's interventions—the extent to which they were traumatic and deviant as compared with helpful and reflective of a sense of unconscious understanding directed toward maintaining the stability of the conditions of treatment (the ground rules or frame). The vertical axis reflects the frequency and overall power of the patient's encoded or derivative communications. The charted points reflect eight random samples, 80 lines each, of the transcribed therapeutic interaction. The points suggest a direct and linear relationship, which has been drawn into the figure. Thus this initial sample suggests that as the derivativewise therapist's interventions were insightful and frame securing, the patient's derivative communication level increased; on the other hand, when the therapist became confronting, uninsightful, and frame deviant, the patient's derivative communications diminished.

There are, of course, many problems in being able to generalize these preliminary observations. No effort of this kind will be made here, except to note the possible lawfulness of this interaction. Perhaps the suggestion of measurable regularity is the most important preliminary finding generated by this research to date. In this light, the possibility that this lawfulness might find a more complex and nonlinear statement is of lesser consequence than the realization that the search for such laws may prove fruitful.

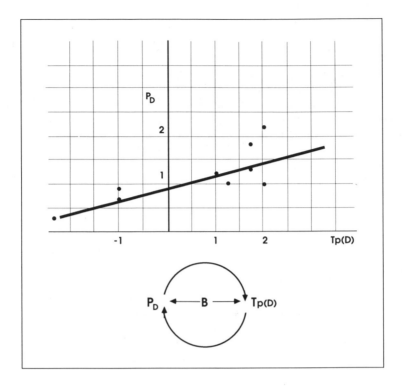

Figure 4. The interaction between the power of the derivative-wise therapist's interventions (Tp(D)) and the patient's derivative production (pD).

Finally, it can be noted a similar consultation carried out by a therapist who tended to disregard both the patient's unconscious communications and the conditions of treatment seems to be pointing toward a rather different curve. In this instance, the traumatic or helpful power of the therapist's interventions produce a surprising "collapse of the function" when the trauma score exceeds a defined intensity. It is here that the shift from encoded images to a direct confrontation of, and attack on, the therapist takes place. Such assaults alternate with highly positive themes that appear to *reflect* a defensive, *unconscious* over-idealization of the excessively hurtful therapist. These findings suggest that a collapse of interactionally symbolic, unconscious derivative functioning plays an important role in the outbreak of direct aggression—a finding that may well generalize to extratherapeutic interactions. It also points to an unconscious reason as to why patients may stay in therapy with very hurtful therapists (see Figure 5).

These initial investigations have concentrated on the patient's production of derivatives, the nature of the therapeutic space, and the attributes of the therapist's interventions as defined by communicative

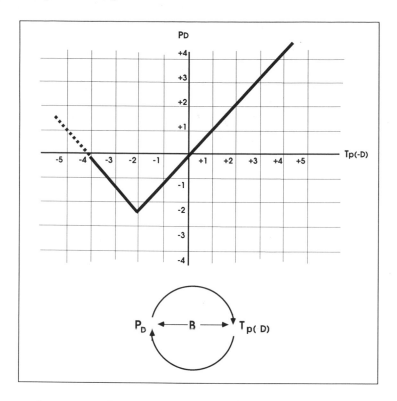

Figure 5. The interaction between the traumatic power of the derivative-ignorant therapist's interventions ($T_{p(-D)}$) and the positive and negative qualities and strength of the patient's derivative responses (P_D) (B = influence of background (frame) factors).

understanding because it appears that these variables are especially vital to the vicissitudes of the therapeutic interaction and are most likely to follow lawful courses that can be captured through mathematical treatments. It is to be understood that these efforts are offered solely as models of ways of approaching the data of the therapeutic interaction in forms that may well generate deep understanding. They are intended to demonstrate the feasibility and possibility of such efforts, rather than to indicate verifiable findings. As we know from comparable projects in biology and physics, a timetable of from 5 to 10 years is rather typical for such undertakings.

Concluding Comments

The complexities of this initial and basic effort to generate mathematical models of the therapeutic interaction are certainly staggering.

However, the present approach differs from that made by Sashin (1985), who began his efforts with a consideration of the psychoanalytic theory of affective responses and attempted to derive a model that could subsequently be applied to the clinical situation. As yet, there appears to be considerable difficulty in making this transition despite the development of impressive mathematical concepts.

In contrast, the present approach to mathematical modeling has begun with the development of a detailed scoring manual that takes into account the vicissitudes of various seemingly meaningful and lawful dimensions of the therapeutic space, the communications of patients and therapists, and of their interaction. The modeling that we are attempting at this time is therefore founded on microscopic clinical measures and on known factors within the therapeutic interaction. The plan is to generate models that have a strong clinical foundation, to study the trajectories that they define, and to then create predictions and understanding that can be applied in turn to fresh clinical data. It is hoped that in this way meaningful models and insight (new theoretical concepts) will be generated and that it will ultimately prove feasible to specifically define the nature of the therapeutic experience. A particular goal is to define as well those interventions of a psychotherapist that are symptom alleviating and those that are symptom exacerbating. It is hoped, too, that these efforts will also eventually define the hidden and often unconscious underlying factors through which such symptom relief or disturbance is generated.

Finally, it is the main purpose of this chapter to ask that psychoanalysts and psychoanalytic researchers be included in the growing community of psychologically minded scientists who are attempting to bring science and predictive meaning to mental processes and transactions. The chapter is furthermore designed to encourage researchers to consider the therapeutic space and unconscious dimensions in developing such models because there is strong indication that these factors are crucial in understanding and predicting symptomatic and behavioral vicissitudes. However crude this beginning may be, there is an unswerving belief that these are endeavors whose time has come and whose future is infinite.

Postscript

Several years have intervened since the writing of this seminal and hopeful chapter. The work that has evolved out of these crude beginnings shows something of the power of transforming theory into empirical research and then into formal mathematical modelling. The prophetic powers of this chapter were far beyond anything I ever dreamed of.

To pick up the story where the chapter leaves off, through a serendipitous act of good fortune I soon met Anthony Badalamenti, Ph.D., a mathematician who quickly became my primary collaborator. Over the ensuing years, we developed a series of mathematical models, most of them drawn from physics, that lay claim to a formal science of psychoanalysis (Langs, 1989; Langs & Badalamenti, 1990; Badalamenti & Langs, 1991; Badalamenti et al., in press; Langs & Badalamenti, in press).

In brief, we used time-series data for six dimensions of human dialogues for these studies: speaker duration; four items that appear to quantify the vicissitudes of unconscious expression (a proposition we were able to validate): Newness of Themes, Use of Narratives versus Intellectualizations, Positively Cast Images, and Negatively Cast Images. Together with a fifth and systemic item—Continuity of Dialogue—these last five items constitute a five-dimensional information particle (IP), whose behavior we have studied in detail. Our data base includes 13 psychotherapy consultations, 2 ongoing psychotherapies, 3 couples in emotional dialogue outside of therapy, and 2 monologues from these couples.

The switching of the speaker role revealed a first-order Box–Jenkins model in most cases, a reflection of deep stability that ran counter to our expectations (as did most of our findings). The histogram of the frequency of utterances of a given duration shows a negative exponential configuration that is a signature of a Poisson process.

The dimensions were used separately to study the cross-correlations between these items within a 10-minute window, 5 minutes before and after a score change, for each party to one of these dialogues, producing a *measure of deep linear influence* that discriminated among the different pairings. The *power spectral density function* (PSD), which measures cyclical power, showed a strong DC (noncyclical) component in all cases, speaking again for strong stability on the level of the deeper nature of mind and communication that this research has investigated. PSD values also meaningfully discriminated among the various individuals under study.

We were able to use the Newtonian model of *work* and *force* to discover a universal deep law, namely, that work is a linear function of time into a session (work increments are everywhere equal), and we also found much evidence for a communicative force field and for strong inductance between members of a dyad. Everywhere we investigated, there were individual differences in the midst of adherence to basic laws.

We have studied many aspects of the *geometry* of the trajectories of the five-dimensional IP, and were quite surprised to discover that all of the points in the 5-cube visited by the IP form an ellipsoidal configuration. We were also able to study shifts from the resting nonderivative space (where there is a relative absence of unconscious meaning) to high derivative space. Indeed, the driving force that has enabled us to formulate these laws of deeper nature appears to involve the vicissitudes of a subject's choice of communicative vehicles; there is a measurable force field that constantly impels the IP to rotate toward new configurations or states while returning as well to prior states. It is the deep power of this psychobiological need that stands behind this science of the mind to this date.

Finally, we investigated the *cumulative entropy* of the IP's trajectory over the course of a session. We found that entropy, calculated with the *Shannon–Boltzmann probabilistic method*, was a logarithmic function of time into a session. In addition, we were able to develop analogues for reversible (and irreversible) heat/energy absorption and temperature (which also increased as a linear function of time); on this basis we were able to carry out measures of the calculus-based *caloric entropy* that confirmed the log curve and agreed with the Shannon values within 5%.

While the implications of these results cannot be discussed here, they are offered as a way of indicating that this chapter, in our opinion, truly adumbrated the entry of psychoanalysis into the family of sciences.

References

Abraham, R. (1983). Dynamical models for physiology. *American Journal of Physiology* 245 (*Regulatory Integrative Comp. Physiol. 14*), R467–R472.

Abraham, R., & Shaw, C. (1982). *Dynamics—The geometry of behavior* (Parts 1, 2, 3). Santa Cruz: Aerial Press.

Badalamenti, A., & Langs, R. (1991). An empirical investigation of human dyadic systems in the time and frequency domains. *Behavioral Science, 36,* 100–114.

Badalamenti, A., Langs, R., & Kessler, M. (in press). Stochastic progression of new states in psychotherapy. *Statistics in Medicine.*

Feinberg, G. (1985). *Solid clues.* New York: Simon & Schuster.

Freud, S. (1895). Project for a scientific psychology. *Standard edition, 1,* 281–397.

Freud, S. (1900). The interpretation of dreams. *Standard edition,* 4–5, 1–625.

Galatzer-Levy, R. (1978). Qualitative change from quantitative change: Mathematical catastrophe theory in relation to psychoanalysis. *Journal of the American Psychoanalytic Association, 26,* 921–935.

Gardner, H. (1985). *The mind's new science.* New York: Basic Books.

Garfinkel, A. (1983). A mathematics for physiology. *American Journal of Physiology 245* (*Regulatory Integrative Comp. Physiol. 14*), R455–R466.

Langs, R. (1982). *Psychotherapy: A basic text.* New York: Jason Aronson.

Langs, R. (1985). *Workbooks for psychotherapists* (Vols. *1, 2, 3*). Emerson, NJ: Newconcept Press.

Langs, R. (1986). Clinical issues arising from a new model of the mind. *Contemporary Psychoanalysis, 22,* 418–444.

Langs, R. (1987). *A primer of psychotherapy.* New York: Gardner Press.

Langs, R. (1989). Models, theory and research strategies: Toward the evolution of new paradigms. *Psychoanalytic Inquiry, 9,* 305–331.

Langs, R, & Badalamenti, A. (1990). Stochastic analysis of the duration of the speaker role in psychotherapy. *Perceptual and Motor Skills, 70,* 675–689.

Langs, R., & Badalamenti, A. (in press). *The physics of the mind.* New York: Ballantine Books.

Prigogine, I., & Stengers, I. (1984). *Order out of chaos.* New York: Bantam Books.

Rapp, P. (1987). Why are so many biological systems periodic? *Progress in Neurobiology, 29,* 261–273.

Sashin, J. (1985). Affect tolerance: A model of affect-response using catastrophe theory. *Journal of Social and Biological Structure, 8,* 175–202.

Thom, R. (1975). *Structural stabilities and morphogenesis.* Reading, MA: Benjamin/ Cummings.

Zeeman, E. (1977). *Catastrophe theory.* Reading, MA: Addison-Wesley.

Author Index

Subject Index